6/94

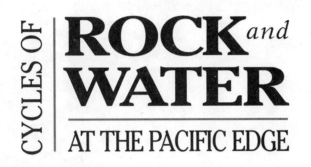

CYCLES OF **ROCK** *and* **WATER**

AT THE PACIFIC EDGE

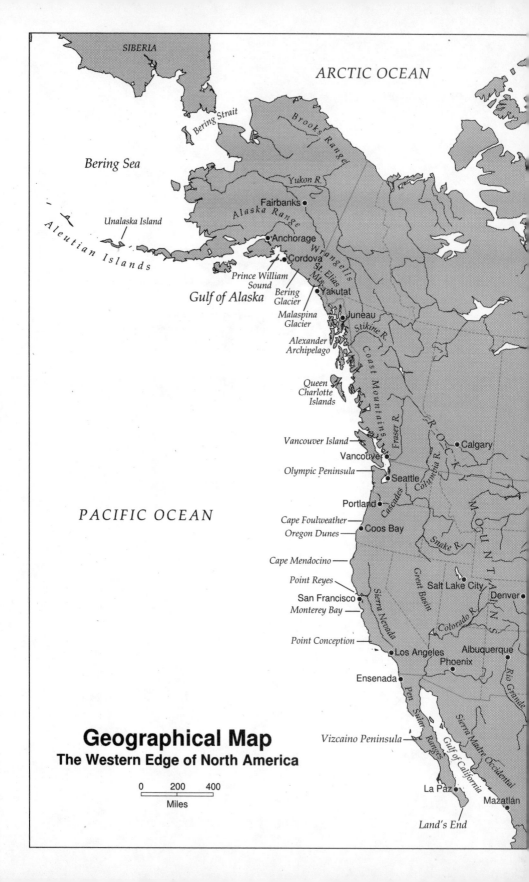

SIBERIA

ARCTIC OCEAN

Bering Strait

Brooks Range

Bering Sea

Yukon R.

Fairbanks •

Alaska Range

Unalaska Island

Aleutian Islands

Anchorage •

Cordova •

Wrangell

St. Elias Mts.

Prince William Sound

Bering Glacier

• Yakutat

Gulf of Alaska

Malaspina Glacier

Juneau •

Alexander Archipelago

Stikine R.

Queen Charlotte Islands

Coast Mountains

Vancouver Island

Calgary •

Fraser R.

Vancouver •

R O C K

Olympic Peninsula

• Seattle

PACIFIC OCEAN

Columbia R.

Cascades

Portland •

Cape Foulweather

Oregon Dunes

• Coos Bay

Snake R.

Cape Mendocino

Great Basin

Salt Lake City •

Point Reyes

Denver •

San Francisco •

Sierra Nevada

M O U N T A I N S

Monterey Bay

Colorado R.

Point Conception

• Los Angeles

Albuquerque •

Phoenix •

Ensenada •

Gulf of California

Rio Grande

Pen. Sierra Ranges

Sierra Madre Occidental

Vizcaino Peninsula

Geographical Map
The Western Edge of North America

0   200   400

Miles

La Paz •

Mazatlán •

Land's End

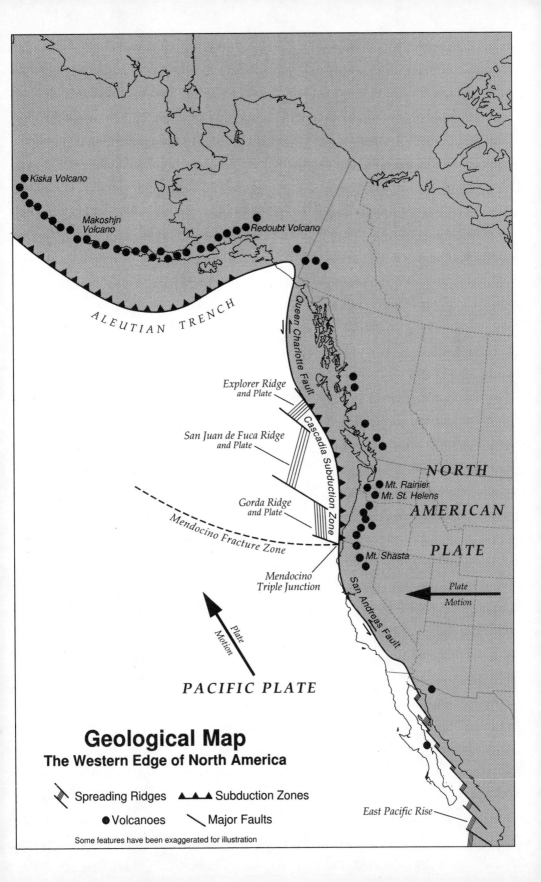

Kiska Volcano

Makoshin Volcano

Redoubt Volcano

ALEUTIAN TRENCH

Queen Charlotte Fault

Explorer Ridge
and Plate

San Juan de Fuca Ridge
and Plate

Cascadia Subduction Zone

Gorda Ridge
and Plate

Mendocino Fracture Zone

Mendocino
Triple Junction

Mt. Rainier
Mt. St. Helens

NORTH

AMERICAN

PLATE

Mt. Shasta

San Andreas Fault

Plate
Motion

Plate
Motion

PACIFIC PLATE

Geological Map
The Western Edge of North America

East Pacific Rise

Spreading Ridges   Subduction Zones

Volcanoes   Major Faults

Some features have been exaggerated for illustration

ALSO BY KENNETH A. BROWN

Inventors at Work

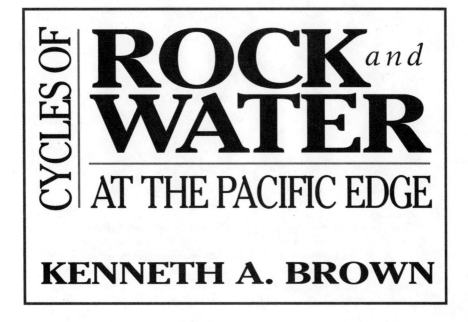

CYCLES OF **ROCK** *and* **WATER**

AT THE PACIFIC EDGE

KENNETH A. BROWN

HarperCollins*Publishers*

HarperCollins books may be purchased for educational, business, or sales promotional use. For information please write: Special Markets Department, HarperCollins Publishers, Inc., 10 East 53rd Street, New York, NY 10022.

FIRST EDITION

Designed by Alma Hochhauser Orenstein

Library of Congress Cataloging-in-Publication Data
Brown, Kenneth A.
    Cycles of rock and water : at the Pacific edge / by Kenneth A.
    Brown. — 1st ed.
        p. cm.
    ISBN 0-06-016056-X
    1. Geology—Pacific Coast (North America) I. Title.
QE71.B76 1993
    557.9—dc20   92-53352

93 94 95 96 97 ❖/HC 10 9 8 7 6 5 4 3 2 1

In memory of my brother MICHAEL,
who dreamed great dreams.

This world is not conclusion
A sequel lies beyond
Invisible as music
As positive as sound. . .

—EMILY DICKINSON

CONTENTS

ACKNOWLEDGMENTS

WRITING IS A SOLITARY DISCIPLINE, BUT IN ONE WAY OR ANOTHER ALL books are the products of the thoughts and ideas of many people. My own work is no exception and after five years of work, I have many debts. Two and a half years of that time were spent traveling up and down the West Coast of North America, gathering the images and material that make up this book. Many of those who helped me are mentioned in the chapters that follow. Others, however, whose comments, help, and insights were no less invaluable are not, and I would like to mention them here. These are the people who helped make my work possible.

For my work in Mexico and Baja California I would like to thank all those who helped open doors for me—to both people and places. Many thanks to Claudia Gonzales Sanchez, Marguerite Brosing, Jose Luis Massa, Jim Crites, and Sergio Flores. The faculty and staff at the University of Baja California Sur provided me with many hours of help and guidance. I would like to thank Carlos Galli-Oliver, Rodolfo Cruz, Juan Guzman, Javier Gaytan, and Oscar Arrizbe. Special thanks too to Bob Douglas at the University of Southern California and Donald Thompson at the University of Arizona.

For the chapters on Los Angeles and southern California I would like to thank the Chevron Oil Company for access to both its personnel and its drill sights in the Los Angeles area; the California Department of Transportation; the South Coast Air Quality Management District; the California Division of Oil and Gas; the Long Beach Public Library; the Southern California Coastal Water Research Project; and the Allan Hancock Foundation at the University of Southern California. I would also like to thank the crew of the *RV Enchanter,* who were kind enough to make room for me in their research and sampling cruise off the coast of southern California. Thanks also to Jack Vedder of the U.S. Geological Survey as well as Stephen Graham and Julie Kennedy of Stanford University who spent hours answering my questions about oil and geology in southern California.

For my work in the Monterey Bay Area I would like to thank the Monterey Bay Research Institute and the Monterey Bay Aquarium for the ready access they offered to both their staffs and facilities, in particular for the time I spent out in the bay on board their research ship the *RV Point Lobos.* My thanks as well to the ship's crew who always made room for me. Support and information for my research were also provided by the Moss Landing Marine Laboratory and the Hopkins Marine Station of Stanford University. Individually I would like to thank Greg Caillet, John Martin, Ken Johnson, and James Nybakken of the Moss Landing Marine Laboratory; Alan Baldridge at the Hopkins Marine Station; Gary Griggs at the University of California Santa Cruz; and Jim Watanabe of the Monterey Bay Aquarium. For my time at sea aboard the *RV Atlantis II,* I would like to thank the Woods Hole Oceanographic Institute and the National Oceanic and Atmospheric Administration. A special thanks to the crews of both the *Atlantis II* and the *Alvin.*

For my work on the San Francisco Bay Area and the Loma Prieta earthquake I would like to thank the U.S. Geological Survey; the American Geophysical Union; the California Academy of Sciences; and the Earth Sciences Library of Stanford University.

For my work in the Point Reyes area I would like to thank the U.S. Park Service and the staff of Point Reyes National Seashore; the Marin County Parks Department; the Vedanta Society; the California Department of Fish and Game; John Finger of the Hog Island Oyster Company; Steve Smith with the University of Hawaii; and Jules Evens of Point Reyes Station, California.

In the Cape Mendocino area my work was helped by Bob Crandall of Humboldt State University and Bob McGlaughlin of the U.S. Geological Survey.

For my work on the Oregon Dunes I would like to thank the staff of the Oregon Dunes National Recreation Area in the Siuslaw National Forest for their help and support, as well as Ralph Hunter of the U.S. Geological Survey and Paul Komar at Oregon State University.

Along the coast of Washington and Oregon my work was helped by the National Oceanic and Atmospheric Administration; the U.S. Fish and Wildlife Service; Sea Lion Caves; the Oregon Department of Fish and Game; and the Hatfield Marine Service Center.

For my work in British Columbia and southeastern Alaska I would like to thank the U.S. Geological Survey's Ice Climate Project and the American Geophysical Union for allowing me to take part in their field trips to the glaciers of southeast Alaska. Help and assistance was also provided by the Canadian Geological Survey; the U.S. Forest Service; the Alaska Department of Fish and Game; and the University of Alaska at Juneau; as well as by Dennis Trabant of the U.S. Geological Survey, Sandy Milner of Anchorage, Alaska, and Terry Quinn of the University of Alaska at Juneau.

For my work in the Aleutian Islands and the Bering Sea I would like to thank the U.S. Geological Survey's branch of Pacific Marine Geology; the University of Alaska at Fairbanks; the Smithsonian Institution; the University of Washington Rare Books Collection; the British Office of Oceanographic Studies; Andy Stevenson of the U.S. Geological Survey; David Hopkins of the University of Alaska at Fairbanks and John Reeder with the state of Alaska's division of geology and geophysical surveys who shared his knowledge of volcanoes with me. Special thanks are also in order to the ship and scientific crew of the *RV Farnella*, who made my extensive time at sea possible and always had time for my questions.

In addition to all the help I received in my travels and search for information, a number of people also opened their homes to me while I traveled, providing me with not just a place to sleep and eat, but companionship and a place to work as well. I would especially like to thank Ann Tape and Phillip Neal for providing me with a base of operations in Seattle for the months I spent working in Alaska and the Pacific Northwest; Edwin and Betty Dunn in Thousand Oaks, California; the Lyman family in Los Angeles, California; Michael Mulcahy and Mr. and Mrs. P. H. Mulcahy in Menlo Park, California; Dan and Janet Farmer of Mountain View, California; and Robert Koenig in Fairbanks, Alaska.

A small group of others helped set this all in motion. I would like to thank my agent John Ware who helped me find such a fine publisher for the book when it saw little more than a loose collection of thoughts in my head. Putting it all together would have been impossible without the steady support of Larry Ashmead at HarperCollins, whose enthusiasm and patience never wore thin, even when what started out as a two-year project swelled into five. In the early stages of the book, a collection of editors at HarperCollins helped guide my thoughts and ideas. In particular I would like to thank John Michel, Keonaona Peterson, and Jon Ewing. As the book finally lumbered into production, Scott Waxman helped me bring it all together and proved to be an invaluable sounding board for all my thoughts and ideas as we strove to hone and polish the final manuscript.

I would like to thank Tracy Vallier with the U.S. Geological Survey for reading the manuscript in its entirety for accuracy and all his helpful comments and insights. Thanks also to my sister Amy who worked as my research assistant tirelessly checking facts and figures and running down references. Her careful reading and proofing of the manuscript and its various drafts helped catch many errors and inconsistencies that would have otherwise gone unnoticed. Most of all to my parents, Margaret Ann and Hubert, for their constant love and support.

The errors of fact and language are entirely my own.

PROLOGUE

THE GOLDEN GATE

THE SAN ANDREAS FAULT IS SOMEWHERE TO THE WEST. ON A CLEAR DAY you could almost trace its path across the Pacific near San Francisco Bay, but it is not by any means a clear day today. It is 7:45 on a Tuesday morning in early February and the fog on the Golden Gate Bridge is so thick that it falls like rain on our windshield. Visibility: twenty feet.

Bright red cables fade in and out of view, slicing the fog into orderly shapes. Street lights spread damp cones of yellow light over clumps of windblown tourists crossing the bridge on foot. Traffic is light in our direction, even though the northbound side has been cut to two lanes for the morning commute. The southbound side is all business: a steady stream of headlights from Marin heading toward San Francisco. Rush hour.

It is not a particularly auspicious day to be heading for the beach, but I am going to Point Reyes National Seashore with Jim Ingle, a professor of geology at Stanford University, to trace the path of the San Andreas Fault. It is one of Ingle's rare days off. Tomorrow he will be "onstage," giving a lecture to the three hundred students in his introductory oceanography class.

In another life, Ingle could have been a successful actor. His lectures are models of content and choreography. While talking about life in the ocean near shore, he leans forward from his ankles like a vaudeville performer and spends the next ten minutes lecturing at a precarious angle. He is talking about sand dollars, and his tilt is intended to mimic their posture as they feed offshore, straining to filter plankton from passing waves. Wearing his best poker face, he stops in midsentence and flashes his eyebrows to make a point. His last lecture of the quarter typically ends with a standing ovation.

Ingle's stage presence is not hampered by the fact that he bears more than a passing resemblance to Douglas Fairbanks. He does not particularly look like a geology professor. An immaculate but conservative dresser, his ties are old-school and carefully knotted. A gold pin holds the wing collar of his pin-point Oxford shirt carefully in place. With a distinguished head of gray hair and a perpetual suntan, he looks like he would be more at home on a Hollywood soundstage than in a university geology department.

Although Ingle grew up in southern California and spent an inordinate amount of time surfing, looks can be deceiving. His suntan is not from successive rounds of golf or tennis at the local country club but from weeks of fieldwork in places as diverse as Peru and the Sea of Japan. And while he can elicit a standing ovation from a class of undergraduates, he is also a first-rate scientist. A former chairman of Stanford's geology department, he is a fellow of the California Academy of Sciences and of the Geological Society of America.

On the north side of the bridge, the Marin Headlands materialize out of the fog. Part of the Coast Ranges, they run all the way to Oregon. Here near the Golden Gate Bridge they rise more than two thousand feet above the freeway. By the time we turn off the highway near Sausalito, the fog turns into a light rain. I pull out my notebook to write down a few notes as we drive.

Ingle spies my notebook and slowly arches his right eyebrow. "Oh no! You brought one of those things! It's not going to be raining at the beach. How could you be such a skeptic?"

One of "those things" is a weatherproof field book, a small five-by-seven-inch notebook filled with thick pages of waterproof paper. A few years ago, a heavy winter rain on the beach turned my notes into a collection of abstract watercolors. Since then I've developed a taste for waterproof notebooks on days like this, but it is an expensive habit—a thirty-two-page notebook costs $5. As we start to wind our way over the mountains toward the coast, I prop my notebook on the dashboard to ward off the rain. "Think of it as weather insurance," I suggest.

By the time we reach the summit, the notebook is clearly beginning to take effect. The rain slows and then stops altogether. When the fog finally burns off we are traveling down the Coast Highway on top of a high cliff overlooking the Pacific. The mountains here run right to the water's edge. The road is barely two lanes wide, precariously cut into cliffs and steep hillsides. On weekends it becomes a proving ground for Porsches and Ferraris from San Francisco and Marin. We would be hopelessly outclassed on a Sunday afternoon in our full-size Chevrolet van, but there is almost no traffic today and we have plenty of room to bump and shudder our way along the coast. The ocean below is deep blue. The surrounding hillsides are green with winter rains, flecked with the first tentative yellow blooms of wild mustard.

These coastal mountains were built out of rocks scraped from the seafloor, Ingle explains. They are less than five million years old. Seen from the perspective of the earth's 4.5-billion-year history, they are fleeting and transitory features. Soft rocks, prone to faulting and sliding, they are flowing back into the ocean almost as quickly as they emerged from it.

"Building a road here is really the ultimate make-work project," Ingle says as we edge our way around a slide of mud and loose rock nearly blocking the highway. "They have to patch it almost daily. These mountains are just melting away. All you have to do is look and you can see that they are just flowing into the ocean."

There is a noticeable flow to things this morning: downhill. Slides and slumps on the road are spaced on a scale of yards, not miles. In places, whole blocks of the mountain have broken free, taking the highway along with them. Repair crews have laid down an impromptu patchwork of asphalt and gravel over the largest breaks, but repairs only a few days old are already laced with a new maze of cracks. "That's what I like about this road," Ingle says. "The pavement's always fresh."

We spot a promising outcrop of rock and park the van at the gate of the aptly named Slide Ranch. Ingle is hot on the trail now and hurries up the road with his rock hammer in hand. We are less than a mile from the San Andreas Fault.

The rocks in this outcrop are cherts, he explains. Composed almost entirely of silica, they were formed in the deep ocean. Near the Golden Gate Bridge these deep-water rocks are divided into orderly layers two and three inches thick. Here on the edge of the fault they have been ground into a pebbly hash, so broken and weathered that at first glance they look almost like topsoil.

In search of a fresh rock face, Ingle begins pounding with his hammer. "Just by looking at these rocks you can" (whack) "see that they've just been pulverized. They've just been" (whack) "mangled. And the closer you get to the fault, the more mangled these rocks get."

Bending over to pick up some small pieces of newly broken rock, he pauses for a moment and begins examining them with his hand lens. "The fault is an enormous thing, but if you look at the rocks here, even the smallest flakes show signs of the fault. Every piece is checkered with slickensides—flat polished surfaces ground into the rocks as they slide past one another. Even the smallest structures here are controlled by the fault."

I pick up my own handful of rock chips. At first they look gray and nondescript. But as I turn them over in my hand, the tiny polished surfaces of the slickensides sparkle in the sun like chips of black volcanic glass. They reveal a kind of hidden instability. Here on the border of the ocean, the edge of the continent is moving.

On the nearby San Andreas Fault, two plates of the earth's crust are actually slipping past one another. While it's regularly rumored that California will fall off into the ocean, movement along the fault is actually north–south. Land on the east side of the fault is part of the stable North American Plate. Land to the west is part of the mobile Pacific Plate. From the Mexican border to Daly City just south of San Francisco, the fault slices most of coastal California from the mainland. San Diego, Los Angeles, and Santa Barbara are all part of that moving slice and they are speeding toward Alaska at the rate of some two inches per year.

Point Reyes is part of the moving slice too. From where we are standing it seems to reach so far out to sea that its tip looks like an island floating offshore. A chevron-shaped wedge of land, it is almost cut from the mainland by the thin watery line of Tomales Bay. The San Andreas Fault actually runs through its base and down the middle of the bay, a drowned valley created by the fault. Point Reyes, Ingle tells me, is made up of displaced rock, what geologists call exotic terrane. Its rocks did not originate here but were formed farther south. Over time they were driven northward along the fault until they collided with the coast. It's suspected that they may have moved as much as one hundred miles.

If you could stand long enough on this cliff at the edge of the Pacific, you would notice that the rocks around you were moving. At first the movements would be subtle and hard to detect. The mountains around you would seem to grow smaller and diminish in size. Points that once appeared to be farther south would seem to move closer. You would be tempted to attribute those movements to hallucinations or the effects of too much sun. With time, however, the Point Reyes Peninsula would break free and begin drifting north. You would have fewer doubts by this time and be more inclined to trust your eyes. In twelve million years, Los Angeles would drift by on its way to Alaska. The mountains around you would disappear into the ocean. You would see cycles of rock. You could trace the movements of continents.

In the ocean offshore you could observe other cycles as well—

cycles of water and life, a kind of fast counterpoint to the slower movements of rocks and continents. Each spring, migrating whales would swim by on their way north to the Arctic from their calving grounds in the desert lagoons of Baja California. In the summer, streams of nutrient-rich water from Alaska would pass by, bringing blooms of plankton and algae. In the still, calm days of autumn you could feel the earth move under your feet.

• • •

Until the early 1960s, continents had a kind of rock-solid stability. They stayed put. While civilizations came and went, rocks and continents endured. All of that changed when the theory of plate tectonics arrived in the sixties. Continents were suddenly mobile and shifty things. They jumped about and wandered across the face of the earth, floating through a sea of denser oceanic crust. The earth's surface was not uniform and unchangeable but divided into plates of oceanic and continental crust. Where they collided these plates built either mountains or deep ocean trenches. Where they split apart they created oceans.

Off the coast of South America the plates move with machinelike precision. New oceanic crust is created at the East Pacific Rise and spreads conveyor-belt fashion toward the coast. Where it collides with South America off the coast of Chile and Peru it is thrust under the edge of the continent through a deep trench in the seafloor. Pushed inward toward the earth's molten interior, the oceanic crust begins to melt. Inland from the coast, this molten rock rises to form volcanoes in the Andes.

In the Atlantic the picture is similarly clear. A rift zone runs right down the ocean's middle, and it is slowly pushing Eurasia and Africa apart from North and South America. Two hundred million years ago, before the rift appeared, there was no Atlantic Ocean. North and South America were joined to Eurasia and Africa as a single supercontinent. Today, the coasts of these separated continents fit together like the pieces of a puzzle.

Off the western coast of North America, however, this well-tuned machine breaks down. The East Pacific Rise runs aground, plowing into the Gulf of California. There is no trench or familiar arc of volcanoes here. Instead the rise is pushing Baja California and the Mexican mainland apart to create a new ocean.

The effects of this motion are more complex in the Pacific than in the Atlantic. Here the rift zone does not run down the middle of a large ocean but along the edge of a continent. The East Pacific Rise represents the boundary between the North American Plate and the Pacific Plate. Spreading along this rift zone is pushing the Pacific Plate slowly northward, creating a complex array of geologic features along the coast. While this motion is creating a new ocean between Baja California and the mainland in Mexico, in California it is driving a thin sliver of the continent northward along the San Andreas Fault, giving rise to earthquakes that level cities and towns. Farther north, in southeastern Alaska and British Columbia, the same drifting plate is driving stray pieces of continent and seafloor into North America to create high mountains and a maze of offshore islands. In the Aleutian Islands on the edge of the Bering Sea, these movements come full circle as the plate is thrust down into the earth through a deep trench on the seafloor. At depth this descending slab of oceanic crust begins to melt, transforming itself into magmas that rise to the surface to form volcanoes in the Aleutian Islands. While rifting inside the Gulf of California represents the beginnings of a new ocean, the volcanic islands being created in the Aleutians represent nothing less than the beginnings of a new continent.

Motion is what West Coast geology is all about. While the East Coast is old and stable, the West Coast is young and dynamic. Its rocks are seldom more than fifty million years old and often less than five million. And while the East Coast is slowly sinking into the ocean, the West Coast is slowly rising out of the Pacific like bread out of a toaster: decades of inactivity punctuated by jumps of several feet. In places its coastal mountains are growing by rates that average as much as two or three inches per year.

Plants and animals do not live apart from these movements of rocks and continents—they are influenced by them immensely. Up and down the coast the moving Pacific Plate has helped shape bays and estuaries as well as deep-sea canyons and offshore islands, a moving stage for a succession of plants and animals.

In turn these plants and animals exert their own subtle influence on the rocks. Trees and plants, for example, often hold soils in place and influence such things as rainfall and erosion. Some plants and animals actually become rocks, leaving fossils behind that provide a glimpse of life in ancient worlds. In the oceans, the skeletons and remains of tiny marine plants and animals build up into thick piles of sediment and debris that are later turned into shales, limestones, and cherts.

Even living plants and animals can provide details about the past. In their spread and distribution they provide clues about changing climate, the rise and fall of sea level, and even the motions of continents. Some plants and animals are almost identical to their ancient relatives that lived millions of years ago. In their form and structure they provide a picture of the past that is often older and more illuminating than those of the rocks themselves.

After my initial trip to Point Reyes with Jim Ingle, I spent the next two years traveling up and down the coast of North America with a collection of geologists, biologists, and archaeologists. As I followed the path of the Pacific Plate, I jumped back and forth across the continent, spending the summer in Alaska and British Columbia and the winter in Mexico. During spring and fall I traveled in California, Washington, and Oregon, filling in the pieces between those two extreme ends of the continent. What follows is a collection of stories and discoveries gathered from a succession of trips both on land and at sea. Motion is what ties the West Coast of North America together.

To understand all of this, however, you must start at the beginning, down in Baja California where the East Pacific Rise is pushing Baja California and the Mexican mainland apart to create a new ocean. Baja California is the key that unlocks the complex geology and biology of the West Coast.

CYCLES OF **ROCK** *and* **WATER**

AT THE PACIFIC EDGE

BAJA CALIFORNIA

DESERT SEAS

THERE HAD BEEN NO RAIN FOR ALMOST TEN WEEKS, BUT THE LATE spring desert outside of town was deceptively green in the early morning light. The sun was only a few degrees above the horizon and the forest of saguaro-like cárdon cacti that rose up through the thorn scrub seemed to glow like the inside of a freshly cut lime in the sharp, low light. Through the tangle of their thorn-covered trunks I could see water, a thin blue line between the desert and the sky. After three months I was still not used to it: a desert that sits on the edge of the sea.

As the bus climbed up the steep grade leading into the mountains and high desert plateaus north of La Paz, the Gulf of California spread out beneath us—a deep blue sea stretching all the way to the coast of Sonora some ninety miles away.

There was no wind that morning and the surface of the gulf was flat calm. Offshore the islands of Espíritu Santo and La Partida seemed to float in midair, suspended in the heat waves that played just above the water's surface. It would be hot that day.

From space, the eight-hundred-mile-long Baja Peninsula is one of the earth's most distinctive geographic features. A narrow finger of land projecting southward into the ocean below California, it is seldom more than fifty miles wide. Here, less than one hundred miles from the peninsula's tip, it is less than five miles wide. As we reached the top of the grade the Pacific came into view: slate-gray and covered by a thin layer of fog, its surface was broken by whitecaps—a counterpoint to the smooth, blue water of the gulf.

For the past four months its surface had been graced with the passage of whales. Each winter the world's entire population of Pacific gray whales gathers to breed and calve in the shallow saltwater lagoons that border the peninsula's Pacific coast. In the spring they head north again to their summer feeding grounds in the Bering Sea and Arctic Ocean. The six-thousand-mile trip takes almost three months.

There are no fixed points here in the desert. Like the whales, this thin sliver of land is also heading north. The surface of the gulf may be smooth, but its floor is scored by an active rift zone, a seam in the earth's crust where lava and magma well to the surface, pushing Baja California and mainland Mexico farther apart. With time the peninsula will reach Alaska. The Gulf of California will become an open ocean.

The effects of this motion are not local but regional, part of the steady northward drift of the Pacific Plate, which is shaping the western edge of North America all the way to Alaska. Here among the bare stones and rocks of the desert you can see how the edge of the continent was first lifted out of the sea and then slowly torn apart to make way for an emerging ocean.

Life on this dry land and in the neighboring desert sea offers its own perspective on the past as well. Among the thorn-covered desert plants you can see the resourcefulness and adaptability of life. In the rich waters of the gulf you can see plants and animals whose forms and structures are the result of several hundred million years of evolution. Some are almost unchanged from their first appearance in ancient seas—a living record of the past.

It was a few minutes past 7:00 on an early morning in April, and I was traveling north out of La Paz with Luis Herrera, a professor at the University of Baja California Sur, and a group of other professors and students. Herrera and I had been trying to get together for several weeks, but we always seemed to miss connections. The night before, he had called to tell me that he was going out to "make a little investigation." Perhaps I might like to come along?

I first met Herrera at his office on campus when I had been looking for someone to talk to about the fossils of prehistoric whales and giant fish found in the desert rocks. His particular specialty is vertebrate paleontology, but his interests are much broader than that. For a time he taught classes in plant biology at the National University of Mexico in Mexico City.

We talked for the better part of three hours about fossils and desert plants. The time passed quickly, punctuated by bursts of activity as Herrera jumped to his feet to check a fact from a book on his well-ordered shelves or combed his files for a reference from a scientific journal. Bones and fossils were dug out of the bottom of filing cabinets and boxes to illustrate a point: the teeth of a giant shark, *Carchardon mexicanus*; the vertebrae of an ancient whale; the leg bone of a desmostylian, a prehistoric relative of the elephant that roamed the coast here a mere twenty million years ago.

Herrera seemed to have an electric enthusiasm for his work, like an explorer who had stumbled across a new continent. Although scientists from the United States and Europe have been drawn here for the better part of a hundred years, logistics and distance have limited most to only a few days or weeks in the field and the area is still very much a scientific frontier. "There are many gaps in the record," Herrera said in a voice that sounded like an equal mixture of resignation and enthusiasm. "There are a lot of mistakes that need to be corrected. We have a lot of work to do."

Outside the windows of our bus the desert was a rolling maze of mesas and plateaus cut by small canyons and broad dry washes. In places, red cinder cones poked through the dry ground, a reminder of the rifting that opened the gulf. Farther north was the start of the Sierra Giganta, part of the Peninsular Ranges, a chain of high granite peaks that runs down the center of the peninsula all the way to southern California. Here, however, the rocks are all marine—a chalk-colored collection of grays and tans. Walk out into these hills and canyons and you will find the fossils of clams and fish. Twenty million years ago these desert rocks were on the floor of the sea.

• • •

By 11:00 we were traveling on foot through the desert following an abandoned mining road outside of Las Pocitas, a small village seventy miles north of La Paz. The road had been laid out ten years earlier by survey crews and geologists from ROFOMEX, the government's phosphate mining consortium, which had long since moved on to more promising sites. We were not looking for minerals but for water. Here in the desert, water is as valuable as gold or oil and often as hard to find. On the surface, the purpose of Herrera's "little investigation" was to find a likely site for a new well for area farmers and ranchers. On a deeper level, however, he was interested in unraveling some of the intricacies of faulting and folding that had shaped this stretch of the peninsula.

Where the road ended on top of a high plateau we paused to look out over the terrain and check directions. There were three others in our group besides Herrera and myself: Cesar Lopez and Ormard Ortiz, two professors of geology from the University in La Paz, and our guide, a local man named Angel who knew the desert well. He picked his way expertly through the cacti and thorns, hiking in plastic flip-flops.

From the top of the plateau the view was expansive. To the east was a series of high ridges and steep hills that led to the Gulf of California just twenty miles away. To the west the view was wider and more open: a broad sloping plain cut by arroyos and dry desert washes. In the distance you could see the Pacific: blue in the midday sun, covered farther offshore by a thick blanket of fog that suggested fresh breezes and cool ocean air.

In the desert, however, the heat seemed to drive things to extremes. The rich, saturated colors of early morning had given way to shades of black and white. The rocks and thorn scrub looked pale and sun-bleached, as if they were covered by a fine layer of ash, while the tall evergreen arms of the cacti seemed almost black, like the charred trunks of trees left by a forest fire. There was no breeze and the ground was noticeably hot underfoot.

We decided to push up the side of a steep brush-covered hill, or *cerro*, on the backside of the plateau. Halfway to the top, its cover of scrub and cacti was broken by a patch of light gray sandstone. Breaking open fist-sized chunks of rock we found fossils: clam shells, sharks' teeth, porpoise bones, and the micalike flakes of fish scales.

A few feet away Ortiz was scratching rocks with the pick end of his hammer and holding them up to his nose.

"What is he doing?" I asked.

"Here, smell," Herrera said, handing me a rock with a few scratch marks on its side. I cupped it in my hand and sniffed it as if it were a flower. It had a faint but pungent smell, like rotten eggs.

"What is it?"

"Phosphate. You can always tell by the smell. That's what **ROFOMEX** was hoping to find here. They did. But not enough to make a prosperous mine."

Used in common products like fertilizers and the coating on match heads, phosphate is an uncommon mineral, deposited only on the outer reaches of the continental shelf, halfway between the deep sea and dry land. Its appearance here locates these rocks as precisely as a street address.

A primary nutrient in the ocean, phosphate is usually absorbed by plants and animals whenever it is available. Where the surface of the ocean is unusually productive, however, the water below can be so low in oxygen that it is almost lifeless. What is bad for marine life is good for phosphate. With no plants or animals to absorb it, phosphate begins to precipitate and collect. Where the continental shelf lies within this low-oxygen zone (usually at a depth of four hundred to five hundred feet), it falls on the sea floor like snow.

The rocks here, Herrera explained, are part of a body of rocks known as the El Cien Formation and are thought to be some twenty-five million years old. What is so important is not just that they came from the sea, but that together with the older and younger rocks that lie above and below them they offer a glimpse of how this area was formed.

In the sides of canyons that cut through the plain below, the dark brown rocks of the underlying Tepetate Formation are exposed. Fossils inside those rocks suggest not only that they are some twenty million years older than those on the hill, but also that they are from water more than three thousand feet deep—not from the continental shelf but from somewhere near its base on the edge of the deep sea. Considered together these two layers suggest that the seafloor became progressively shallower as time went on—an idea that is all but confirmed by the bands of red rock covering the top of the hill.

The red rocks above are part of the Comandu Formation, the eroded remains of a chain of volcanic mountains or highlands built up by activity that was a prelude to the rifting that opened the Gulf of California. The rocks are conglomerates, composed of volcanic sands laced with pebbles and stones. By the time they were deposited, these rocks were well above sea level. Their hard surface has kept the softer marine rocks of the hill from being eroded as flat as the plain below.

• • •

The forces that lifted these rocks out of the sea were not local and self-contained but global in nature. This series of rocks in the desert can be interpreted as a lesson not just on how Baja California was built up but on how the whole West Coast of North America was formed.

Two hundred million years ago the Pacific coast of North America lay somewhere near what is now central Nevada. On the other side of the continent the Atlantic was just beginning to open along a rift zone known today as the Mid-Atlantic Ridge. The spreading not only opened the Atlantic, it also pushed the continent westward, driving it through the Pacific like the blade of a plow, building up the edge of the continent with rocks scraped from the seafloor.

At the same time, the edge of the continent was growing from within. The Pacific was scored by rift zones of its own, but instead of pushing continents apart, the crust from these spreading ridges spread outward and then slid under the edge of the continent. At depth these sinking slabs melted, then rose to the surface to form mountain ranges of granitic and volcanic rock. Here in Baja California that history is recorded in the high granite peaks of the Peninsular Ranges that run down its center. Farther north, molten rock from the subducting plate built up the Sierra Nevada of California, the Sawtooth Mountains of Idaho, and the Coast Ranges of British Columbia. While the continent moved westward, a series of now all-but-vanished pieces of oceanic crust—the Kula and Farallon Plates—danced across the floor of the Pacific with movements of their own.

Twenty-five million years ago this period of building came to an end when the continent rode over the East Pacific Rise. Farther to the north this collision is still taking place, but here in Baja California lava rose to the surface as the rise worked its way into the continent, building volcanic highlands and mountains that were later eroded.

Five million years ago rifting began in earnest, splitting Baja California from the Mexican mainland and opening the Gulf of California. It was a sequence of events that would be repeated over and over again up and down the West Coast: after building this edge of the continent up, plate tectonics would begin tearing it apart, sending small slivers and pieces of continent sliding northward along the coast.

••••

The road had been flat and level for miles, a narrow dusty track through the desert that peeled off the Carretera Transpeninsular, near Santiago. At first, small handmade signs had pointed the way down side roads to pueblos and ranchos farther out in the desert: El Torote, La Soledad, El Refugio, Rancho Nuevo. Now, however, there were no signs, only fresh tire tracks and well-worn ruts showing that others had been this way recently.

It was late March and the desert was still green, but there had been no rain for several weeks and the ground was very dry. The cacti were evergreen, but the trees of this dry landscape—mesquite, palo blanco, and torote—were beginning to show signs of wear. Their small tough leaves had begun to shrivel and fall off, as though they were preparing themselves for a long, hot summer.

I was traveling with Jim Ingle and a group of graduate students from Stanford University. Ironically we were traveling in the same full-size Chevrolet van Ingle and I had taken to Point Reyes nine months earlier. Since then, Ingle had been working in Peru and Japan. I had been up in the Aleutian Islands off the coast of Alaska and Siberia, exploring the spot where, in some two hundred million years or so, this dry edge of the continent will end up.

We were just fifty miles from the peninsula's tip at Cabo San Lucas, traveling across a broad flat plain toward the Gulf of California. You could see approaching cars and pickups for miles. They threw up clouds of dust more than a half-mile long that rose up into the air in low graceful arcs. Cacti and plants alongside the road were covered with a patina of fine white dust. There was no hint of the gulf or ocean here, just a broad, flat tableland broken by an occasional line of low cliffs. In the rearview mirror, I could see the mile-high granite peaks of the Sierra de la Laguna.

We were traveling toward the coast not to see the gulf but its opening. After a morning of fast driving we were still roughly five million years too late, but that was largely irrelevant to a group of geologists. Here in the desert the opening of the gulf is recorded in layers of rock. You can read them like a book.

The road was good and our two-van caravan made good time— thirty to forty miles per hour in the straight stretches between washes and potholes. A half-hour off the highway it all came to an end. At the top of a gentle rise, the flat landscape suddenly opened up in front of us. Both the flat plain and the road we had been traveling on came to an abrupt end at the edge of a five-hundred-foot-deep canyon cut into the soft rocks of the tableland. The canyon was broad and headed east in a soft, gentle curve that led into a range of dry, red hills that would have been called mountains almost anywhere else. Between their peaks you could see the gulf—a dark blue line on the horizon marking the spot where those red hills meet the coast near Punta Colorado.

When we had turned off the highway an hour earlier, Ingle had happily shouted over the collective roar of an assortment of rock hammers and several hundred rock samples rattling on the floor of the van: "Finally! A real Baja road! That pavement was just Baja for wimps. This. . . This is the real Baja!" As I looked over the edge of the canyon it occurred to me that perhaps we had at last stumbled onto the true essence of Baja: impassable.

Two white tracks—roads in this part of the world—followed the dry floor of the canyon. The problem was how to get down there. The road we had been traveling on continued on ahead almost straight down the side of the canyon. At one time it had been the main road through the area for local ranchers running their herds in the desert near La Trinidad and Cerro Colorado—a quick, steep route into the canyon below. Lately it had fallen into disuse, its edges eroded by landslides and heavy traffic. In places it was less than a half-lane wide, the other half being some sixty or seventy feet down slope. The grade was improbably steep and scored with ruts two and three feet deep. Passable to some vehicles perhaps, but not to a pair of overloaded vans with no four-wheel-drive, no high-lift jacks, and no winch.

Below us we could see an old pickup truck speeding across the canyon floor with a precarious load of thin, near-wild cattle in back. What was back country for us was a backyard for others. There had to be a way down.

We decided to have lunch and think things over. When in doubt, eat. Afterward, Ingle and I walked back to a promising side road—narrower than the first, but also newer. After winding through a forest of cacti and twisted desert trees, it cut downward toward the canyon bottom past low cliffs of gray shale.

"Deep-water rocks," Ingle announced, pausing to examine some pieces of rock pried from an outcrop alongside the road. Ingle had seen these rocks before, both in the field and in the laboratory on the stage end of a microscope. A few years back one of his graduate students had done the research for her dissertation in the area. Fossil diatoms—tiny floating bits of microscopic algae with a two-piece silica shell shaped like a hat box—found inside these rocks suggest that they were deposited at a depth of more than four thousand feet.

The grade steepened but it was still passable. We decided it was worth a try and walked back to the vans, where Ingle announced to the group: "I'm sure we can make it down, but getting back up will be difficult. You may have to get out and push. We better get going."

Thoughts of pushing a seven-thousand-pound van up a steep canyon road ran through our minds. "Notice how he says *we* may have to push the van uphill," said a not-so-quiet voice from the back of the van. "I thought this was supposed to be kind of a vacation. I don't remember reading anything about pushing vans uphill in the field trip guide."

"Sure you do," Ingle replied. "That's what I meant when I said fun and adventure."

True to prediction, we made it down without a hitch. The floor of the canyon was wide and sandy, more than one hundred yards from side to side and covered with a head-high blanket of thorn scrub. The road was split into two well-traveled parallel tracks, a kind of desert expressway.

• • •

All rocks have stories to tell. In most cases they are incomplete stories or merely short ones: the path of an ancient streambed or a thirty-five-million-year-old lake inhabited by snapping turtles. Perhaps the eruption of a volcano. Every so often, however, you come across a series of rocks that tell a larger story, one that pulls all the small pieces together. The rock walls of this desert canyon are one of those stories.

The rocks here are stacked up like a layer cake: red on top, white on the bottom. The pattern is striking even to the most casual observer. To a geologist, however, these layers of rock are much more than just admirable scenery.

The almost-white rock below is granite. Composed of translucent grains of quartz and pale gray crystals of feldspar and peppered with black flecks of biotite and horneblende, it is not unlike the granite that makes up the high peaks of the Sierra de la Laguna some thirty miles away. The upper layer is a dull red sandstone. Unlike the granite, it is not uniform or well ordered, but shot through with channels of gravel and poorly sorted stones—like the dough of a fruitcake that hasn't been completely mixed.

The granite, Ingle explained, is not marine but continental—part of the granite spine that runs down the center of the Baja Peninsula. Twenty to thirty million years ago it was not on the floor of a canyon where sands and gravels were deposited, but high in the air.

The red rocks on top are alluvial deposits: the leavings of desert streambeds and sediment fans, not unlike the broad, sloping fans of rock and sand that run out of the Sierra de la Laguna today. They were deposited on land and exposed to air. Their red color is due to nothing more sublime than rust. That these red rocks lie on top of this granite suggests that the granite sank or subsided at some point in time, forming a basin deep enough for them to collect in but not yet deep enough to let in the sea. That came later.

Farther up the canyon, these white granites and red conglomerates are overlain by the gray marine shales we drove down through to reach this spot in the floor of the canyon. The transition of rocks from desert streams to deep sea is abrupt. What is so interesting, Ingle says, is not that this area sank, but the speed at which it happened. In the space of probably less than a million years this piece of coast went from being above sea level to nearly a mile deep in the sea.

Geologists have no definitive age for either the granite or the overlying red beds, but the marine rocks above them have been well dated and are about five million years old. This is an interesting age to geologists because the Gulf of California is roughly five million years old as well. These layers of rock in the canyon wall record not just the creation of deep ocean basin but the actual opening of the Gulf of California. When rifting opened the gulf, the earth's crust was pulled apart, creating a series of deep basins on the seafloor. These rocks went through a kind of geologic free-fall. Once part of the desert, they were suddenly on the floor of the sea. Later they would be lifted up again to become part of this desert canyon—a record of the past written in stone.

■ ■ ■

For all the speed with which this area subsided, it took several million years to set the stage. Twenty-five million years ago when the North American Plate rode over the flanks of the East Pacific Rise, the edge of the continent began to stretch and thin. Inland from the coast, the crust of the continent was broken into blocks that tilted and sank, like the panels of a sidewalk underlain by the roots of a large tree. Here, however, the panels were not a few square feet in size but involved whole mountain ranges. Between what is now Baja California and the mainland, this stretching and tilting created a mountainous interior basin. From the south, the Pacific invaded, creating an inland sea that stretched as far north as Arizona. Geologists call this ancient sea the proto-gulf.

By twelve million years ago this proto-gulf had dried up and disappeared. Its floor, however, was a zone of weakness, a way into the continent. Seven million years later, when rifting began to open the gulf, it would follow the path of this ancient sea almost exactly.

Unlike the proto-gulf that had preceded it, the floor of this new sea was not made up of continental crust, but oceanic crust—built up by dense lavas rich in iron and magnesium. As they rose to the surface, particles inside them sensitive to magnetic fields aligned themselves with the earth's own magnetic field. As the rocks cooled and spread outward, this orientation was locked into the rocks.

For geologists this ability to record magnetic orientation is an invaluable tool for unraveling plate tectonics because the earth's magnetic field periodically reverses itself—north becomes south and south becomes north. Over time these irregularly spaced reversals (sometimes every few hundred thousand years and sometimes every few million years) coupled with seafloor spreading have created a pattern of magnetic stripes on the seafloor.

Inside the gulf this pattern of stripes marches outward from the ridge in orderly rows, each side a mirror image of the other. Each stripe records the gradual opening of the gulf, like a series of time-lapse photographs. The same is true of the patterns found on the floor of the Atlantic radiating outward from the Mid-Atlantic Ridge.

Near the mouth of the gulf, however, the pattern of stripes is irregular and chaotic. "It's as if the rise played around down there," Ingle said, "trying to find its way into the continent."

• • •

The processes that opened the Gulf of California are still at work today. Each year Baja California moves some two inches away from the mainland—an amount that roughly correlates with movement along the San Andreas Fault. That correlation is more than just coincidence.

In the northern end of the gulf the East Pacific Rise disappears under the continent. Directly to the north of it the San Andreas Fault appears, running diagonally across California to the Pacific only to meet up with another spreading ridge on the seafloor not far from the California–Oregon border.

With time the Gulf of California will become not only wider but longer, as rifting reaches farther into the continent. In spite of widespread rumors that California will fall off into the sea, in all likelihood that new rifting will not be along the San Andreas Fault but farther inland. As Baja California slides northward it is taking a thin sliver of the coast along with it. The San Andreas Fault is simply taking up slack, the border between that thin slice and the more stable edge of the continent.

In all likelihood the new rift will run northward through central Nevada or eastward into New Mexico through the Rio Grande Rift Zone. Seams of volcanic rocks running through Nevada and the Rio Grande Rift have a chemical composition almost identical to that found on midocean ridges—as if they came almost directly from the earth's mantle. The earth's crust under both areas is thinning, as if it were preparing for the arrival of a new sea. Geologists believe that the opening of this new sea will take place not sometime in the next few hundred million years but sometime in the next few hundred thousand. When it does, the East Pacific Rise will be not just slicing off a sliver but cutting right to the continent's core.

At the moment, the rise is simply making up its mind. It may go east through New Mexico or north through Nevada. The future is unpredictable. In time, central Nevada may have something it has been missing for the last 180,000,000 years since this edge of the continent began forming: a coastline.

■   ■   ■

The air was cool and smelled like rain. The sky was clear over the desert six thousand feet below, but up in the mountains a small raft of clouds floated over the highest peaks of the Sierra de la Laguna. I was sitting on a ledge of bare granite watching the early morning light play across the dry landscape below. The sun was not yet above the highest peaks, but stray beams of sunlight shot through passes and gaps in the mountain front, illuminating the desert with long triangles of light that were first orange and then yellow.

Beyond the desert below you could see two oceans. To the west was the Pacific. From the high ridge of rock where I sat, its surface seemed to cut across the horizon in a gentle arc—as if you could actually see the earth's curve. To the north was the Bay of La Paz on the Gulf of California. The semicircular sweep of the bay reached so far inland that it seemed to all but cut this mountainous tip of land from the peninsula. To the south and east the view was cut off by mountains, an endless series of peaks and high ridges that ran all the way to the peninsula's tip at Cabo San Lucas.

I had left for the mountains the day before, heading out of Todos Santos in a thick morning fog with enigmatic directions: "Take every left except the last one." I had spent the morning looking for the trail, driving down the desert roads that paralleled the mountain front. Finally two vaqueros on horseback pointed me in the right direction. By 1:00 I started walking, following a trail up a small side canyon whose walls reflected the heat with mirrorlike intensity. My small thermometer registered 120 degrees, the strongest heat of the day.

Once out of the canyon the trail headed straight up the mountain, picking its way across successive ridges and ledges. Cacti and short desert trees overhung the trail but offered little shade. Traveling solo and heavy, with food and water for a three-day trek, I made poor time. An hour into the hike I was climbing the mountain in five-minute segments: short bursts of walking broken by ten minutes of rest crouched in the meager shade of an ocotillo or an organ-pipe cactus.

There were few signs of life in the midday heat. Lizards scurried from rock to rock in search of life-preserving shade. Stopping to rest, I could hear the soft musical ring of bells tied to cattle ranging through the desert scrub in the canyon below.

The desert changed slowly as I climbed. Cacti gave way to spindly trees. Halfway up the mountain, scattered live oaks appeared, their evergreen leaves burned to a coppery brown by the heat and several weeks of no rain.

I watched the sunset from a flat parklike area in a saddle between two ridges, then pushed on through the twilight until it was too dark to see. There was no flat ground and I ended up sleeping right in the middle of the trail. Taking off my shirt and folding it up to make a small pillow, I noticed that it was brittle with salt. Live oaks bordered the trail. Lying on my back I could peer through their branches and watch the spring stars wheeling through the desert sky. Waves of cool night air washed down the sides of the mountain toward the desert below, rustling the leaves of the trees, making a sound like running water.

I was up by 4:30, and two hours later I was on top of the mountain looking out over the dry landscape below—a carpet of thorns and cacti that ran all the way to the water's edge.

It was no desert here on top but a dense forest of pine, live oak, and madrone. Thick mats of lichen and moss clung to the sides of trees. Clumps of mistletoe and Spanish moss hung from the branches overhead. Only solitary yuccas and agaves scattered among the trees served as a reminder that this forest was surrounded on all sides by desert.

Geologically and biologically, these high peaks on the southern tip of Baja California are an island. While the peninsula as a whole has been moving northward for roughly five million years, this mountainous tip has engaged in some extracurricular motion of its own. While the granite rocks of these mountains resemble those running down the center of the peninsula farther north, they are not identical to them. Instead, they are nearly the same as the granites found farther south on the coast of mainland Mexico, not far from Puerto Vallarta in the present-day state of Jalisco. Geologists believe they may have drifted northward on their own until they were added to the Baja Peninsula—like the dot at the bottom of an exclamation point.

The forest of trees found on top of these mountains support this notion of a separate origin as well. Walk into the hills of Jalisco and you will find a forest of pine, oak, and madrone much like this one. Several million years of separation have brought few changes—the needles on the pines (*Pinus cembroides*), for example, are slightly shorter than those on the mainland, a possible adaptation to the drier climate, but aside from those kinds of small changes, one could almost believe that this forest had been uprooted and transplanted intact from the mainland to the tops of these desert peaks. Drifting pieces of continent may have carried not just rocks, but plants as well.

While this mountainous tip of the peninsula is no longer a geographic island, the tops of its high peaks are still very much a biological one. The pine and oak forest that covers its top is surrounded on all sides by desert. Similar forests are found on the mainland to the north and south, but here in the mountains, altitude replaces latitude, lifting the forest up into the cool thin air while the wall of high peaks squeezes moisture out of scarce rain clouds. While summer temperatures in the desert below frequently climb above one hundred degrees, here in the mountains they rarely climb above eighty. And while rainfall in the desert is rarely more than five inches per year, here in the mountains it is regularly more than thirty.

But although the desert isolates this mountain forest, it is also subtly linked to it. Biologists believe that the cacti and thorn scrub evolved in part out of ancient forests much like this one on top of the Sierra de la Laguna. The desert and this mountaintop forest are separated not just by several thousand feet of elevation but by several thousand years of evolution. Over time, hard outer bark became a green waxy skin. Leaves became a protective mantle of thorns. When the desert below is in full bloom, that link is easy to see. When the rains return, the desert becomes as green as this forest of pine and oak on top of the mountains.

• • •

I first came to Baja California in the winter of 1989, driving down the peninsula solo on the edge of a cold front that swept down from Alaska and brought freezing temperatures as far south as Los Angeles. I had spent the dispersed equivalent of several years working and traveling in the deserts of North America—the Mojave, the Great Basin, the Colorado Plateau, even a few months around the edges of the Sahara desert in Africa. Somehow, however, I had never made it to the deserts of Baja California. Nothing prepared me for how green it would be.

It took me more than a week to make it down the peninsula. The diversity of shapes and sizes of plants left me numb. The winter rains had been good and the desert was in full bloom. It was as if I had stumbled onto a new world. I would park my truck by the edge of the highway and wander off through the desert with a collection of guide-books, notebooks, and cameras, trying to identify different plants and flowers to get a feel for this new landscape.

In other North American deserts like the Mojave and the Great Basin, you can walk for miles and see no plants higher than your knee. The land is open and dominated by solitary, well-spaced plants like sagebrush and snakeweed. Baja California is different, part of what biologists refer to as the Sonoran Desert. Plants here come in all shapes and sizes. Cacti are the first to catch the eye: tall, saguaro-like cárdons, which rise forty to fifty feet into the air, and others like the pin-cushion and hedgehog cacti, which rise only a few inches above the desert floor. In between are scores of others with fantastic names and shapes: organ-pipe, beavertail, prickly pear, teddy bear cholla, jumping cholla, candelabra, old man, creeping devil.

There are other striking plants as well: tall, whiplike stands of ocotillo topped with bright red flowers, the waxy green bayonet-like leaves of magueys and aloes. In places, tough desert trees cover the dry ground with a canopy of leaves: green-trunked palo verdes, thin palo blancos with straight white trunks, twisted elephant trees covered with peeling bark, or the whispy branches of an accacia or smoke tree.

In the spring this collection of plants comes into full bloom. The trees are full and green with thousands of tiny leaves that set off the bright magenta, yellow, and white flowers of the cacti. Bare ground gives rise to a garden of flowers: fields of mallow, lupine, brittlebush, and sand verbena that run between the trees and cacti.

Two basic reasons lie behind this desert's richness and variety, both of which have to do with location. The Sonoran Desert is located farther south than any other North American desert. It stretches from Sonora and Baja California northward into southern Arizona. In general, tropical and subtropical regions have a greater diversity of species than those in temperate and polar regions. Here near the Sierra de la Laguna it sits almost squarely on top of the Tropic of Cancer. At the same time the Sonoran Desert's more southerly location also puts it within a different weather system.

While rainfall in northern deserts like the Mojave and Great Basin is generally limited to winter months, rainfall in the Sonoran is almost evenly split between winter and summer. Even though the land is exceptionally dry (less than five inches of rain falls per year in much of the peninsula—half as much as commonly falls in the cooler deserts to the north), the desert can bloom twice a year, supporting both winter and summer annuals.

••••

For all the desert's richness and diversity, however, survival still means adjusting to a harsh environment. Some areas can go for years without rain. Away from the cooling effects of ocean breezes, daytime temperatures often climb well above one hundred degrees.

Survival means adapting to a limited supply of water. For annuals, like flowers, the solution is simple. When the desert becomes too hot, they die off, leaving seeds behind that will sprout when the rains return. Perennials like the cacti and trees, however, have no such option. Instead, they have developed a variety of strategies for coping with limited rainfall.

Plants like the creosote bush are drought-resistant, with thick waxy leaves that help reduce water loss. Other treelike plants, such as the torote and the palo verde, are more opportunistic, putting out a fresh crop of leaves whenever the rain arrives. They respond to heat the same way deciduous trees in a northern forest respond to cold weather and the onset of winter: they drop their leaves to wait out the long dry spells between rains.

Cacti are perhaps the most ingenious of all desert plants. While other plants hibernate or die off when the rains end, cacti are water savers, capable of soaking up enough water to make it through the long dry spells between rains. Their waxy outer skin encloses a fleshy pith where food and water are stored. A forty-foot-tall cárdon may hold as much as a ton of water.

Such a storehouse of water is sure to attract predators. In cacti, leaves have evolved into a protective mantle of thorns that wards off predators and casts a small halo of shade over the plant. With what were once leaves put to other uses, photosynthesis is carried out by chlorophyll in their green outer skin. The cacti's waxy surface helps reduce water loss.

Like all plants, cacti depend on photosynthesis, the transformation of carbon dioxide and water into sugars in the presence of sunlight. Most plants take in the carbon dioxide necessary for this process by opening their pores, or stomates, during the day when sunlight is available, but cacti go through the process piecemeal, reducing water loss due to evaporation by opening their stomates only at night. The carbon dioxide they take in at night is stored as malic acid and then used during the day to complete the complex chemical reaction of photosynthesis. This specialized plant chemistry is called crassulacean acid metabolism, or CAM. It is not limited to cacti but is also used by yuccas and agaves.

••• •

Newcomers to the desert are often overwhelmed by its sparseness. It seems barren and hostile. They turn away angry and not infrequently bored. So much land with so little on it. Once the subtlety of the landscape is pointed out, however, they become more interested, looking for the flowers that explode over the landscape after the first hint of rain or admiring the cacti and their hidden storehouses of water.

The economy and resourcefulness of the desert is often mistaken for a sign of wizened old age. Yet while it seems old, in reality the Sonoran Desert is very young. Deserts have existed for several million years, but biologists believe that the Sonoran may be no more than ten thousand years old, one of the youngest ecosystems on earth.

In the Pleistocene, 1.8 million years ago, when the climate was wetter and cooler, tropical pine–oak forests similar to those on top of the Sierra de la Laguna covered much of Baja California and may have even extended as far north as Oregon and Washington. As Pleistocene glaciation drew to a close, the climate became drier. Pines, madrones, and live oaks survived in the high mountains, but in the lowlands these trees slowly evolved into desert trees and cacti. Plants from the cooler and older deserts to the north moved in as well to complete the creation of this rich and sometimes green desert.

Walking up to the top of the Sierra de la Laguna is a walk through evolutionary history. In the forested tops of the mountains you can see the origins of the desert below. The desert is not old, but remarkably young, the product of more than 3.5 billion years of evolution, filled with plants and animals that have only recently learned to live without water. Throughout time evolution has developed increasingly sophisticated forms of life capable of living farther and farther away from water. To see the beginnings of life on earth, you should not look in the desert or the mountain forest, but in the sea, where life began.

• • •

Both the beach and the desert were quiet that morning. A light wind was blowing, sending ripples through the water offshore—the only sign of movement or life. It was two hours past sunrise, and the sun, while not quite high in the sky, illuminated the desert with a bright, white light that set off rocks and cacti along the shore in sharp relief.

Desert hills, still green from the heavy rains brought by a late summer hurricane, defined the western horizon. Behind them were the high, forested peaks of the Sierra de la Laguna.

Here on the beach, however, the eye was not drawn to the hills or the higher peaks behind them but to the blue, open water of the Gulf of California. There was a broad open curve to the shoreline. It led northward past the fishing village of Cabo Pulmo to the high, dome-shaped headland of Cabo Pulmo itself, for which the area is named.

Bars of coral in the water offshore spread outward from the head-land like the ribs of a Japanese fan, running along the coast for the better part of three miles. You could see the reef in the subtle changes of color and hue that marked the water's surface: bands of purple amid the solid blue of the gulf. It was the reef that had brought us here.

The gulf is only five million years old, but its waters are rich, filled with more than five hundred species of fish and more than a thousand different species of invertebrates—animals like snails, clams, crabs, and starfish. Nowhere else in the gulf is that life as concentrated as it is at Cabo Pulmo Reef. While the rifting that opened the Gulf of California offers a glimpse of the beginning of oceans, the reef offers a glimpse of the beginning of life. Its rich flora and fauna is a living record of evolution, the product of more than six hundred million years of change and development.

It was late fall and I was preparing for a dive on Cabo Pulmo Reef with Hector Reyes, a student at the University of Baja California Sur. It was my fourth trip to the reef and roughly Hector's fortieth. For the past four years he had been studying the reef as part of his thesis. In a few weeks he would be receiving his *licenciado*, the equivalent of a bachelor's degree in the United States, although it requires almost as much research as a master's degree. After graduation he would be going to a biological research station near Cancún to study corals in the Caribbean. The day's dive was sort of a last farewell.

"I don't know if I will ever get back here," he told me as we slipped on our tanks. "I grew up in Mexico City and at the time I didn't even know there was a reef in Baja California. I first came here during my second semester with one of my ecology classes and I fell in love with the place. Pretty soon, I started coming to the reef every chance I could get. People say that they know the reef, that Cabo Pulmo is very small. But I swear that in all the times I've been here, I've never been to the same place twice. The reef is always different."

• • •

Entering the water was like entering another world. The still quiet of the desert was replaced by a strident chorus of snapping and popping sounds from literally thousands of fish. Sound travels roughly four times faster in water than in air—so fast that human hearing, split stereophonically between left and right ears, cannot tell where it is coming from. Instead, it seems to come from everywhere at once, filling your head with a loud collection of sounds: the rasp of fish teeth scraping algae off of a piece of rock or coral, the popping sound of a claw grasping for food or warding off a predator, the clicks of a dolphin echolocating offshore, the soft rolling sound of the surf breaking over a high rock. You could hear for miles.

We snorkeled out from the beach with our regulators trailing behind us. The near shore was sandy and scattered with rocks. Small schools of perchlike sergeant majors with thick, horizontal black and yellow stripes flitted from rock to rock. In the pockets of sand between rocks, spine-covered balloon fish hovered like dirigibles, their pectoral fins whirring like tiny propellers.

A few hundred yards from shore we reached deeper water and the innermost bars of the reef. Pausing for one final check at the surface, we put our regulators in and dropped quietly to the bottom, fifteen feet below. Underneath, the water was clear and seemed to glow with a soft, blue light. Ripples on the surface broke the sunlight into beams that spun through the water below, lighting up a world of coral and fish. Visibility was more than one hundred feet. Looking up, I could see our bubbles of air rising to the surface like tiny beads of quicksilver. In front of us, the corals were everywhere—a rich carpet that seemed to stretch outward in all directions.

The first few minutes of a dive are always the same: the eyes are drawn to large shapes and things that move. Thirty feet ahead, a school of yellow surgeonfish moved through the water in a vertical plane. Moon-shaped fish with speckles of black on their silver sides, their wedge-shaped yellow tails caught the light as they swam by—first right, then left, dancing in unison to some hidden choreography. Overhead, a school of several hundred scissortail damselfish swam through the water, casting a speckled shadow over the bottom.

With time, however, the eyes begin to focus, and other facets of the landscape emerge. Rounded heads of coral, *Pocillopora elegans*, seemed to grow right out of the rock like small, lush trees, their branches covered with a layer of brown-green fuzz. Sea fans and gorgonians spread delicate, fan-shaped lattices of red and yellow through the water. Fernlike feathery hydroids weaved and danced in the currents. Anemones coated the surfaces of rocks with flowerlike blooms of red, gold, green, and brown.

Appearances, however, are deceiving. What look like plants are really animals. The corals, gorgonians, hydroids, and anemones of the reef are not trees or flowers but animals known as invertebrates—animals whose scientific classification suggests what they most prominently lack: a backbone, or spinal column. What seem to be petals or branches are actually a collection of filters and tentacles used to gather and filter food from the water. The corals, hydroids, and anemones of the reef feed on microscopic plants and animals floating in the water.

These immobile animals in turn are food for a host of predators. Others use the structure and shelter they provide for their own purpose—everything from a place to hide to a point of reference in open water. Every surface of the reef seems covered by layer upon layer of life.

•   •   •

Stopping over a sand pool surrounded by corals, Hector slowly turned in a circle, pointing out solitary cortez damselfish positioned like sentries over individual heads of coral. Fiercely territorial, each one defends a small patch of reef, warding off fish several times their own size. Nearby, a solitary cortez angelfish, neon blue with bright orange piping, hovered near a sea fan.

Swimming up to a crevice in the rock base of the reef, we were surrounded by dozens of curious fish. Inside was a cluster of sea urchins with long, stiff, iridescent purple spines. Nearby, another held an octopus curled into a tight knot. Barnacles, clams, and limpets covered the exposed surfaces of rocks, their shells flecked with green bits of algae and the tiny pink plates of coralline algae.

Even the coral heads themselves were filled with life. Snails living inside them had strung their branches with feeding webs that looked almost like spiders' webs. Breaking off a loose branch of coral, Hector pointed out the velvety form of a red sponge, growing on its inner reaches like moss. The branch itself was a cluster of life. Holding it a few inches from my mask, the layer of brown-green fuzz that covered it became a carpet of tiny flowers. When I touched them with my fingers they folded up in a flash. What appeared to be flowers were actually individual corals, tiny animals known as polyps. What look like petals are actually tentacles tipped with nematocysts, stinging cells used to gather prey. Symbiotic algae known as zooxanthellae living inside the polyps gives them their green or brown color.

The large coral heads of the reef, which are so noticeable, are actually a kind of communal skeleton created by thousands of tiny coral polyps. As I turned the branch of coral over in my hand, a group of rainbow wrasses with bright red, yellow, and black stripes circled around it, taking quick tentative bites.

•   •   •

The density and diversity of life on the reef is almost overwhelming. In the space of a few feet there are representatives of every living phylum in the animal kingdom. Few places on earth can match the concentration and variety of life found here, and those that do usually involve water: a desert spring, a freshwater lake filled with migratory waterfowl, or a tropical rain forest. Of these three, the rain forest comes closest to matching the reef. Like the rain forest, the reef manages to thrive in a nutrient-poor environment.

As the thousands of acres of rain forest cleared each year to make way for farming and cattle grazing sadly prove over and over again, tropical soils are actually nutrient-poor. When the trees are stripped away, heavy rains quickly wash the soil away, reducing land that once supported a lush forest to barren, open ground. The rain forest's phenomenal productivity does not lie in the soil but inside the forest itself, where nutrients, food, and energy are locked up and held in place by its intricate layers of plants and animals. The biological structure of the coral reef in some ways is almost identical.

Tropical waters are typically nutrient poor. Bright year-round sunlight fosters so much productivity in the upper layers of the ocean that nutrients are quickly used up, making it almost sterile. At the same time, the sun also heats the surface water, creating a thermal cap that keeps the colder, nutrient-rich water below from rising to the surface.

Like the rain forest, the reef thrives by locking nutrients up inside itself. The secret of its richness and diversity is not in the surrounding water but inside its successive layers of plants and animals. It is not merely a collection of individuals, but a collection of individuals that behaves like a single organism. The reef feeds and grows off of itself— a magical source of life.

•••

It is no accident that the reef contains such a complete picture of life. Life on earth began in the sea, evolving from simple cells into plants and animals as complex as cacti and whales. While that progression is remarkable, it is no less remarkable that some forms of life have remained almost unchanged for hundreds of millions of years. At Cabo Pulmo you can see the past in the form and structure of the plants and animals of the reef.

The oldest known fossils are single cells of blue-green algae, or cyanobacteria, that are believed to be some three and a half billion years old. For nearly two billion years they were virtually the only form of life on earth. They grew in large mats on the seafloor known as stromatolites.

Forms of blue-green algae still survive and can even be found at places like Cabo Pulmo Reef. Today, however, it no longer dominates the ocean or the reef but is forced to compete with more modern forms of algae and is food for a host of tiny animals and fish. Its structure is so primitive that it has neither nucleus nor chromosomes. But for all its primitiveness, it still occupies a niche in the marine world—a living record of the past.

As for how these simple cells evolved into more complex forms of life, the same competitive life of the modern reef and ocean that holds them in check offers some insight. Based on the study of primitive living cells, biologists have come to believe that the first structured cells came about when larger cells consumed smaller ones and failed to completely digest them. Over time these absorbed cells took on specialized functions as centers of cell respiration (mitochondria), photosynthesis (chloroplasts), and reproduction (nuclei). These structured cells were the building blocks for more advanced forms of life. Once they appeared, changes came quickly.

The fossil record suggests that the first cells with nuclei and other advanced internal features appeared about 1.4 billion years ago. By eight hundred million years ago the first multicellular animals appeared. For cyanobacteria it was the beginning of the end as these primitive animals began to graze and feed on them, opening the way for new forms of life. Like seedlings that sprout after a forest fire, new plants and animals appeared almost instantaneously.

Six hundred million years ago representatives of every living phyla of marine invertebrates appeared in the space of only a few million years. Their arrival marked the beginning of a division of geologic time known as the Paleozoic era—"the age of ancient life." Their appearance was not only sudden but worldwide, marked by fossils that were large and readily observable.

Ever since Charles Darwin published his theories of evolution, this sudden appearance of invertebrate life has been problematic. Darwin himself went so far as to say that this explosion of life in the Cambrian refuted his theories of evolution. Until recently, however, the cellular forms of life that preceded this sudden surge of invertebrates were unknown. Armed with this knowledge, modern scientists are able to see this so-called explosion of life as part of an ongoing process.

While traditional evolutionary theory placed its faith in slow and steady change, modern evolutionary theorists like Harvard's Stephen J. Gould have proposed that evolution is actually punctuated: periods of slow and steady change are broken by bursts of radical creativity and innovation, like the sudden appearance of marine invertebrates that filled a gap left by the decline of blue-green algae.

Since this almost instantaneous arrival, invertebrate evolution has been characterized largely by slow and steady change. The fossils of ancient clams and mollusks so strongly resemble those of modern animals that their relationship is obvious to even an untrained observer. At the same time, other invertebrates, like starfish and sea urchins, have changed so much from the forms of their ancient relatives like the crinoid (sea lily) that, externally at least, they are almost unrecognizable. Internally, however, the details of structure and symmetry reveal their links to the past.

This is an important evolutionary point to understand: the development of radically new forms of life is relatively rare. Most of the evolutionary changes center around minor modifications to a limited set of designs—the established kingdoms and phyla of life. In the form and structure of the modern plants and animals that make up the reef, you can see the past.

• • •

Not only do ancient plants and animals resemble those of the modern reef, they also helped build reefs. Instead of corals, however, these ancient reefs were dominated by primitive sponges and other animals. Over time the composition of these ancient reefs changed as different types of plants and animals rose and fell from prominence in the ocean. The plants and animals that make up reefs today were put together piece by piece as time went on. In the rich life of places like Cabo Pulmo Reef you can see a living record of the progression of life itself.

Five hundred million years ago primitive spongelike animals known as archaeosyathids were building reefs on the seafloor. They were small, not much more than low mounds of life. Primitive corals–horn-shaped rugoses and flat-bladed tabulates—were present but played an insignificant role in reef building.

Three hundred fifty million years ago, about the time that the first plants were appearing on land, primitive corals and spongelike animals known as stromatoporoids were already building large reefs on the seafloor. In places, they were as much as thirty-five feet high and two miles long with atolls and central lagoons similar to large reefs found today in the South Pacific. Primitive fish were also present by this time, but still relatively rare. They had scales that barely overlapped one another and asymmetrical tails that resembled those of modern sharks. Some were armored with bony plates. Others had no jaws.

This era of "ancient life" drew to a close 245 million years ago with a massive extinction that wiped out 90 percent of the existing species of marine invertebrates. Like the arrival of cellular predators, this extinction cleared the way for the development of new forms of life. The period that followed is known as the Mesozoic, the period of "middle life."

By two hundred million years ago the first modern reef-building corals appeared. As the Atlantic began to open they developed a symbiotic relationship with algae.

The Mesozoic is also known as the age of reptiles. On land, life was dominated by dinosaurs—literally, "terrible lizards"—like the brontosaurus and stegosaurus. At the same time, the sea had its own terrible lizards—fantastic animals with elongated necks and flippers and equally fantastic names: placodont, nothosaur, ichthyosaur, plesiosaur.

While these sea monsters were roaming the ocean, the first modern fish also appeared there. Known as teleosts, they had overlapping scales and symmetrical tails like the fish that weave over Cabo Pulmo Reef today.

By seventy-five million years ago, about the same time that flowering plants appeared on land, another type of bloom appeared in the ocean in the form of tiny floating plants—the diatoms, calcareous nannoplankton, and other modern forms of plankton that are the base of the marine food chain today.

The Mesozoic closed the same way the Paleozoic did: with a massive extinction. This one, possibly caused by the impact of a large meteorite, killed off the dinosaurs and cleared the way for new forms of life on land. In the sea, however, most plants and animals survived intact. Today their descendents fill the modern ocean.

The new era of geologic time runs right up to the present and is known as the Cenozoic, the era of "modern life." It is also known as the age of mammals, but although the rise of mammals would change the complexion of life on land, life in the sea already had a surprisingly modern look by the start of Cenozoic time.

By twenty-five million years ago when the East Pacific Rise collided with North America, the first baleen whales appeared. Ancestors of the gray whales that gather in the lagoons of Baja California today, they evolved from land animals that fed in the shallow water near shore. Other facets of the marine world were already set: 90 percent of the species of clams and snails alive in those ancient seas are still alive today.

By five million years ago when the Gulf of California began to open, North America was not yet joined to South America and the Pacific was linked to the Atlantic via an open seaway. Fish from the Atlantic and Caribbean worked their way up the coast and into the emerging Gulf of California. Today the sea bass, grunts, snappers, damselfish, and gobies that swirl over Cabo Pulmo Reef are near-perfect images of their relatives in the Caribbean—distinguished only by minor changes in behavior and form like twins separated at birth— what biologists call geminate pairs or species pairs.

By twenty-five thousand years ago, about the same time that early humans began to migrate across the Bering land bridge, corals were growing here at Cabo Pulmo, building reefs in the shallow water near shore. It had taken them a long time to get here, more than three billion years of evolution to develop their intricate forms, but you could read nothing of that early history in the more solid rocks of the desert. Their memories were too short; their history, too young. But when the modern reef arrived, it carried with it a living record of the earliest forms of life on earth. In the rich flora and fauna of the reef you can see the past.

• ▪ •

I am quite sick indeed of this place, every day seems a long year, and every one with the same monotonous desolation around me, not affording the least pleasure, variety or enjoyment of any kind. I have no reading matter what ever. . . . I am now of Gods grace nearly two years perched on this sandbeach, a laughingstock probably of the Pelicans & Turkey buzzards the only signs of life around me. To the E & SE the terminally smoky Gulf, to all other points of the compass the sandy desert, covered with white salina and ornamented with cactuses in every form, sticking out like candlesticks on a white cloth. . . . There is not a blade of grass in the country, & not a green leaf. Sand, salt, trunks of shrubs, Rock & the like everywhere and covered everywhere with bleached bones of cattle mules horses etc, died by the thousands lately of starvation.

—NATURALIST JOHN XANTUS
LETTER TO SPENCER BAIRD
DIRECTOR OF THE SMITHSONIAN
FROM BAJA CALIFORNIA, 1860

The fig tree grew right out of the canyon wall. Its tangle of white roots seemed to spring right out of the rock. Underneath its widespreading branches was a circle of shade, a shelter from the heat of the late afternoon. It was a good spot. We were not alone in this assessment. Higher up on the canyon wall behind the fig tree was a group of small stick figures, petroglyphs left by the Pericu Indians, an all-but-vanished tribe that came here several hundred years before the first Europeans arrived.

While the Aztec and Maya built great cities and temples, the Pericu lived from hand to mouth. When the desert was ripe, they gorged themselves on the pear-shaped fruit of the organ-pipe cacti, *pitaya dulce,* the sweet cacti. In the long dry spells between rains they virtually starved to death. They made no pots and had few tools or, for that matter, clothes. They left little behind them except these scattered paintings and writings on the desert rocks. In places, the canyon walls are covered with vast panels of figures several hundred feet in length with images of whales and manta rays. On the face of it, life for the Pericu was bleak. But these pictures and rock drawings suggest something more, as if their primitive struggle for survival did not preclude a sense of wonder and appreciation for the stark world around them.

After a morning spent swimming in the gulf, John Hans Bickel and I were following a small canyon back from the coast up into the mountains. In the 1940s and 1950s, Bickel was a photographer and reporter for the *Saturday Evening Post,* with assignments that took him around the world. In 1950 he did the magazine's first article on Baja California. Twenty years later, at an age when most people opt for a soft and retired life in Florida or southern Arizona, Bickel moved to Baja California to an empty stretch of ground on the coast not far from Cabo Pulmo Reef and began building his house by hand.

Now in his early seventies, he could pass for someone half his age—a tribute to the clean desert air and his daily half-mile swims in the open waters of the gulf.

Ostensibly we were looking for flat stones to use in a new patio on the side of Bickel's house. In reality, however, that was a flexible goal, one that gave us almost endless opportunities to enjoy the desert rocks and cacti in the sharp late afternoon light. As we walked up the canyon, my own appreciation of the scene was tempered by an awareness that this desert peninsula has broken many. The letters of clerics and early explorers who came here are filled with despair: like those of men and women sent off in exile. They could never come to terms with the desert. It was not just that the struggle for survival was keen, but that their perceptions of the land were shaped by the expectations of their time and culture. The desert was not what it seemed.

•••

The word *peninsula* is derived from the Latin *paene insula:* "almost an island." For the better part of two centuries the Spanish clerics and soldiers who came here believed that Baja California was an island. Many assumed they had found the mythical island of California described by García Oredóñez de Montalva in his novel *Las Sergas de Eplandian.* Popular with sailors at the time, the book told of an island called California located "very near the terrestrial paradise" bordered by steep cliffs and inhabited by a race of beautiful Amazons who carried weapons of pure gold.

From the beginning, however, things went badly. In 1532 the remains of a mutinous crew of Spanish sailors sent out by Hernando Cortés to follow the coastline northward from central Mexico sailed into a broad bay on the Gulf of California, which they named *La Paz* ("the Peace") for the seemingly peaceful natives they encountered there. At first they were content to trade with the Indians for their rich supply of pearls gathered from oysters in the water offshore. Later their interests ran to women as well. The women, however, were not Amazons but mothers, sisters, and wives. Enraged, the Indians killed twenty of the sailors, including their leader, Ordoño Jiménez. The remainder fled back to Mexico. Their stories of a sea filled with pearls encouraged others.

Three years later Cortés sent another expedition north to establish a colony on the same site. By the time he returned with supplies from the mainland twelve months later, twenty-three of his men had starved to death. These brutal, early failures set the tone for the successive waves of would-be colonists and missionaries who followed. Nothing seemed to take root here. Jesuits, Franciscans, and Dominicans came and went, leaving behind a string of mission churches, but not much more. It rained too little or too much. Fields and gardens dried up and blew away, while flash floods washed away missions and churches. The natives themselves were a source of unending problems. Unwilling or at best indifferent converts, they could match the Spaniards in brutality when it suited them, murdering scores of priests and soldiers in bloody massacres. Dreams of an earthly paradise vanished into thin air like a desert mirage.

Neither the desert fathers nor their soldier garrisons left any more mark on the land than did the Indians who had arrived here some seven thousand years before them. Although the Indians were looked down upon as crude and bestial, they had managed to survive here for several thousand years while the Spaniards—with their imagined advantages of culture and grace—merely starved to death. In the end it was not their superior culture or weapons that defeated the Indians, but their superior diseases: smallpox, measles, and syphilis.

The only newcomers who seemed to thrive here were English pirates like Thomas Cavendish and Woodes Rogers who used the rocky, mountainous tip of the peninsula near Cabo San Lucas as a lookout to watch for treasure-laden Spanish galleons wallowing across the Pacific from Manila to Acapulco. For the others who were drawn here, there was enough wealth to make life promising but never enough to make it truly possible. The sea was rich, but the land was poor. Pearls from Cabo Pulmo Reef and Espíritu Santo Island near La Paz graced the crowns of Spain and England, but there was scarcely enough water on the land to support anything more than a few scattered small towns and a network of isolated ranchos. In places five and six years passed between rains. In the entire eight-hundred-mile length of the peninsula there were just two year-round streams. So much water evaporated from the surface of the gulf that it was 30 percent more saline than the ocean.

By the late 1700s, navigators had discovered that Baja California was not an island but a peninsula. A thousand miles north of La Paz a new settlement called Monterey was established on the shores of Monterey Bay in Alta, or Upper, California. The land there was greener and more promising, and Spanish interests in the region began shifting slowly northward. One by one the missions closed up. The departing fathers cheered like schoolboys let out unexpectedly for recess. The peninsula and its vast deserts sank into roadless obscurity.

• • •

Two hundred years later, the desert has not changed much. The rough path that led northward through a string of missions to Alta California is now a paved two-lane highway, the Carretera Transpeninsular, and towns around the peninsula's tip like Cabo San Lucas and San Jose are now bordered by a thin veneer of tourist hotels. But turn off the highway or travel past the edge of town and you are in the midst of a desert that seems timeless and untouched.

Although the physical landscape has not changed much, the psychological landscape has changed immensely. What was once a wasteland is now a landscape to be savored and contemplated. Places where missionaries and colonists struggled mentally and physically to survive now attract tourists and travelers.

Visitors from around the world come to fish and dive in the rich waters of the gulf. Others travel to remote desert lagoons in the spring to watch gray whales. Local families flock to remote beaches during *Semana Santa*, Easter Week, for a few days of camping and visiting. Students from Mexico and the United States come to study marine life in the gulf or travel through the desert studying rocks and plants. Others come to see the desert in bloom or travel down back roads looking for pictographs and pictures on the rocks or the fossils of whales and turtles. We see the world through different eyes.

While the first missionaries and settlers sought domination over the landscape, we seek communion with it. In an industrial society that increasingly isolates us from the natural around us, we strive to find our place in things, to get in touch with our roots. Unlike the settled world of routines and regulations we inhabit, the desert is free and open. You can take time to admire the play of light on the thorns of a cactus or sit under the shade of a fig tree studying the pictographs that decorate a rock wall. There is no one here to second-guess your thoughts or ask you if your time was well spent.

• • •

When the sun fell below the canyon rim we started walking back. Currents of cool air began to flow down the canyon, carrying with them the smell of water and desert flowers. The day's heat was no longer overhead but seemed to rise up from the stones underfoot. As we walked, Hans pointed out other sites: the small caves higher up on the canyon walls where he gathers honey, the well that supplies his home with water—two hundred feet deep and dug entirely by hand. He seems to have taken root here. Living apart like this is something many contemplate but seldom choose; the force of routine and habit is too strong. On the way back to the canyon mouth we talked about the appeal of living here, back and forth, turning the idea over in our heads like a stone held in the palm of the hand.

"I spent most of my life in the Pacific Northwest where it is very green and lush," Bickel told me, stopping now and then to catch my eye. "I loved it at the time, but now I really prefer it here in the desert. Things are much sharper here.

"When I first came here, there was nothing here. I remember driving up to where my house is now and getting out of my car and thinking, 'Well. . . here I am. . . .' There was almost no one to talk to. For the first eight months, I thought I would go crazy. But it has been the chance to watch things, watching the seasons come and go that has kept me here.

"You've heard about happenings? You know the idea that some people have that there are places like a particular mountain or a natural event like the summer solstice that have a certain power? Well, I don't really believe in happenings, but I have had a few things that I would call *events* since I have been living here.

"One time I was sitting at a table outside my house reading a book when a hummingbird suddenly flew by and stopped a few feet in front of my face. He just stood their hovering for about twenty minutes, moving from side to side and flying around my head. I wasn't wearing red or any bright clothing, so he wasn't attracted to me for that reason. I had the uneasy feeling that this hummingbird wanted to communicate with me, that he was trying to tell me something. I've never forgotten that.

"Another time I was sitting inside with all the windows open writing a letter to my daughters and suddenly the whole room was filled with butterflies. Thousands of them! You should have seen it: yellow, blue, and white ones, moving in a beautiful sinuous path. It was like an open invitation, as if they were inviting you to dance with them. I wanted to fly off with them. I didn't want to be a butterfly. I just wanted to travel with them. I've never seen anything like it before or since. After that happened I began to realize that I could make it here. It's been that kind of thing that has kept me here. Nature has been my saving grace."

• • •

Land's end. Cabo San Lucas. The southern tip of Baja California. This is the way the continent ends: a narrow fin of granite sticking out between two oceans, a framing arch, a few small islands of rock, and then nothing but ocean all the way to Antarctica.

It is a stark and simple landscape, nothing but rock and water, the building blocks of continents and oceans. From where I was sitting halfway up the granite cliffs I could not even see so much as a solitary cárdon or ocotillo. Spires of white granite rose right up from the water's edge into the bright desert sky. Waves and sea spray had carved them into rounded, surrealistic shapes and fluted columns shot through with holes. High waves from the Pacific bent around the point and pounded into the rocks, sending clouds of spray twenty and thirty feet into the air. With my back against the rock I could feel each one break.

It was late in the day and in the west over the Pacific the sun was already low on the horizon. The clear sky above it was a pale, golden yellow. In the east over the gulf was a bank of purple and gray fog several miles offshore. It marked the point where the warm water of the gulf met the cold water of the Pacific.

Somewhere beyond it on the floor of the sea was the East Pacific Rise, the undersea rift zone that pushes Baja California and the Mexican mainland apart. Its spreading lavas, however, are responsible for more than just the opening of the gulf. Spreading, both inside the gulf and elsewhere along the East Pacific Rise, is the engine that drives the Pacific Plate northward. The forces at work on the rift zone that lies on the floor of the gulf are shaping the coastline all the way to the Aleutian Islands on the edge of the Bering Sea. They cause large earthquakes in California. Farther north they build high mountains along the coast of Alaska. Motion is what ties this edge of the continent together. With time this narrow desert peninsula will reach Alaska.

At sunset the water began to take on the color of sky: first orange and then red and pink. The tide was rising but in the water near shore a pair of dark mounds rose up out of the sea. At first they looked almost like sandbars, but then dove back toward the bottom only to rise again—a pair of gray whales, a mother and calf, heading north to their feeding grounds in the Arctic, hugging the coast to avoid the killer whales lurking in the deeper water offshore.

Like the mobile Pacific Plate, these migrating whales tie the two ends of the continent together. Each year they travel back and forth between their breeding grounds in Baja California and the rich waters of the Arctic.

Like us, they are warm-blooded mammals who nurse their young on milk. Fossil evidence suggests that they evolved from land animals that fed in the shallow water near shore. Inside their massive flippers are an array of bones whose shape and architecture resembles that of a human hand.

For the ancient tribes who lived here, whales were the guardians of the world. Mythical creatures come to life, they guarded the son of the creator who had sided with evil and been banished to a sea cave. Farther north in Alaska and British Columbia those same whales were a valuable source of food. A single whale could supply a village with literally a mountain of meat. Both the whale and its hunters were objects of mystery and power.

Our own culture's relationship to whales has been more varied. From the early 1800s until the turn of the century the gray whale was hunted to the verge of extinction. Its habit of traveling in coastal waters made it an easy target for whalers, and for a time they were even hunted in the shallow coastal lagoons of Baja California.

Today the same coastal habits that nearly brought about their extinction have proved to be their saving grace. Easily seen from shore and in coastal waters, their migrations are watched by literally tens of thousands of people. As a result they are perhaps the best-known and best-loved whales in the world—which has made it easier to ensure their preservation. Protected by international treaties, their population today is an estimated sixteen thousand and growing by as much as 2.5 percent per year. With their massive, fluid forms, they are as capable of exciting awe and wonder in our own world as they were in the past.

• • •

I stayed up on the rocks until well past sunset watching the moon rise over the gulf, finding my way down the rocks and across the beach in the moonlight. By the time I started driving back to La Paz, the moon was already high in the sky. I headed north on Highway 19, a straight, narrow road that skirts the edge of the Pacific on the broad coastal plain that lies on the flanks of the Sierra de la Laguna.

There was no traffic outside of town and the desert was lit by a soft blue light of a full moon. The black shapes of cacti rose up from the desert floor like bare trees. The land smelled rich and green in the cool night air.

Offshore in the Pacific the scalloped edges of waves glistened like pearls in the moonlight. On such nights their illusory wealth must have seemed almost real to the Spanish sailors and explorers who were drawn here by dreams of gold and treasure. I kept watching the water, waiting for the black shapes of whales rising to the surface in the moonlight, traveling north in the wake of the continent.

LOS ANGELES

A RIVER OF OIL

THE DESERT SEEMS TO GO ON FOREVER NORTH OF LA PAZ. IN PLACES IT IS more than one hundred miles between towns. You can drive for hours without passing a single car. This should not be taken to mean that it is boring. Desert plants are rich and varied: forests of cárdon and organ-pipe cacti, whiplike stands of flowering ocotillo, and low thorny thickets of mesquite and torote. The land itself is sharp and irregular, bordered by a blue, tropical sea. The high granite peaks that run down the center of the peninsula are edged by broad valleys and high desert mesas made up of gently folded and faulted rock that give the desert a look that is at once both open and mountainous. In the clear, dry air you can stand on flat ground and see hills and peaks sixty and seventy miles away. At night the stars extend right down to the horizon, casting a soft, blue light over the desert. On clear, moonless nights you can pick your way across the desert by starlight—following the sweet, almost narcotic smell of night-blooming cacti that travels for miles in the cool night air.

The changes come slowly as you head north. Towns and pueblos become more frequent. Desert plants give way to coastal scrub. By the time you reach El Rosario, two hundred miles below the U.S. border, the cárdon, organ-pipe, and prickly pear cacti of the desert have been replaced by live oak and chaparral. Instead of twisted desert thorn scrub, open hillsides are covered with buckwheat and cheat grass. The change in flora is a result of changing climate: in northern Baja California the tropical climate with its evenly proportioned winter and summer rains gives way to a Mediterranean one of long, dry summers punctuated by two or three months of light winter rains.

This change in plants and climate coincides with a change in geology as well. The terrain is still rugged and untouched, but the valleys are small and narrow, surrounded by bowls of steep-sided mountains. Flat ground is valuable in the nearly vertical landscape, and ranches and vineyards spread across valley floors in a lush, green carpet.

Like the hills and peaks farther south, the mountains here have been built up from rocks that were once part of the seafloor. Here, however, they are not gently folded and faulted but twisted and shattered. The difference in the style and shape of the landscape is due to nothing less than a change in the forces that are shaping this thin edge of the continent. In the northern end of the Gulf of California the East Pacific Rise is transformed into the San Andreas Fault. Sliding replaces spreading as the driving force shaping the edge of the continent. The Pacific Plate begins slipping past the stable North American Plate, pushing small pieces of the earth's crust northward.

Geologists are uncertain just how or where this transformation takes place. In the northern end of the gulf the spreading rise is buried by several thousand feet of silt and sediment deposited by the Colorado River, which empties into the head of the gulf. (Dig a Grand Canyon and you have to dump the sand and rock somewhere.) To the north, the San Andreas Fault appears, cutting across southern California in a path as clear and straightforward as the furrow left by a plow. Across the dry, inland valleys of southern California its path is marked by a trough ten to fifteen feet deep. As it heads down the backside of the San Bernardino Mountains toward San Francisco and the coast, it passes within sixty miles of downtown Los Angeles.

Above Ensenada the two-lane highway becomes a four-lane parkway skirting the edge of the ocean. In the high cliffs that border the coast you can see the details of the shattered piles of marine rock that make up the coast.

Tijuana. The border. A three-hour wait at Customs. The landscape changes again, but this time the changes are almost entirely manmade. Crossing the border is like entering another world. The differences are due not to biology or geology but to culture and economics. Nowhere in the world do two so very different countries share a common border: the Spanish and Indian world of Mexico and the Anglo and industrial world of the United States.

The broad details of the landscape, however, are still the same—a knotted fabric of high coastal mountains broken by small narrow valleys—but the appearance is entirely different. The valleys are carpeted not with vineyards and ranches but with shopping malls and subdivisions. Metropolitan Los Angeles is the largest urban area in the world:

not in terms of population, but size—a sprawling network of cities and suburbs spread over some six thousand square miles. It is home to some fifteen million people—one out of sixteen Americans.

In spite of its size, the overall impression is one of not space but motion. Eight- and ten-lane freeways loaded with traffic head off in all directions with cars bound for San Diego, San Bernardino, Los Angeles, Orange County, Riverside, and the San Fernando Valley. The air is not sharp and clear but filled with a soft, brown layer of smog that all but hides the pattern of high mountains and deep valleys. At night the stars are replaced by the landing lights of planes bound for LAX, Orange County, Burbank, and Ontario, while the freeways are filled with rivers of red and white light.

The foundation for all this motion is oil—not just in the engines of cars and planes but underground as well. Underneath Los Angeles's steep topography of basins and ranges is a series of deeper ones—not topographic basins but geologic ones—fifteen and twenty thousand feet deep and filled with oil-rich sediments. Los Angeles sits on top of some of the most productive oil fields in the world: a string of billion-barrel oil fields stretching from Ventura to Huntington Beach, a Prudhoe Bay with palm trees and orange groves. In Baja California the moving Pacific Plate created the Gulf of California; here in southern California the same moving plate has created deep basins filled with oil. Literally and figuratively, Los Angeles is a city that runs on oil.

••••

Ernie Severn leaned back in his chair and looked at the ceiling and then let out a sigh. "Go ahead. Put in the sprinkler."

"But he didn't even have one originally," said Leroy Dohrman. "We had to do some work on the water pipe that runs in front of his house and while the ground was torn up he decided he wanted us to install a lawn sprinkling system for him. He also wants us to pull out a palm tree near the sidewalk that's bothering him."

"Put in the sprinkler system," said Severn. "It's not worth the hassle. But don't touch the palm tree. The palm tree's not even on his property. It's on the city's. If you want to cut down a palm tree in the city you've got to get a permit. I don't want to get in trouble for that. That he can take care of if he wants to."

We were sitting in the control room of Chevron's Packard Drill Site on Pico Boulevard in central Los Angeles. Dohrman is the drill site foreman. Severn is the company's area supervisor. In theory their jobs are about oil—keeping the pumps and pipelines in this small part of Chevron's oil empire running safely and smoothly twenty-four hours per day. In practice they seem to spend as much time worrying about landscaping and general contracting as they do about oil. Running an oil field in Los Angeles is not quite the same as running an oil field in West Texas or Louisiana. Their drill site is surrounded not by dry desert plains or cypress-studded bayous but by storefront businesses and trim two- and three-bedroom homes with white stucco walls and red tile roofs. Working on the pipelines that carry oil from the field to the refinery or water back to the wells to be pumped into the ground means tearing up city streets, sidewalks, and lawns.

Today's problem was brought on by a leaking water pipe on a nearby side street that flooded one homeowner's lawn and left a six-inch-deep pool of water standing in the street. Over the years, Severn tells me, they have paved streets and driveways, rebuilt sidewalks and curbs, and resodded lawns and planted gardens. "If we tear it up, it's our responsibility," he explains. "We try to get things fixed up as quickly as we can. We want to keep our neighbors happy."

If keeping the neighbors happy also involves installing a lawn sprinkling system or paving a driveway, that is fine too. It is part of the cost of doing business in Los Angeles and business has been very good. Since the site went into production in 1964, Chevron has pumped more than fifty million barrels of oil from under this small piece of central Los Angeles.

From the outside the drill site looks like a ten-story office building, so much so that for the first few years after its construction the company routinely got telephone calls for information "about leasing some office space in that new building you've got on Pico Boulevard." It is not an office building, however, but an oil rig—a hollow, ten-story facade built over some eighty wells that pump oil from underneath some of the most expensive real estate in Los Angeles.

Slant-drilled wells radiate outward from the site like the spokes on a wheel. Canted as much as eighty degrees, they reach outward over a half-mile or more toward La Cienega Boulevard on the edge of Beverly Hills and the Central Wilshire district before plunging five or ten thousand feet straight down to the oil-rich rocks below.

Inside, field hands in hard hats tend to wells and pumps. Well casing and steel pipe lie stacked up in the yard. Lift pumps whine, pulling oil up from the ground, while others send it on its way through pipelines that lead to the company's processing plant at Inglewood ten miles away. Inside the hollow ten-story tower you can watch the well crews at work on a skid-mounted drilling rig that slides over the orderly rows of wells below. There is no roof to the building: you can look straight up and see the sky. Down in the production cellars you can walk through row after row of well heads painted a bright apple green, studded with gauges and dials and connected by a maze of pipes. There is a faint smell of oil—not the sharp penetrating smell of gasoline but the thick, sweet smell of crude oil. The air is hot and thick: the oil comes up out of the ground at 180 degrees.

The oil wells pump twenty-four hours per day, 365 days per year. Outside you will not hear the whine of pumps or smell any oil. There are no signs to tell you that this nondescript but well-maintained building is an oil field. Insofar as is humanly possible, it is all but invisible. Although it is socially acceptable to pump gasoline into your car in Los Angeles, it is not socially acceptable to pump oil out of the ground. The key to success in the oil business in southern California is to remain out of sight and out of mind.

• • •

According to the California Division of Oil and Gas, there are more than five thousand producing oil wells in Los Angeles County, making them only slightly less plentiful than gas stations. Driving or walking around town, however, you would be hard pressed to find more than a dozen. Oil is one of Los Angeles's best kept secrets.

There are oil wells in Beverly Hills and Century City hidden behind manicured hedges and stuccoed walls. There are drill sites in West Los Angeles and the Central Wilshire district disguised as office buildings, still others in Venice Beach and Palo Verdes. Long Beach Harbor contains more than twelve hundred oil wells—hidden just a few hundred yards from shore on artificial islands complete with waterfalls and palm trees. At night you can drive north on the Ventura Freeway toward Santa Barbara and see the oil platforms offshore lit up by work lights like Christmas trees.

Los Angeles is home to as many major oil companies as is Houston. They are here not for the climate or the beaches but for the oil. In the 1920s southern California's oil fields were the most productive in the world. While they have since been overshadowed by newer and larger reserves in the Middle East, the fields are far from dry. In 1988 more than forty million barrels of oil were pumped out of the ground under the cities and suburbs that make up Los Angeles by companies like Shell, Arco, Chevron, and Exxon and a host of smaller companies like the Just Barely Oil Company, Koch Oil, and Manley Oil.

There is so much oil under Los Angeles that it seeps out of the ground and collects in pools. On the sides of steep hills and cliffs you can see it trickling out of the ground in thick black drops. It rises up under sidewalks and streets, staining the concrete and pavement above with dark crude oil. Offshore at places like Coal Oil Point near Santa Barbara, it seeps out of the seafloor and rises to the surface, creating a near-permanent oil slick.

In the heart of the Central Wilshire district you can walk out the front door of the Los Angeles County Museum of Art and catch the rich, sweet, sulfurous smell of crude oil wafting through the air from the neighboring La Brea Tar Pits. Eighteen thousand years ago these pools of tar and oil were a trap for the saber-toothed tigers, dire wolves, and mastodons that roamed over what is now Los Angeles. Today their well-preserved bones fill a nearby museum. The largest pit has been so extensively excavated for its treasure trove of fossils that it has become a small lake. The changes, however, have not stopped the oil. From the top of the overlook in front of the George M. Page Museum of La Brea Discoveries you can watch it rise to the surface of the lake in thin streams accompanied by bubbles of methane gas—like a pot about ready to boil. Oil is persistent and unstoppable, directed by a singular purpose. Warm and fluid, lighter than the rocks that surround it, oil rises up through the ground under its own power. Oil is nondiscriminatory and equal opportunity. It is as fond of Chinatown as it is of Beverly Hills. It has no respect for property or position. It does what it wants. And what it wants in Los Angeles is to get to the surface.

• • •

Several hundred years before the first Europeans arrived, the coastal Indians who inhabited southern California were using tar and oil from seeps to caulk their boats and make waterproof baskets. In 1769 the Spanish explorer Gaspar de Portola traveled through what is now Los Angeles while on his way north to Monterey from Baja California. In his diary Portola described passing "swamps of bitumen and asphalt." Terrified by a series of earthquakes that rocked the area as they passed through, he and his men speculated that the tremors were somehow linked to the oil and tar that flowed out of the ground. His route roughly followed that of present-day Wilshire Boulevard and skirted the edge of the La Brea (*brea* is the Spanish word for "tar") Tar Pits.

Twelve years later when El Pueblo de la Reina de Los Angeles was founded, residents used the oil and tar from the tar pits and other local seeps to waterproof their roofs and burned it like coal in place of scarce firewood. On the broad plains west of town, the La Brea Tar Pits became a landmark for travelers.

One hundred years later Los Angeles had passed from Spanish to Mexican and finally American hands, but little had changed with respect to oil. City residents still collected tar and pitch from the seeps by the wagon load and burned it as fuel or used it to caulk boats and waterproof roofs.

In 1892 this relationship with oil changed completely when a nearly bankrupt gold miner from Wisconsin by the name of Edward L. Doheney realized that the tar and pitch being collected by Los Angeles residents was really oil that had congealed on contact with air. With a partner, Doheney purchased a lot in downtown Los Angeles and began looking for oil—not with a drilling rig but with a pick and shovel—digging a four-by-six-foot shaft by hand. At seven feet he struck oil, not enough to come gushing up out of the ground but enough to come seeping out of the sides of his hand-dug shaft like water filling a well. Encouraged, he kept right on digging. At 155 feet he was almost overcome by the fumes of oil and gas rising up through the shaft. Fashioning a crude drill out of the trunk of a sixty-foot eucalyptus tree, he pushed down deeper.

On the fortieth day of digging he hit a pocket of natural gas that nearly blew his homemade drilling rig apart. Then the oil started flowing: seven barrels per day, the first oil well in Los Angeles. Doheney, who had a hustler's instinct for self-promotion, proclaimed a new day in the city's economic life. Twenty years later, in 1920, he reenacted the scene for the benefit of cameras, standing in front of a towering wooden oil derrick in a straw boater and black suit with his arms held high over a worshipful crowd of onlookers like a tent show revivalist. The oil boom was on.

Doheney's first well was not on open ground but right in the middle of the city. Speculators rushed to the area and bought up city lots while homeowners sunk exploratory wells in their own yards. Five years after Doheney's initial discovery there were more than five hundred wells in a ten-block area bordered by Figueroa, First, Union, and Temple in downtown Los Angeles. Derricks and pumps vied for space with two- and three-bedroom bungalows and Queen Anne Victorians. By 1900 there were more than 140 oil companies operating within the city limits.

The oil boom created far more money and more millionaires than did the Gold Rush of 1849. Doheney quickly became one of the city's earliest and most eccentric millionaires, lending his name and money to everything from the Doheney Library at the University of Southern California to Doheney Drive in Beverly Hills. After a lifetime on the fringes of both society and the law (before discovering oil Doheney had worked his way west on a string of failures that included careers as a mule skinner, fruit packer, singing waiter, and part-time pimp) he built a private chapel at his mansion, Chester Place, and developed a taste for monocles and British tailoring. His wife purchased a title from the Vatican and took to calling herself Countess Estelle Doheney. Doheney would eventually go down in disgrace, playing a central role in the Teapot Dome Scandal, which brought down the Harding administration.

About the same time that Doheney was discovering oil in downtown Los Angeles, others near Santa Barbara were discovering that the oil and gas under southern California extended offshore as well. There was no technology for offshore oil drilling at the time (that would not be developed until after the Second World War), so they simply built piers out into the surf and began drilling, using the same techniques they had been using on land. By 1900 there were more than fifteen oil piers stretching into the Pacific at Summerland near Santa Barbara. Railroads running through the area promoted the oil wells, that crowded the coast as a tourist attraction.

By the 1920s the oil boom was in full swing. Demand was growing by leaps and bounds as the automobile became more popular. For the oil business in southern California the timing could not have been better. This growing demand for oil coincided with the discovery of even larger oil fields near Long Beach.

In March 1921 Shell Oil began drilling east of downtown Long Beach on top of a mound-shaped hill known as Signal Hill. Two months later the well reached a seventy-foot-deep layer of oil-rich sand. Good news travels quickly, and within days workers were erecting barricades around the drill site to keep sightseers away. On June 23, they hit oil at 3,114 feet—a gusher that blew to the surface with such force that it destroyed the oil rig and sent a fountain of oil 114 feet into the air. Two days later, when the well was finally brought under control at 4:00 A.M., there were five hundred people on hand.

In a matter of days everyone in Long Beach seemed to be getting into the oil business. "Population Oil-Crazed!" local headlines proclaimed. Well logs, normally the stuff of dry, technical reports, became front-page news. Books on oil and gas disappeared from the shelves of the Long Beach Public Library. In less than a year there were 850 oil wells spread out across the top of Signal Hill in a swath five miles long.

By 1923 Signal Hill was the largest oil field in the world. In the same year, California replaced Texas as the nation's leading oil producer and became the only state in the Union to produce more oil than it consumed. Part owner of the land that lay on top of the Signal Hill oil fields, the city of Long Beach raked in more than $7,000 per day in royalties. The sudden surge of income left city officials giddy: for a time they even debated the merits of leveling the downtown to make way for more oil wells. The phenomenal finds at Signal Hill were soon followed by others at Huntington Beach, Santa Fe Springs, and Seal Beach. A particularly euphoric piece in the Sunday magazine of the *Long Beach Press* seemed to catch the heady mood of the times in a single paragraph:

> Four million three hundred and forty thousand gallons of petroleum with every revolution around the sun! What does it

mean to Long Beach? First: great fuel supply. Second: it spells wealth—riches—luxurious homes—wondrously dressed women—great fortunes.

•   •   •

While the tastes and tenor of the times were different from our own, it would be misleading to suggest that oil was met with universal and unquestioning approval in southern California. From the start, its discovery and development was problematic. Early oil operations were often messy and dangerous. In the Old Los Angeles City Field, oil leaked out of holding tanks and flooded lawns and streets. While some homeowners struck it rich, others grew tired of wheezing, steam-driven pumps that worked around the clock. In 1907 Echo Lake on the edge of the oil field became so fouled with oil that it caught fire and burned for several days. Less than ten years after Doheney's initial oil well, the city was already moving to restrict oil drilling inside the city limits.

In Long Beach similar problems soon followed. There was so much oil underground that new wells were often explosive, heralded by gushers that spewed oil over city streets and nearby neighborhoods for days at a time. Others hit pockets of natural gas, igniting fires that burned out of control for days at a time.

Even the area's newfound prosperity was problematic: while the booming oil fields on the edge of town brought money and jobs, they also gave rise to a boomtown full of bootleggers (Prohibition was in full swing in the 1920s), whores, and cardsharps.

Less than five years after oil was discovered at Signal Hill, the romance was growing cold. While oil made the city wealthy, once the oil wells began to push into wealthy neighborhoods, opposition began to rise. In 1926 a citizen's group led by the city attorney pushed for an ordinance prohibiting oil drilling inside the city limits. Put on the ballot it passed by a nearly 3-to-1 margin. Afterwards no one said anything more about tearing down the city center to make way for new oil wells. Three years later, the state legislature postponed indefinitely the sale of oil gas permits on state-owned tidelands and coastal areas. Both onshore and off, oil fell out of favor.

Sixty years later the battle over oil still goes on, driven by contradictions and self-deception. While the city, with its sprawling suburbs and clogged freeways, is unquestionably addicted to oil, it is vehemently opposed to oil wells. Adding fuel to this already hot fire, the region sits on top of some of the most productive oil fields in the world. There is no escaping the oil that lies underground. It is a product of the area's dynamic geology and the biology of the ocean offshore.

• • •

The brick walls were high and topped with razor wire. The alarm bell went off as soon as I drove through the gate. A tall man in a straw Stetson hat and wraparound mirrored sunglasses appeared from behind an oil tank in the plant yard and walked up to the side of my pickup.

"Can I help you?" asked a clear, clipped voice from behind the sunglasses and hat. It was a hot, late fall day in Los Angeles. Ninety-four degrees, with Santa Ana winds blowing in off the desert. I could feel a fine line of sweat that had been collecting between my shoulder blades begin to run down the center of my back.

"I have an appointment to see Bruce Manley," I said, studying myself and the plant in miniature in the silvered-turquoise lenses of the man's sunglasses.

"Your name?"

"Ken Brown."

The face broke into a smile and a hand reached in through the window to shake mine. "I'm Bruce Manley. Nice to see you. Why don't you come inside?"

• • •

The office was cool and dark. A wooden hat rack loaded with Stetsons and hard hats stood by a sofa. Manley took off his sunglasses and hat and laid them carefully on his desk, revealing a pair of light-blue eyes and a soft round face. He was not dressed in a three-piece suit but wore a pair of dark-blue work pants and a light-gray work shirt. On his feet were a pair of work boots, not hand-tooled cowboy boots. On the wall next to his desk was a sign: I'M NOT DEAF / I'M IGNORING YOU.

Nearby was a bulletin board with a collection of more than a dozen permits from an assortment of city, state, and federal agencies certifying that Bruce Manley and the Manley Oil Company in particular have been duly inspected, licensed, authorized, sanctioned, and permitted to pump, transport, and process oil in the City of Los Angeles.

At thirty-six years old, Bruce Manley is the head of Manley Oil, a fifth-generation, family-owned oil company in the heart of downtown Los Angeles. With a little work, he admits, one might be able to build the family into something like the Ewings from the television series "Dallas." "We had plenty of back-stabbing and screams and yells and so forth," he admitted. But that is about as far as the comparison goes.

"We were always a hat-in-hand working group. Ninety-five cents of every dollar we ever made got put back into the company," he said. "The Manley Oil Company, if there ever was such a thing, consisted of two brothers slugging it out doing all the work by hand. We didn't even have an employee to speak of until the late 1950s. It's a misconception to think that everyone in the oil business is rich and famous and that all they've got to do is sit on their rear ends and collect their royalty checks. I guarantee you that kind of oil man disappeared back in the 1940s with oil wells that gushed up out of the ground in the middle of nowhere. Most of the time it's just hard work."

As head of Manley Oil, Bruce's job entails minding not only the front office but the oil fields as well. He is both chief executive and chief field hand. It is not a job that lends itself to extended vacations or early nights off. The wells pump oil seven days a week, 365 days per year. The company has just six employees. His mother balances the books.

The company's wells are part of the Old City Field, which stretches from Chinatown and Dodger Stadium through the Temple-Beaudry district to the edge of Elysian Park. Most of their wells are more than ninety years old. A few of them were drilled by Edward L. Doheny himself in the heyday of the Los Angeles oil boom. In the early 1900s there were more than fifteen hundred oil wells in it. Today there are just fifty-three, all but ten of which are managed by Manley Oil. The company functions as a kind of cooperative, gathering oil from its own wells as well as those of a host of private owners ranging from a ninety-two-year-old grandmother to the United Cerebral Palsy Foundation. Together they produce about three thousand barrels of oil per month. While most people think of oil in terms of large multinational corporations like Arco, Exxon, and Shell, 40 percent of the oil produced in the United States comes from small producers like Manley Oil.

The company was started in 1894 by Bruce Manley's great-great-grandfather Marcellus Manley. He was not a "wildcatter" who drilled his own wells but a buyer and seller of properties. The company was pieced together from a collection of existing wells. Like most everyone else operating in the Old City Field at the time, he was not an absentee owner but lived right in the midst of it. He believed that if the oil wells were good enough to earn money for them they were also good enough to live next to.

Today the neighborhood is part of the combat zone for the city's violent street gangs. Bruce and his wife live in suburban La Cañada, but the company is still very much downtown—eight blocks east of Library Park on the industrial backside of the city that borders the concrete channel of the Los Angeles River. You can stand in the midst of the pipes and tanks of the plant yard and see the First Interstate Tower and the World Trade Center. It is, if you have a strong arm, within rock-throwing distance of City Hall—and throwing rocks at City Hall is something Manley has thought of doing on more than one occasion. If Bruce Manley has a siege mentality (the plant's high concrete walls topped with razor wire do give one that feeling) it is because he feels that his company is slowly being squeezed out of existence by the city's unstoppable growth and an increasing thicket of regulations. Los Angeles may run on oil, but oil wells do not seem to fit into anyone's vision of modern Los Angeles.

• • •

Three days later I went back to Manley Oil to spend the morning with Bruce as he made his rounds of the oil field. Two minutes after leaving the plant we were downtown, rolling down Alvarado past the white expanse of City Hall in Bruce's pickup truck, a two-tone Ford with a bumper sticker on the back for the *Million Dollar Cowboy Bar* in Jackson Hole, Wyoming.

All but forgotten in the sixties and seventies by the explosive growth around it, lately downtown Los Angeles has been going through some explosive growth of its own: upward. Tall glass-and-steel skyscrapers rise above old four- and five-story buildings of stucco and brick.

By 9:30 A.M. the white-collar rush hour was in full swing. Traffic was bumper to bumper on the streets and shoulder to shoulder on the sidewalk.

Bruce shifted into neutral as we waited at a light. "If you look at the oil and gas division's maps of this area, the oil wells seem to run in a line. Originally that wasn't because it was the only place where people drilled for oil. The line is there because that's where the oil lies. Most of the oil is centered on this fault zone that runs right through the center of the city all the way out to Beverly Hills. When people drilled on top of it they found oil. When they drilled off to the sides they didn't. I'm not an oil geologist, but it's kind of like an underground river of oil. It runs from right here in downtown out through the Central Wilshire district and the La Brea Tar Pits to Beverly Hills. I've lived here all my life, but when I try to tell people that there's lots of oil underneath Los Angeles and that the name La Brea means oil, they say, 'Oh no. Can't be.'"

Passing under the Hollywood Freeway we continued up Alvarado and into the oil field. There was no maze of pipes covering the ground and no forest of oil derricks marking the horizon. Instead the streets were lined with stucco bungalows and Queen Anne Victorians. The oil field was not so much in the neighborhood as part of it. As we drove around the steep, narrow streets, Bruce pointed out the wells hidden in backyards and in fenced-in vacant lots. In places, they sat in front yards like motorized lawn ornaments.

Relatively shallow as oil wells go, the average well is just one hundred feet deep and produces between one and two barrels per day. But while production is slight, it has also been steady. Over the course of a few years the steady trickle of oil from even a half-barrel-per-day well can turn into a respectable profit. Red tape and the high costs of drilling, however, would make it almost impossible to drill these wells today, and they cannot be turned on and off like a faucet on your kitchen sink. The trick is keeping them running.

"It's like losing your hair," Manley said, taking off his hat and pointing to his own receding hairline. "You only lose one or two hairs at a time and you don't think much of it. But all of a sudden you're going bald and you wish you had those one or two hairs per day back on your head."

That analogy is also true of the oil industry in general. After supplying the country with a river of oil through the forties and fifties, U.S. oil fields are slowly running dry. While the average oil well in the Middle East produces nearly four thousand barrels of oil per day, the average well in the United States produces just fourteen. Although the United States uses more than 25 percent of the world's oil each year, it has less than 4 percent of the world's oil reserves. If it were forced to depend solely on its own supplies, the country would run out of oil by the year 2000.

Those figures have not been helped by the lack of a coherent national energy policy. Since 1980, in fact, official U.S. policy has been one of consumption, not conservation: low taxes on oil and gas and cutbacks in federally mandated fuel economy standards for automobiles that have caused oil consumption to skyrocket. After the Gulf War, there is no question that the U.S. dependency on oil is a deadly serious problem.

■   ■   ■

We spent the morning driving around the oil field, stopping to listen to pumps for telltale sounds of trouble and checking pipes and tanks for leaks. At a well site across the street from the offices of the Metropolitan Water District, I climbed to the top of a holding tank. From on top, beyond the hedge of oleander and bougainvillea that surrounded the lot, you could see downtown. I opened the hatch and peered inside. Where the sunlight caught the incoming flow of oil from the well, it looked as clear as tap water. When I let the incoming flow wash over my hands, my fingers were quickly flecked with tiny black spots of oil.

"The fire department likes to tell me I have a bunch of water wells with a little bit of crude oil floating on top," Manley said.

Although it is axiomatic that oil and water do not mix, the "oil" that comes out of the ground in the Old City Field is a mixture of 95 percent water and just 5 percent oil.

Between stops Bruce was the consummate tour guide, pointing

out sites both past and present: the site of the first gas well in Los Angeles on Glendale Boulevard, now a bare pipe sticking up through some tall weeds on a vacant lot; the home where his parents spent their honeymoon, with oil wells in the yards both back and front; a rusting sign for the long-gone Clampitt Oil Company (no relation to the Clampitt family from the television series "The Beverly Hillbillies," who found oil on their farm in Tennessee and moved to Beverly Hills; in reality, they would have been more likely to find oil on the grounds of their Beverly Hills estate).

Los Angeles is one of the most ethnically diverse cities in the United States. In most parts of the city those ethnic groups are separated by sprawl. Here in the Old City Field, however, those groups all seem to come together. Neighborhood residents are black, white, Spanish, and Asian.

The area is economically mixed as well. Some blocks are clean and solidly middle class, with well-trimmed hedges. Others are filled with homes and apartment buildings that have been in need of new paint and roofs for more than a decade—owned by slumlords who are more inclined to raise the rent than fix leaks. The tenants, often refugees from Central and South America, are in no position to complain or move elsewhere. Vacant lots are strewn with the debris of life on the bottom: dead TV sets, rotting mattresses, and the stripped carcasses of worn-out cars. Buildings and walls are peppered with graffiti and gang symbols. On more than one occasion Manley and his workers have found dead bodies dumped at the well sites. The week before my visit, two workers painting over graffiti on a holding tank had to dodge the bullets from a drive-by shooting.

All of this is visibly upsetting to Manley, who lived here as recently as 1984 and still thinks of this area as home. He talks about it with a kind of resigned despair. Most depressing of all is not the decay but that some parts of the neighborhood are gone for good—bulldozed flat to make room for a group of four high-rise office buildings planned by a Japanese development group.

"I'm going to show you something you probably haven't seen before in a city this size—seventy-six acres of vacant land," Manley said as we headed up a steep side street. The view from on top was panoramic, made even more so by the blocks of houses and apartments that had been cleared to make room for the impending high rises. In the pickup we sat and talked, watching a bulldozer flatten a block of houses a few hundred yards away.

"I can fight a lot of things, but I can't fight this," Manley said. "Right now City Hall is going around with the developer of this project. They want them to put in eight thousand units of low-income housing and pay the city sixteen dollars per square foot for development rights. If the city and the developer end up sleeping in the same bed, I'm dead. They don't want me here. They'd like to pretend I don't exist. We've been in business here for ninety years, but I will be surprised if this oil field lasts another ten years.

"In my lifetime I've seen probably twenty-five or thirty wells abandoned here. They weren't abandoned because the oil petered out. Some of them were producing ten barrels of oil per day when they were closed up. They took the money and ran. They left because it became too hard to do business in Los Angeles. If this field goes much beyond the nineties, I will be very much surprised. Not because the oil is disappearing, or because it's not economical to run it anymore. But just because the cost of doing business, the cost of keeping everybody happy and all the permits right, is just impossible."

■ ■ ■

From the turn of the century to the end of the Second World War, the oil industry was a key part of the southern California economy and pumped literally billions of dollars into the area. Oil fields were as profitable as they were productive. Today, however, they are often valued as much for the land that lies on top of them as the oil that lies beneath them. Cities and suburbs have sprung up on top of what once were oil and gas fields, but that has not stopped oil and gas from finding its way to the surface.

In 1983 oil rising from an improperly capped well in Newport Beach cracked the foundation of a suburban home, lifted it more than two feet off the ground, and flooded the kitchen with oil. The area is so rich in natural gas that neighborhood residents have been known to make tiki torches by simply sticking a pipe into the ground and lighting the top.

In the Central Wilshire district the La Brea Tar Pits are not the only site where oil and gas seep to the surface. They are simply more visible there. Oil seeps into the basements and elevator shafts of the sleek high-rise office buildings and apartment houses that cluster around Wilshire Boulevard. It comes up through sidewalks and streets, collecting in small rivulets. Not all of these seeps are small. A few produce more than one hundred gallons of oil per month. In the 1920s the area was the site of the Salt Lake City Field, one of the most productive oil and gas fields in Los Angeles. Much of the Central Wilshire district and the nearby Fairfax shopping district were simply built on top of it.

In 1981 construction crews drilling an elevator shaft for a new office building for the Farmers Insurance Group on Wilshire Boulevard struck a pressurized pocket of natural gas that almost blew their drilling rig out of the ground. After assessing the risks, they decided to incorporate a natural gas well into the building's design. Today the building is heated and cooled by natural gas pumped out of the ground beneath it.

Not all problems are so happily resolved. In 1987 a clothing store in the Fairfax shopping district filled up with methane gas seeping up through the foundation and exploded. The two-day firestorm that followed looked like something concocted on a back lot at nearby Studio City. Sheets of fire and flame shot out of cracks in the sidewalks and streets, fed by methane gas seeping out of the ground. Fire fighters were able to bring the blaze under control only by drilling a relief well to vent off the excess gas. A four-block area of stores was closed for four days while fire and safety officials waited for gas levels to drop off.

Miraculously, no one was killed in the Fairfax blast, but twenty-two people were injured, some of them seriously. At first authorities attributed the explosion to a leaking gas main. Later it was decided that the source was entirely natural—the gas was simply seeping up through the ground. Two weeks after the blast the Southern California Rapid Transit District confidently announced in the *Los Angeles Times* that the explosion would have "no impact at all" on their plans to build the city's first subway system right through the middle of the area underneath Wilshire Boulevard.

Based on these and other incidents, oil and gas experts believe that the area's oil and gas fields are slowly repressurizing. With no wells to tap into this naturally occurring stream of oil and gas, it is slowly seeping to the surface again as it had been doing for thousands of years.

With development edging into the Old City Field, these problems are very much on Manley's mind. There has been some informal talk with city officials about running a series of "relief wells" if the planned high-rise development goes through, but it is not something he cares to contemplate. The same oil he now sells to Union Oil's refinery would be classified as hazardous waste, and becoming a hazardous waste hauler is not something he finds particularly appealing. "I've looked at the regulations and I don't even want to think about it," Manley said.

• • •

We got out of the truck and began walking through the empty lots strewn with trash and broken glass toward the top of a small rise overlooking the city. It was a clear day with hot Santa Ana winds that had blown the smog offshore and sent the temperature soaring into the nineties. Oil wells pumped away on the top of a small bluff across the street, but Manley was not looking at the wells, but out over the city and the soaring heights of downtown—the gleaming towers of fifty-, sixty-, and seventy-story office buildings that had sprung up like weeds over the past ten years. At our feet the Hollywood Freeway was filled with bumper to bumper traffic. A sermon was coming on. He had been building up to it all day:

"We run a clean business here and it just galls the hell out of me every time I pick up a paper and read about these 'goddamn oil companies.' As far as I'm concerned, it's just a lot of people being hypocritical. Here we are living in a city that eats, sleeps, and breathes automobiles and the products produced by oil—plastics, asphalts, petrochemicals, and synthetic fabrics—and they're out there screaming and yelling that they want to save the world from oil production. Now, granted, oil is a messy thing to deal with and there have been problems in the past. History shows it. But if we're going to do away with oil we're going to have to do away with cars and fabrics like nylon and Gore-tex and all of the modern advantages that oil produces. There are certain risks in anything.

"Fine. You don't want oil. What do you want? Nuclear? Nuclear power entails an incredible waste problem and it's more of a risk than the oil industry. I'm sure Chernobyl and Three Mile Island have proven that to just about everyone.

"You want wind power? Well this is the seventeenth of November in 1989 and I think if you do some checking down at City Hall you will find out that the county supervisors said no today to a wind power project located in the city for 'environmental reasons.' Those environmental reasons really boil down to a question of aesthetics.

"Solar energy? Solar energy is wonderful! I hope someday it works. But right now it's not even 12 to 15 percent effective. I believe everyone should put solar collectors on their houses to help conserve energy.

"But it all boils down to one thing: the average person does not want to be bothered with recycling or saving energy. They aren't willing to go the least bit out of their way as far as being comfortable is concerned. They don't want to make sacrifices. They don't want oil production in their own backyard, but they want all the advantages that oil produces."

• • •

The last few pages have been almost exclusively about oil and the oil business. Oil, however, is what Los Angeles geology is all about, an intrinsic and unavoidable part of the landscape. What follows is another story about oil, but it is different from the others in that it comes full circle, heading back toward the biological and geological forces that created it.

• • •

G. Allan Hancock was born in Los Angeles in 1892. He grew up on the 4,444-acre Rancho La Brea, named for the La Brea Tar Pits that sat in its midst. His father had been one of the original surveyors of Los Angeles, and the city, which was land rich but cash poor at the time, had paid him in land—one lot for every eight surveyed. At the time Hancock was growing up, his father's holdings stretched from the Hollywood Hills to Santa Monica and included most of the Central Wilshire district.

At the turn of the century Los Angeles was still something of a frontier town, and Hancock was sent to private school in San Francisco. He spent his summers in Los Angeles, however, working on the ranch and driving the family's six-team wagon to downtown Los Angeles with loads of tar. The fourteen-mile round trip took all day, for which his mother paid him $1.25. Black-and-white photos from the period show a Los Angeles as wide open as the African savanna. West Los Angeles, Hollywood, and the Central Wilshire district were open ranch land. Sheep herders ran their flocks in Beverly Hills. Western Avenue was a long dusty track leading off to the foot of the Santa Monica Mountains. The city would not reach out to Rancho La Brea until the 1920s.

In 1900 the family issued its first oil lease on the ranch to the Salt Lake City Oil Company. Hancock found the oil business interesting and began working with the company. By 1907 he was ready to start looking for oil on his own and asked his mother for a $10,000 loan. A little less than a year later he struck oil at 1,391 feet. When it went into production, that first well produced three hundred barrels of oil per day. Over the next few years Hancock drilled seventy-one more wells and never hit a dry hole. At its peak the Rancho La Brea Oil Field produced more than four and a half million barrels of oil per year. He had the magic touch.

His oil company later became part of Mobil Oil. With the money he made from oil Hancock started the Hibernia Bank, now part of First Interstate Bank. When the population of Los Angeles began to swell, Hancock's holdings grew right along with it. He subdivided Rancho La Brea and created Hancock Park, donating the La Brea Tar Pits to the city. In San Fernando Valley he started a small development known as Sherman Oaks. Farther north he branched out into farming, buying land in the Santa Maria Valley north of Santa Barbara to grow fruit and produce as well as raise livestock and poultry for the growing Los Angeles market. Eggs and produce from his Rosemary Farms can still be found in southern California supermarkets. Access to the area was difficult, so Hancock built his own railroad out to the main line. In Mexico he bought a seven-thousand-acre hacienda near Mazatlán to grow winter vegetables. Rather than haggle with shipping companies, he bought his own freighter to bring them to market.

Hancock's hobbies and personal interests were as in tune with the city's future as with his business interests. He was the owner of the second car in Los Angeles (a 1894 Milwaukee Steamer) and was a founding member of the Automobile Club of Southern California. An accomplished cellist, he was also a founding member of the Los Angeles Symphony Orchestra—now the Los Angeles Philharmonic. Interested in film and television, he bought the prototype television equipment displayed at the 1939 World's Fair and set up the first television studio in Los Angeles at the University of Southern California.

His interests were catholic and intense at the same time. A train buff, he became a licensed engineer and often drove the locomotives on his Santa Maria Railroad. Interested in sailing and navigation, he held a master's license and was qualified to sail anything from a small fishing boat to a full-size ocean liner. (He accumulated the needed hours of ship time for his master's license by sailing back and forth from Mazatlán with the crew on his freighter.)

Science, however, was Hancock's consuming passion, and in Los Angeles he became a kind of one-man National Geographic Society, sponsoring oceanographic expeditions on his private yacht to places as far away as Alaska and the southern tip of South America. Most of those expeditions took place on Hancock's private yacht, the *Velero III*. Two hundred feet long with individual staterooms for the scientists he invited to travel along with him, and with amenities like a music room with a grand piano, the ship was more of a private ocean liner than a yacht. The guests "did science" during the day and held music recitals in the evening, with Hancock and the members of his trio (Hancock traveled with his private chef and a cadre of musicians) playing works by Mozart, Mendelssohn, and Hayden.

Trips to places like the Galápagos Islands could last for six weeks or more, and Hancock picked up the entire tab. On board he was both the ship's captain and the chief scientist—he was at the helm of the *Velero III* when they were under way, and operated the winch when they dredged or trawled for samples. When they sent parties ashore to collect plants and animals he piloted the launch. The only complaint scientists had was that he was constantly on the go, recalls marine biologist John Garth. "He couldn't stay in one place for more than three days. The only way we got to spend more time in an area was by coming back the following year."

• • •

After the Second World War, Hancock created the Hancock Foundation at the University of Southern California to sponsor research in marine science. He also bought a new ship, the *Velero IV,* equipped more along the lines of a traditional research ship. Instead of funding and organizing private expeditions to exotic locations in Mexico and South America, Hancock poured his time and money into his new foundation and USC. Using his money and ship, scientists and the university and the foundation began studying the geology and biology of the nearby ocean in detail—all at Hancock's urging. The seafloor off southern California, they found, was not flat, but dotted with deep basins. In addition to rocks, plants and animals, they also collected data on a host of chemical and physical properties in the ocean: its salinity, temperature, and density as well as oxygen content and nutrient levels.

The research Hancock funded and the institute he founded helped train a generation of marine biologists and geologists. It also helped make the sea off southern California one of the best understood ocean regions in the world. Today the information that Hancock urged scientists to collect is invaluable for making decisions about everything from managing local fisheries to controlling pollution.

"He saw needs where other people didn't see them in the study of local flora and fauna—for knowledge of the local oceanography, which he felt would be someday essential to managing the resources of the whole area," said Dorothy Seoul, a marine biologist who heads the Harbors Environmental Project at the Hancock Foundation. "He felt that if you didn't put it all together you weren't doing it. This was the hardest thing to get across to people. That you needed all this data: oxygen, salinity, temperature, and density as well as a wide collection of animals to understand an area of the ocean."

It is here that this particular part of the story comes full circle. The same broad scientific approach that Hancock was so insistent on pursuing is also the key to understanding oil. The oil under Los Angeles is a product of the area's unique geology, biology, and oceanography. In Baja California the Pacific Plate is creating a new ocean. Here along the coast of southern California it has created a series of deep basins filled with oil.

• • •

Oil is a fossil fuel, formed from the decayed remains of tiny marine plants and animals. It requires active biology and relatively inactive geology: thick piles of organic material must be built up and buried and then slowly cooked to make oil. The process is slow and precarious: it typically takes several million years to complete and occurs only within a narrow range of temperature and pressure. Too little heat and the oil never forms. Too much and it gives way to hard, crystalline rock. While stability is important during these formative stages, eventually some faulting and folding is needed to help pool the oil underground. Here again, balance is critical: if the rocks become too fractured and folded, the oil simply leaks out and rises to the surface.

It should come as no surprise then that most of the world's oil reserves are concentrated in areas where the earth is relatively stable. Nearly two-thirds of the earth's estimated nine hundred billion barrels of oil lie under the Arabian Peninsula and Russian Platform, two areas that have remained geologically stable for nearly six hundred million years.

Southern California, however, is as precocious geologically as it is socially. Although it is one of the youngest and most unstable areas on earth, it contains some of the most productive oil fields in the world. Ten billion barrels of oil lie under a few square miles of land. While its reserves are only a fraction of those in the Middle East, they are significant in terms of U.S. production.

Geologically speaking, three things are required to make oil: source rocks, reservoir rocks, and trapping structures—or to be more direct something to make oil from, something to store the oil in, and a lid to keep the oil from leaking out. In southern California those features all come together.

• • •

Twenty-five million years ago the East Pacific Rise collided with the edge of the continent. Prior to that collision the continent was bordered by an oceanic plate that was being slowly thrust underneath it. Afterward it was bordered by a plate that was slowly moving northward—the present-day Pacific Plate. In southern California the faulting and folding associated with this new direction of movement pulled the seafloor apart, creating a series of deep basins. By five million years ago, about the time that the Gulf of California began to open, some of these basins were more than a mile deep, positioned like pots on the seafloor. While those closest to shore were quickly filled in with rock and sand eroding off the nearby coast, those farther away were filled by a steady rain of organic debris—the remains of plants and animals—falling out of the waters above.

Organic material was not only deposited in these deep basins, it was also preserved once it reached bottom. Productivity near the surface was so high that the water below was anoxic, or low in oxygen, and almost devoid of life. As plant and animal debris falls to the bottom, it is typically used and reused by a host of scavengers and predators. In the anoxic and lifeless waters of these deep basins this organic debris could remain undisturbed for thousands of years—a near-perfect condition for the formation of oil. Today the rocks from these basins are known as the Monterey Shales. They are the so-called source rocks of the southern California oil fields.

However, there is more behind these low-oxygen zones on the seafloor than just productivity in the ocean above. The oceanography of the Pacific—the large-scale current patterns inside it—has also played a role.

■ ■ ■

Below the surface the ocean is not uniform but stratified into layers of different temperatures, salinities, and densities. In simplified terms, one can think of the ocean as divided into three layers: surface water, intermediate water, and bottom water. In southern California the intermediate water in the North Pacific plays a critical role in creating conditions that preserve organic material on the bottom and ultimately lead to the creation of oil.

Surface waters are those in the uppermost three hundred feet of the ocean. This upper layer is sunlit and stirred by the wind, and most of the ocean's productivity occurs here. Below this three-hundred-foot threshold, water movements are driven not by the wind but by differences in temperature, salinity, and density.

Bottom water is the coldest and densest water in the ocean. More saline than water at the surface and only a few degrees above freezing, it largely fills the ocean below one thousand meters. Formed in the Antarctic Ocean where surface water cools to almost freezing, it sinks to the bottom and slowly heads northward. It can take this newly formed bottom more than one thousand years to reach the equator.

Intermediate water, warmer and generally less saline than bottom water from the Antarctic (but still colder and more saline than the surface water), is formed by cold water in the Arctic Ocean that sinks to the seafloor. Once on the bottom it begins moving southward and eventually up and over the Antarctic bottom water to fill the ocean between three hundred and one thousand feet. Movement is relatively quick, a few knots in places, and it takes this intermediate water only a few hundred years to reach the equator.

In the Atlantic this intermediate water flows southward relatively unimpeded. Because it stirs the water below, anoxic water rarely forms in the Atlantic and organic debris is swept away.

In the Pacific, however, the flow of intermediate water from the Arctic is impeded by the Aleutian Islands, which stretch across the North Pacific like a chain, sealing the Arctic Ocean and Bering Sea off from the Pacific. As a result, the flow of intermediate waters in the North Pacific is slow and sluggish—almost stagnant in places. Because of this almost dead calm below the surface, anoxic bottom water develops with relative ease in the northern half of the Pacific. In southern California not only was the water below the surface anoxic, but the tops of the deep basins that dotted the continental borderland lay within this low-oxygen zone as well. As a result, the water above these basins and within them was anoxic. It ensured not only that organic debris would be deposited on the bottom but that it would be preserved there as well.

Geology created these deep basins off southern California, while the region's biology and oceanography ensured that they would be filled with organic-rich sediments—source rocks that could be slowly transformed into oil. Geology would play a role in the next phase as well—the formation of reservoir rocks that could hold and store this oil once it was formed.

• • •

Like blocks of broken ice on a frozen river tilted by the current below, the same faulting and fracturing that dropped some rocks down to make basins lifted other areas up to create mountains and high ridges. It was not a one-time event but an ongoing process. As the basins grew deeper, the mountains around them grew higher. And while faulting and folding pushed the mountains upward, erosion was wearing them down; sending a steady stream of sand, stone, and rock into basins below.

Offshore, this sand and debris did not flow into the basins in a continuous stream but in periodic slumps or underwater land slides, which geologists call turbidity currents. Piles of sand and sediment would build up to some critical level of instability and then break loose, sending a cloud of sand and debris into the basin below. After the dust settled, the rain of organic debris would begin again.

Over time, these alternating deposits of sand and organic debris created successive layers of source rocks and reservoir rocks—oil rich rocks stacked up like nested pots. Instead of having just one shot at finding oil from a given well, oil companies and independent operators often had more than a dozen. Men like Allan Hancock could drill seventy-one wells and never hit a dry hole.

• • •

A final bit of geologic activity was needed to complete the creation of the rich oil fields that lie under southern California. If the basins had simply filled up, there would have been oil but it would have been so dispersed that it would have been of little practical value. A trapping structure or lid was still needed to keep the oil in place.

Roughly three million years ago the Pacific Plate made a slight change of direction. Instead of moving smoothly, it began pushing inland as it moved north. Along the coast, the edge of the continent began to buckle like a throw rug caught by the underside of a door. Geologists see evidence for this change in motion up and down the West Coast as well as a pronounced kink in the trend of the Hawaiian Islands halfway across the Pacific. In southern California this folding is expressed not only in the rise of the Coast Ranges but in folded layers of buried rock and thrust faults—ramplike faults where layers of rocks have been thrust up and over others.

These faults and folds tilted the rocks below, freeing the oil to flow toward the surface. In places like the La Brea Tar Pits it actually reached open air. Elsewhere, however, its progress was checked by layers of impermeable rock and sometimes faults—trapping structures—with the oil building up behind them like water behind a dam. In essence, these trapping structures acted like a lid that kept the oil below from leaking out.

■   ■   ■

The processes that created these rich basins of oil in southern California are still at work today. Unlike the passive East Coast, which is bordered by a broad, flat continental shelf, there is almost no continental shelf on the West Coast and in most places the deep ocean is less than fifty miles from shore.

In southern California, however, the coast is bordered by a broad "continental borderland"—a submerged terrain of deep basins separated by high submarine ridges and mountains that occasionally break the surface as islands. It stretches from Santa Barbara to Ensenada and the water within its basins is almost anoxic. In places it is so low in oxygen that it is almost lifeless. Schools of fish have been known to swim into these pools of airless water and die en masse. Scientists exploring the area in research submarines report that the boundary between this low oxygen basin water and overlying sea water is easily seen, expressed by a sudden change in water clarity. Organic debris is still being deposited in these basins today in finely detailed layers that offer a yearly record of the past, like the growth rings on a tree. The layers vary with the seasons: a light gray during the spring and summer plankton blooms, followed by a darker gray in the fall and winter. In the past few decades other markers have arrived on the bottom as well: DDT and PCBs from factories and cities along the shore and a thin layer of oil from the 1969 Santa Barbara oil spill.

In turn the processes that created these basins still make themselves felt on land. Faulting and folding have not stopped. The mountains that border Los Angeles are still growing. This growth is not painless and the area is regularly rocked by earthquakes. Most are small and almost imperceptible, but some are large enough to level homes and freeways. They occur not only on the San Andreas Fault east of town but also on the web of side faults and thrusts that run through the city and play such a critical role in the appearance of oil.

On land, Los Angeles is a terrain of basins and ranges, a mirror image of those on the seafloor. Those on land, however, are not a trap for oil but for the smog and pollution produced by its use. Like the oil underground, this pollution is one of Los Angeles's inescapable problems, a part of the landscape as intrinsic as its mountains and beaches.

•   •   •

Before sunrise the air offshore had seemed sharp and clear. In the soft blue light that preceded the sun you could see the black shapes of hills and peaks: Saddleback Mountain and the rise of the Palos Verdes Peninsula along the coast and the higher thrust of the San Bernadino and San Gabriel Mountains farther inland. Overhead the sky was still dark, the stars a mirror for the lights of towns along the shore. By midmorning, however, it was all hidden by a dense cloud of smog that rose up into the sky like a high brown wall.

I had been out in the ocean since five in the morning with scientists from the Southern California Coastal Water Research Project on one of their weekly sampling cruises. More sewage than rainwater runs off the land here, but thanks to a careful monitoring program and aggressive pollution control efforts, the water quality offshore is very good—far better, in fact, than the air overhead.

We were only a few miles from shore, and looking over the side of the boat I could follow the yellow line of the marker buoy we had lowered over the side down through the water for more than eighty feet. Small ripples on the surface sent beams of light spinning through the water below. The wind had not yet arrived. The sea was almost flat calm, broken only by a soft rolling swell. Its smooth surface gave no hint of the extreme topography below.

A few miles away the water was more than two thousand feet deep. If you could drain off the water from this stretch of the sea, you could see a terrain of deep basins and high mountains that mimics the landscape onshore. Like those in the sea, however, those basins on land are often hidden from view. At times the air is almost as anoxic as the water in the deep basins offshore.

•   •   •

In places like Westwood or Beverly Hills, Los Angeles seems like a lush tropical garden. Morning fogs keep the days cool while onshore breezes keep the air relatively clear and smog free. Inland, from places like the roof of the South Coast Air Quality Management District's (SCAQMD's) monitoring station in Azusa, you can see Los Angeles for what it is—a dry semidesert surrounded by mountains.

The monitoring station sits in the midst of a neighborhood of warehouses and trailer homes spread out along the front range of the San Gabriels. They are not covered with oleander and exotic palm trees but with bare stones and tough desert plants. Smog from the factories and freeways that lie between here and the sea are blown by prevailing winds inland, where it collides with these high mountains and stacks up, giving Azusa and other nearby towns one of the worst air pollution problems in southern California. From an array of instruments scattered on the building's roof the agency monitors pollution levels around the clock. Similar stations scattered throughout the region also keep track of air pollution, and the numbers they gather are not encouraging.

Los Angeles has the worst air quality of any major city in the United States. Levels of ozone and carbon monoxide are three times higher than the minimum federal safety standards. Levels of particulate matter are two times higher, and in 1989 it was the only city in the United States to violate the standards for nitrogen oxides.

The health effects of these pollutants are varied and serious. The most dangerous is ozone, which damages cells in the lung's airways and makes respiratory infections more likely. It also damages the body's immune system and causes coughing, shortness of breath, and wheezing. Carbon monoxide aggravates heart disease and a number of blood disorders because it bonds with hemoglobin in the blood instead of oxygen. Fine particulate matter penetrates deeply into the lungs, where it causes irritation and damage. Studies suggest that these fine particles may be responsible for as much as 10 percent of the noncigarette lung cancer in the region.

Smog in the area is so corrosive that it eats away at the stone surfaces of buildings and monuments and corrodes paint and rubber on automobiles. Even the plants and trees are not immune. The smog discolors their leaves and kills some outright.

While long-term residents suggest that one develops an immunity to smog, medical evidence suggest otherwise. Lung damage and scarring continue long after the body has learned to turn off the pain caused by air pollution. Most disturbing of all is its impact on children. Studies at the University of Southern California suggest that children growing up in the area have a 10 to 15 percent decrease in lung capacity compared with those growing up in areas where the air is cleaner.

• • •

The coastal Indians who inhabited southern California before the first Europeans arrived called the Los Angeles Basin "the Valley of Smokes." The bowl of mountains that surrounded the city acted as a natural trap for fog off the ocean and the smoke from campfires and forest fires blown inland by the onshore flow of air. That circulation pattern has not changed over the past few thousand years and over the past few decades the problem has become even more noticeable. Los Angeles today is home to more than ten million people and six million cars.

"We have the most adverse weather of any of the ten largest urban areas in the United States," said Margaret Hoggan of the SCAQMD. Weather is not something Los Angeles residents usually complain about, but Hoggan, a senior scientist with the region's pollution control agency, is talking about weather in terms of air pollution, not in terms of tennis, golf, or a backyard barbecue.

"It is the most pollution-prone area in the United States," she explained. "There's not much wind here so things move slowly up toward the San Gabriels. Our nice friendly weather gives us a pollution problem."

The ocean off Los Angeles is cool year round, hovering between fifty and sixty degrees. Inland, the daytime sun heats up the ground, causing the air to rise. This rising pocket of warm air pulls the cool, foggy marine air inland, where it blankets the city and acts like a lid— trapping the fog as well as the smoke from cars and factories below to create smog. Winds in the area are light—typically less than five miles per hour. As a result the air below becomes stagnant—trapped by the mountains like seawater in the deep basins offshore.

As early as the 1940s doctors and scientists in Los Angeles began noticing that the area had an unusually serious air pollution problem. Residents regularly complained about sore throats and eyes. Rubber seemed to develop a peculiar form of cracking. Leaves on plants and trees seemed to develop spots and discolorations for no apparent reason; sometimes they dried up prematurely and dropped off altogether. In 1947 the Los Angeles County Pollution Control District was formed—the first pollution control district in California.

By the early 1950s the agency's scientists had discovered the source of the area's unique pollution problem: ozone. Ozone is not produced directly by cars or factories but is the result of a complex chemical reaction involving nitrogen oxides (which are produced primarily by cars, trucks, and buses) in the presence of sunlight. Like oil, the formation of ozone depends on a sequence of natural events—and in Los Angeles those conditions all come together.

The nature of the chemical reaction that forms ozone helps explain the unusual pattern of the region's pollution problems. While most of the emissions occur in the western and central areas of the city, the most serious pollution problems are found on its eastern side.

Nitrogen oxides are spewed out by automobiles during the morning rush hour and then drift eastward with the flow of cool marine air. As the day goes on sunlight becomes brighter and more intense, slowly converting these nitrogen oxides to ozone. Because of its dependence on sunlight, the peak period for ozone pollution occurs in the early afternoon, just after the peak period of sunlight. Because the formation of ozone needs sunlight, there is no corresponding peak of ozone pollution after the evening rush hour.

In spite of increasing population growth, pollution levels have been slowly declining in the Los Angeles area since the early 1970s. Much of that improvement can be attributed to the installation of catalytic converters on automobiles and the fact that most "smokestack" industries have left the area. While that has helped clear the air, Hoggan believes that improvements in the future will be harder to come by. "The easy solutions were taken long ago," she said. "It's easier to regulate one factory than ten thousand cars." The outlook for the future is not encouraging. Since 1980, car use has been increasing twice as fast as the population.

"We hope to be in full compliance with federal standards by 2007," Hoggan continued. "We know how much emissions have to be reduced by, but we don't know how to get there. The technology we need isn't even on line yet. It's not going to be a simple problem. There are hydrocarbon emissions in almost everything people do. It may mean doing things like telling people they can't use their favorite hair spray. You're impacting the individual, but when you have ten million people living in one place, you have to start talking about that kind of thing."

The automobile is still the key to the pollution problem in southern California, but no one, least of all those in the SCAQMD, believes southern California residents will cut back on their driving anytime soon. Idealistically, the agency supports car pooling and mass transit. Realistically, the major task on its agenda is replacing not the automobile but the fuel it runs on.

By the year 2000, regulations currently in place call for 70 percent of the trucks and buses and 40 percent of the cars in the area to be run on methanol. Liquid fuel derived from natural gas, methanol would cut automobile emissions roughly in half. The air in Los Angeles would still be dirtier than almost anywhere else in the United States, but it would be a step in the right direction.

• • •

It was a clear, late fall day. Hot Santa Ana winds off the desert had blown the smog offshore again and the air was unusually clean. Downtown, the view up Spring Street from Wally Rothbart's office window at the California Department of Transportation (Caltrans) gave one the feeling of being at the bottom of a large well. That effect was due not so much to the high office buildings scattered around downtown but to the high peaks of the San Gabriel Mountains, which seemed to rise right up from the edge of downtown. In reality their ten-thousand-foot-high peaks are more than fifty miles away. Surrounded by high buildings and mountains it was easy to see why the area is so regularly covered with smog. With the department's figures on cars and traffic in hand it is also easy to see where all that smog comes from: at present there are more than six million registered cars and trucks in the Los Angeles area and on an average day those vehicles travel more than eighty-one million miles on area freeways, filling the air with a poisonous collection of fumes and gases.

It was 4:30 in the afternoon and all was quiet across the street at the *Los Angeles Times*. The paper's loading dock faces Rothbart's office, and the delivery trucks were lined up in a row with their blue snouts facing outward. On the street, however, the evening rush hour was beginning to take shape as the estimated ninety thousand cars that squeeze into downtown Los Angeles each day were beginning to make their way home to Ventura, the San Fernando Valley, Orange County, and Riverside. Several thousand others were also beginning to pass through the city on their way home from offices in Inglewood, Century City, and West Los Angeles, adding to the slow creep of traffic on the surrounding loop of freeways. On an average day more than three hundred thousand cars pass through the freeway interchanges near downtown.

Inside Rothbart's office, that congestion was easy to visualize. Propped up on a bookcase is a map of the Los Angeles Freeway System. The routes are marked by thick black lines and most of those lines are bordered by strips of color: yellow for average speeds under forty-five miles per hour, orange for speeds under thirty-five miles per hour, and red for average speeds under twenty-five miles per hour. Nearly half of the map is orange. A good part of it is red.

As head of project studies for Caltrans' District Seven, which covers Los Angeles and Ventura Counties, Rothbart is intimately aware of the traffic level on area freeways. Professionally, it is something of a constant problem. Personally, however, it is less of a worry. On most days the region's chief of project studies walks to work.

"I grew up in New York using New York subways," Rothbart said. "When I first came to Los Angeles I was kind of a wild and crazy guy. I used to say, 'We should build a subway in Los Angeles! I'd use it. It's the only way to travel!' But after living in California for twenty-four years I'm a confirmed drive-by-yourself kind of guy."

• • •

In the early 1900s Los Angeles had a population of less than 150,000 and one of the most extensive urban railroad systems in the United States. The Southern Pacific's Red Cars, as the urban commuter trains were called, reached as far inland as Irvine, and the city itself was criss-crossed by the streetcar lines of the Los Angeles Railroad. Today Los Angeles has a population of more than ten million, and those rail lines have been replaced by freeways. It is a city built more for cars than for people, a far-flung network of suburbs and shopping malls that sprawls over more than six thousand square miles.

Part of that sprawl has to do with timing. The city underwent its period of most explosive growth after the automobile had become an integral part of the American life-style. Local geology played a role as well: the area was (and still is) regularly rocked by damaging earthquakes, a fact that tended to discourage the construction of high-rise buildings. Instead of reaching upward, the city spread outward.

In 1940 the first freeway in Los Angeles, the Arroyo Seco Parkway, was opened. Known today as the Pasadena Freeway, it connected the city of Pasadena with downtown Los Angeles. By the early 1950s construction was under way on segments of the Hollywood and Harbor Freeways near downtown Los Angeles. Elsewhere, tracks and trolley lines throughout the region were being torn up to make room for additional lanes of traffic on city streets.

In 1958 plans for the California Freeway and Expressway System adopted by the state assembly called for the construction of more than fifteen hundred miles of freeway in Los Angeles, Ventura, and Orange Counties. By the late 1960s, however, freeways had fallen out of favor. Area residents had grown disillusioned with the disruption and environmental damage caused by freeway construction. In the space of a few years public opposition brought construction to a standstill. Today less than half of the freeway mileage planned for southern California in the late 1950s has been built.

When it is completed in 1993 the seventeen-mile-long Century Freeway will be the first new freeway in the Los Angeles area in ten years. At a cost of more than $100 million per mile, it will be one of the most expensive highways in the United States. Its $2 billion price tag includes not just the cost of construction but also the costs of nine years of legal battles with citizen and environmental groups and a complex resettlement program for displaced residents, as well as the cost of building some 4,300 units of low-income housing.

Southern Californians' relationship with freeways is nearly identical to their relationship with oil: while they love using them, they are vehemently opposed to their construction or expansion.

• • •

When plans for the Southern California Freeway System were drawn up in the 1950s, the area had a population of less than five million people and less than three million cars. Today those figures have doubled. Although population growth is slowing, since 1980 car use has been growing twice as fast as the population. Not only are area residents driving more, they are also driving alone. The average car carries just 1.2 people, a depressing figure that becomes even more depressing when one realizes that it includes buses and car pools. While most of the talk about traffic centers around the need to get to work, more than 70 percent of the trips on area freeways are "discretionary," according to Caltrans' figures, involving trips to the mall or beach or simply a drive around town. It is one of the subtle ironies of life in southern California. In an area with perhaps the best weather in the United States for walking, almost everyone drives.

If present trends continue, the Southern California Association of Governments predicts that by 1997 the average speed on area freeways will be just nineteen miles per hour, with sections near downtown moving at just seven to nine miles per hour. Rothbart, however, believes the situation will never get that bad because the system is, in essence, self-regulating. As traffic gets worse, more and more people will begin looking for alternatives to driving—everything from car pooling to changing jobs to avoid a long commute.

"Traffic today is much greater than we ever anticipated twenty years ago," Rothbart said. "I won't say the system has more traffic than it can handle, but we are running at congestion." One of the reasons the system still works, Rothbart believes, is that local residents have such a high tolerance for those traffic problems. "We've got people here in southern California who will drive two and three hours per day to live where they want to live and work where they have to work."

Since the 1980s, freeway expansion has been largely restricted to what Rothbart calls "squeezing more capacity out of existing highways." Lanes, median strips, and shoulders have been narrowed to create new lanes of traffic. Today that option is all but played out. In the future, says Rothbart, "the best we can hope to do is maintain congestion at present levels."

Over the next decade Caltrans' goal in southern California is to squeeze additional capacity from area freeways by squeezing more people into fewer cars, by making some existing lanes into car-pool, or high-occupancy-vehicle (HOV), lanes.

At the moment, however, that idea is not going over well. Plans to include an HOV lane as part of the widening of the Hollywood Freeway from Topanga Canyon to downtown created a political firestorm as local residents and politicians objected to the idea largely on the grounds that their right to drive in any lane at any time might be restricted. (Six lanes of the eight-lane freeway would have remained open to all traffic.) Eventually, plans for the special lane were scrapped. Plans to install HOV lanes on Orange County freeways have met with similar opposition.

While the city is in the process of building its first-ever subway and a commuter rail line between Los Angeles and Long Beach, Rothbart, like officials at the region's air quality district, does not expect mass transit to play a major role in the area any time soon. No small part of this is due to the natural aversion that area residents have toward mass transit. Los Angeles already has the second largest bus system in the United States, yet it carries just 5 percent of the traffic. All of this is bad news for traffic and pollution in the city.

"People are always going to want to drive because of the convenience of it," Rothbart said. "It's amazing when you talk to people. The very same people who tell you that the answer isn't more construction and that more people have to take the bus and car pool are the same people who don't do either. When you ask them: 'How about you?' the answer is: 'Well, I can't. . . my job. . . I need my car.' Everyone seems to think that taking the bus is a wonderful idea—for somebody else."

• • •

On the day before Thanksgiving I went back to the California Department of Transportation to see its Traffic Operations Center (TOC) in action. On Thanksgiving eve Los Angeles freeways are typically clogged with the heaviest traffic of the year, and on my way into town radio announcers had been suggesting that those driving to the L.A. airport "add another hour to that extra hour on top of an extra hour that you've allotted for getting to the airport."

Traffic in Los Angeles often seems to have a mind of its own, but the TOC tries to impart some logic to the flow of things on the freeways outside. Computers track the flow of traffic on area freeways through a network of sensors embedded in the roadway. The information they pick up triggers small lights on the large wall map that looms over the control room: green for speeds faster than thirty-five miles per hour, yellow for speeds of twenty to thirty-five miles per hour, and red for traffic that is moving at twenty miles per hour or less.

In spite of the impending holiday, things were moving smoothly. Traffic was heavy near the airport and there were some red lights on the interchanges near downtown, but that was all expected. Operators sat at desks watching a bank of monitors carrying pictures from closed-circuit cameras positioned along the Santa Monica Freeway and around key downtown interchanges.

They were not watching for speeders or unlicensed cars but for traffic problems. Observers at the TOC keep their eyes open for accidents and breakdowns and pass the information on to the highway patrol and area police departments in an effort to keep things moving. They try to get help to places where it is needed and help steer traffic away from problem spots by feeding information about accidents and back-ups to area TV and radio stations. Along some of the busiest sections of freeway there are remotely controlled electronic signs that the TOC can use to warn motorists of problems ahead.

A few days earlier I had been reading in the *Los Angeles Times* about the trial for the Alaskan oil spill. Buried inside the newspaper was an ad taken out by the environmental organization Greenpeace. It featured a photograph of Joseph J. Hazelwood, the captain of the *Exxon Valdez*, which had fouled Prince William Sound with oil. Underneath it was a caption that read: "It wasn't his drunk driving that caused the Alaskan oil spill. It was yours." An often overlooked fact in the justifiable outrage that followed the spill was that the *Exxon Valdez* had been bound for refineries in southern California when it ran aground on Bligh Reef—it was the demand for oil here that had sent the ship and its cargo in motion.

As I watched the TV monitors at the TOC flip from camera to camera with images of endless bumper-to-bumper traffic, that ad was suddenly very much on my mind.

Traffic problems and air pollution are front-page news almost every day in Los Angeles. They are a staple of everyday conversation and a source of regular complaint. But while the volume of complaint is very high, the level of action is very low.

Through a series of carefully contrived arguments, the problems of pollution, automobiles, and oil are typically traced back to corporate greed and government incompetence. In the end, however, we have no one but ourselves to blame. We are unwilling to see the link between private actions and public problems, that these problems are inextricably tied to our own addiction to the use of oil and automobiles. We are unable to recognize the limits of the natural world around us and live within them. We are worried about damage to the environment, but we are unwilling to commit ourselves to causes that require anything more than self-righteous indignation.

Los Angeles is an easy target in this regard. Sprawling and oversized, its car culture and conspicuous consumption make it a ready scapegoat. In reality, however, the problems faced by the city and its surrounding network of suburbs are merely symptomatic of those faced by the country at large. You can find the same sprawling growth, the same clogged freeways, and the same foul air with increasing frequency in places like Seattle, Denver, Chicago, Dallas, and Washington, D.C.

Southern California is different only in the sense that the problems are more concentrated. In Los Angeles all the threads come together: oil, air pollution, and automobiles. Two of those—oil and air pollution—are the result of geology and geography, the natural setting of the area.

On the way back to my hotel I got stuck in traffic on the 405 behind an immaculate four-door Mercedes-Benz with tinted windows and a cellular phone—a car so elegantly inefficient that it gets less than twelve miles per gallon in town and carries a gas-guzzler tax of more than $2,000. On the rear of the car was a bumper sticker with the silhouette of an oil well framed by a red circle with a slash through it and the message: "No Offshore Oil Drilling." It seemed like the latest sign of hypocrisy.

■   ■   ■

We had been circling a small patch of ocean for more than an hour, anchoring instruments on the seafloor and collecting water samples as part of an ongoing research project by scientists at the Southern California Coastal Water Research Project. It had been a long day that began well before sunrise. The last of the marker buoys were in place. We were ready to head back.

We were less than ten miles from shore, but it felt much farther away. The break between land and sea is sharp and distinct—a meeting of different worlds. Here it was made even sharper by the fact that a city of more than ten million lay only a few miles away.

At 4:00 A.M., when I had driven down to Newport to meet the boat, traffic on the freeway was already heavy, part of a rush hour in southern California that increasingly seems to last all day long. Offshore, however, there was a sense of openness and space. We had seen only two boats all day. Two seals swam lazily off our starboard bow. A pelican floated off the stern, pausing for a moment in hopes that we might pull up something more promising than scientific instruments and water samples.

The settled world of freeways and towns onshore gives one the illusion of living apart from nature, suggesting that our society has somehow mastered the earth. For most, that illusion is broken only occasionally, by natural disasters like earthquakes, fires, and droughts.

In a small boat offshore, however, it is easier to keep things in perspective, to sense how small our own world is in relation to the larger natural world around us. Out here it is easier to imagine how the ocean influences the weather on shore, sending a lid of cool air over the city that acts as a trap for smog. It is easier to imagine the steady rain of geologic debris from these rich surface waters that falls on the seafloor and leads to the creation of oil. In the distance you can see the mountainous shapes of islands floating offshore, suggesting not only the possibility of deep basins on the floor of the sea, but that the land here has been lifted out of the sea.

• • •

You can find this same sense of detachment on land as well. Los Angeles is unique among major U.S. cities in that its extreme topography of high mountains and deep basins has preserved pockets of wild land too steep or unstable for development right in the center of the city. Side streets off of Sunset Boulevard dead end into the flanks of the Hollywood Hills. Six blocks away from the nightclubs and movie theaters you can leave the pavement and walk up a side canyon into the mountains, following the tracks of coyotes and deer. There is no sound or sight of the city, only dry grass and steep hills dotted with live oaks and occasional rows of eucalyptus. Climbing up the steep hills you can find seeps of oil and patches of loose broken rock—buff-colored, light-gray marine rocks that were once deep in the sea. On a clear day, from high points like the top of Mount Hollywood in Griffith Park, you can see the spread of the cities below: Los Angeles, Santa Monica, Beverly Hills, Long Beach, Wilmington, and Inglewood. From its backside you can look down into the suburbs that fill the San Fernando Valley and the cluster of towns that run north and south from Pasadena along the front range of the San Gabriels. Farther south you can see the endless sprawl of Orange County and Riverside.

In the end, however, the most lasting impression is not one of cities but one of mountains and hills. They seem to rise up on all sides, running parallel to the coast with a pronounced northwest grain: the Santa Monicas, closest to the shore along the northern edge of town, and the San Gabriels, farther inland along the edge of the desert. In between are others: the Verdugo Hills, the Hollywood Hills, the Baldwin Hills, and still others that run like a welt from Newport and Inglewood along the eastern edge of downtown toward the Central Wilshire district and Beverly Hills. Like the banks and islands on the seafloor offshore, you can see how these chains of hills and mountains divide the city into a land of basins and ranges.

These basins and ranges are merely an expression of the ongoing geologic forces that have shaped this section of the coast. The mountains and hills here have been built out of large blocks of rocks that have been faulted and folded and lifted out of the sea. Underneath all this are deep basins filled with oil. Farther inland, along the backside of the San Gabriel Mountains, is the San Andreas Fault. Like Baja California, this whole region—the surrounding bowl of mountains as well as the city and the oil beneath it—is slowly heading north toward Alaska.

MONTEREY BAY

A CANYON IN THE SEA

T O HEAD NORTH ALONG THE COAST FROM LOS ANGELES YOU MUST FIRST go west toward Santa Barbara along the front of the Transverse Ranges and around the tip of Point Conception where these mountains meet the sea. From Baja California to southeastern Alaska the coastal mountains that border the western edge of North America trend northwest like the San Andreas Fault. North of Los Angeles, however, the Transverse Ranges run east-west. They are some of the fastest-growing mountains in the world—shooting upward in places by as much as two inches per year. Their location coincides with a major bend in the San Andreas Fault. After angling up from the Gulf of California through the Mojave Desert for some 150 miles, the fault makes an abrupt left turn and heads straight for the Pacific. It does not quite reach the coast but runs due west for some fifty miles before turning northwest again and heading toward San Francisco.

In terms of plate tectonics, the bend is problematic. While the trace of the fault, which marks the boundary between the North American Plate and the Pacific Plate, changes direction, the Pacific Plate is still heading north. Along the length of the fifty-mile bend it is not sliding past the North American Plate but heading right into it. This head-on collision, geologists believe, has given rise to the Transverse Ranges, with the two plates squeezing the rocks between them like a vice, driving them upward into the air.

The mountains are a dividing line, separating southern and northern California. The boundary is no less striking in the ocean offshore—north of Point Conception both the geology and biology of the ocean changes as well. The Continental Borderland, with its deep basins and high ridges, is replaced by a narrow continental shelf. The water is no longer warm and clear but cold and cloudy. Warm-water fish like sea bass give way to cold-water species like salmon and halibut. The winds are no longer light but strong and gusty, giving rise to rough, choppy seas.

One hundred and fifty miles above Los Angeles at San Luis Obispo, the Coast Highway peels off of Highway 101 and runs right along the water's edge through an area known as the Big Sur. The shoreline here is bordered not by cities or suburbs but by high mountains. At times the road seems to defy both gravity and common sense, built on the sides of mountains so steep that construction crews did not lay out the route by coming overland but came ashore in rowboats, laying dynamite charges to blast out the roadbed.

The mountains are young and built largely from marine rocks that are often soft and unstable. During the winter rainy seasons the Coast Highway here can be closed by slides for weeks at a time. The forces that created these high mountains are the same as those that created the rich pools of oil in southern California: the steady northward drift of the Pacific Plate coupled with a slight east-west compression.

Rainfall increases as you head north as well. The landscape becomes greener. The dry coastal grasslands and semidesert of southern California give way to an increasingly thicker cover of plants. The sides of mountains and hills are often covered with thick forests of live oak, pine, and madrone. Redwoods appear in the dark shade of mountain canyons.

At Monterey the wall of mountains fronting the coast is broken by a broad coastal plain quilted with fields of artichokes, strawberries, and lettuce. The plain seems to be surrounded by a bowl of mountains, but that impression is an illusion. Inland from the coast the mountains are lined up in rows, like ships tied up in port—the Santa Mountains, the Gabilans, the Santa Lucias, and the Sierra de Salinas. Near Monterey the northwest grain of the mountains is cut by the northward trend of the coast, opening the interior valleys to the coast. The two ends of the bay, Point Piños and Santa Cruz, are anchored by hard rock. In between, waves and currents have carved the edge of the bay into a smooth, semi-circular curve.

North of Watsonville, not far from the center of the bay, you can drive through fields of broccoli and lettuce that run right to the water's edge. On clear fall days when there is no fog and the irrigation sprinklers send silver arcs of water over the fields, you can stand on the high cliffs and see the bay in its entirety. To the north and south the two mountainous points of land that mark the ends of the bay stretch out into the water like encircling arms, reaching so far offshore that their tips seem almost like islands. Inland are endless ridges of high mountains, their tops covered by dark forests of pine and redwood broken by meadows of grass burned to gold by the late fall sun. The view is so broad and open that the ocean offshore seems almost rounded. You can almost feel the earth's curve beneath your feet.

And yet the most striking feature of this landscape is one you cannot see. Running westward down the center of the bay is Monterey Canyon—a submarine Grand Canyon that splits the bay almost perfectly in half. It starts just a few feet from shore, opening outward into the ocean like a funnel. At its head, less than a quarter mile from shore, the canyon is a half-mile wide and five hundred feet deep. Ten miles farther out it is thirty miles wide and more than ten thousand feet deep—a Grand Canyon in the sea.

More than the encircling arms of the shoreline or the surrounding bowl of mountains, the canyon defines the bay's character. It provides a pathway between the deep sea and the shallow waters of the bay and helps direct the flow of nutrients and currents. Like the young ocean and deep basins farther south, it too has been shaped by the moving Pacific Plate.

•••

We were several miles offshore, due west of Monterey. The sky had been gray and overcast all day, but the winds were light and the ocean was almost flat calm. The *RV Atlantis II* was quiet, engines off—as still and calm in the water as if it were sitting in dry dock. We were listening and waiting. Ten thousand feet below, the research submersible *Alvin* was winding through middle reaches of Monterey Canyon.

On board the *Atlantis,* computers and crew members tracked the submersible's movements. Occasionally, messages from its three-person crew came to the surface with information on battery levels and bottom conditions. For the most part, however, the dive was uneventful for those on the surface. The excitement would come later, when the sub returned with a collection of rocks and animals plucked from the floor of the sea. For the *Alvin* support crew it was time to grab some much needed sleep from the all-night job of prepping the sub for its daily trip to the floor of the sea.

I was sitting in one of the main deck scientists' cabins with Gary Greene, a senior geologist with the U.S. Geological Survey. In an average year Greene racks up more flight miles than most airline pilots, jetting between his office at the survey's branch of Pacific Marine Geology in Palo Alto and projects and meetings in Europe, Asia, the South Pacific, and South America. Like most successful geologists, he built his reputation on fieldwork and by getting out and seeing things. And as with most successful geologists, his success has catapulted him into the middle of tasks that are increasingly bureaucratic and political. "I keep finding myself locked up in rooms where I can't even open a window," he had confided to me a few months ago.

At the moment, however, Greene was enjoying a rare and relatively undisturbed week at sea. Yesterday he had been down in the *Alvin* himself, exploring a series of meanders in the outer reaches of Monterey Canyon. Today he was plotting their course through the canyon onto a series of charts and geologic maps that littered the top of his desk and bunk.

For the past twenty years Greene has been studying Monterey Canyon, compiling charts and geologic maps and thinking about the geologic processes that created it. He has spent more time both in and over the canyon than anyone else, traveling through it in research submarines and over it in surface ships with sonar and seismic gear. He is, without question, the world's authority on the Monterey Canyon. For all that, understanding and describing the canyon is still a formidable task.

Monterey Canyon is one of the largest submarine canyons in the world, but you cannot stand on its edge and peer at the canyon floor a mile below or fly from rim to rim in a plane and capture its expanse in a single glance. It it is several thousand feet deep in the ocean, and its depths, except for the occasional lights of a research submarine, are perpetually dark.

"I had the feeling I could very well be flying through a portion of the Grand Canyon," Greene said of the previous day's dive. "The problem is that it is like flying through the Grand Canyon at night or in a fog. The view is limited from the submarine, but you can see cliffs and they drop off quite rapidly. The relief is just as steep and exaggerated as it is in the Grand Canyon. The only thing that's different is that it's softened by sands and silts. It's like going into an old house that hasn't been dusted for years. You know there's furniture in the house, but you don't know if it's pine or mahogany or what have you. Your imagination is going all the time trying to picture what the big thing looks like."

• • •

By the early 1700s, Spanish explorers had noted the existence of a large submarine canyon in Monterey Bay. Like other submarine canyons in the Pacific, it was considered a navigation hazard at the time. Early sailors often navigated along the coast by depth, dropping a leaded line over the side to measure depth and then using that figure to judge how far they were from shore. Near submarine canyons, however, the bottom dropped suddenly and unexpectedly out of reach, leaving them without a point of reference.

Monterey Canyon was also noted by early sailors because it was a site of fresh water in the ocean. Like canyons on land whose walls sliced though aquifers to create springs, the walls of Monterey Canyon also created springs. These springs, however, were not in air but in salt water. The fresh water from these submarine springs, instead of cascading down the canyon walls, rose upward through the denser salt water to create freshwater boils. Ships could gather water at these ocean springs by simply lowering a cask over the side.

In the 1920s, surveyors for the U.S. Coast and Geodetic Survey noted the existence of a "fine and picturesque submarine valley" in Monterey Bay. The first detailed surveys of the canyon, however, did not come until after the Second World War and the development of sonar. The same sonar gear that detected submarines could also be used to detect the seafloor, revealing such features as submarine mountains and canyons. In the 1950s and 1960s, sonar was used to document the worldwide occurrence of submarine canyons, and some cursory surveys were done near the head of Monterey Canyon. In the 1970s, when Gary Greene began to study Monterey Canyon as part of his doctorate at Stanford, it was still terra incognita.

• • •

Gary Greene grew up on a farm in California's Central Valley more than one hundred miles from the Pacific. Later when the family moved to southern California he got his first glimpse of the ocean on weekend fishing trips with his father and fell in love with ships and the sea. Out of high school he joined the coast guard for three years. When his enlistment was up he made plans to join the merchant marine. To make it into the California Maritime Academy he needed to brush up on his science and math, and so he enrolled at Glendale Community College for some courses. Out of curiosity he decided to take a class in geology. "At the time," he says, "I didn't even know what geology was. I knew what dirt was, but I didn't know what a rock was."

Within a year he had scuttled his plans to attend the maritime academy and was enrolled at Long Beach State and majoring in geology. He worked his way through school, in part by using his skipper's license from the coast guard to pilot private yachts out of Long Beach Harbor.

In 1966, when he started graduate school at San Jose State, the science of marine geology was beginning to come into its own. Scientists had discovered a globe-girdling chain of submarine mountains and patterns of magnetic stripes on the seafloor that had opened the door to the development of the theory of plate tectonics. In nearby East Palo Alto the U.S. Geological Survey had created a new Pacific Marine Geology branch. Greene began working at this new branch of the survey as a technician to help cover the costs of graduate school, using the research he did there on Arctic beaches for his master's thesis, while his wife worked as a social worker to support them. By the time he started to work on his Ph.D. at Stanford, everything began to come together, centered around Monterey Canyon.

By the late 1960s, farmers in the Monterey Bay area and nearby Salinas Valley were pumping so much water out of the ground to irrigate their crops that salt water from offshore was beginning to seep into underground aquifers, making them unfit for either farming or drinking. Two hundred years ago the rocks underground had been so full of water that they created boils of fresh water at the surface for Spanish sailors. Now, with the aquifers pumped almost dry, the flow was in the other direction: salt water from the sea was beginning to flow inland and contaminate underground water supplies on land.

By the early 1970s the problem was so bad that the California Water Resources Board was proposing to spend $50,000 to study local water supplies. Greene applied for the money and won the grant, using it to fund more than seven years of research on the local geology and water resources. At the center of all this was Monterey Canyon, and for Greene it was the most intriguing problem of all. The canyon, he found, was not simply a product of erosion but was intimately related to plate tectonics and the faults that sliced through the area. Like so many other things on the West Coast, it was slowly heading north on the Pacific Plate.

• • •

Conventional scientific wisdom had it that submarine canyons were no more than a few hundred thousand years old. It was believed that they were formed during the ice ages, when so much water was locked up in ice that sea level was as much as three hundred feet lower than it is today. Such a drop, theorists reasoned, would have exposed broad stretches of the continental shelf. Streams and rivers flowing across this newly exposed land would have cut channels and gorges that would later become submarine canyons when they were submerged by rising sea level.

That theory worked quite well on the East Coast where submarine canyons extended outward from the mouths of existing rivers, but it did not work at all in Monterey Bay.

While Monterey Canyon was the same size and scale as the Grand Canyon, there was no corresponding Colorado River. Instead the Monterey Canyon spread outward from the mouth of the Salinas River, a stream so marginal that it often ceased to exist during the dry California summers. Imaginary geology and geography solved the problem by rerouting streams and drainages to channel large rivers into the head of the canyon. One popular theory suggested that San Francisco Bay had been blocked off several hundred thousand years ago and that the Sacramento River (which drains California's Central Valley and a good portion of the Sierras) shot through a break in the Coast Ranges known as the Pajaro Gap to empty into Monterey Bay. There was, however, nothing but speculation to support this idea. In the Pajaro Gap and in the head of the canyon there were no sediments from the Central Valley and no river gravels from the Sierras.

Instead of a link to San Francisco Bay, what Greene found when he began exploring the area inland from the head of Monterey Canyon was an ancient submarine canyon several thousand feet deep. Studying the data from a series of coastal wells, Greene discovered that the soft marine rocks found near the surface were underlain with hard granite. To either side of the head of Monterey Canyon those rocks were only a few hundred feet below the surface. Immediately behind it, however, they were more than seven thousand feet below the surface. The present-day canyon seemed to be excavating an ancient one that had been filled in with sediment.

That ancient canyon, Greene found, had been filled in with not just any sediments, but with marine rocks from the deep sea that were more than twenty million years old. Monterey Canyon, it seemed, was not merely a few hundred thousand years old but several million. Suddenly it was possible to see the origin of the canyon in an entirely new light.

"If you have millions of years instead of hundreds of thousands of years you don't need a big river," Greene explains. "You can erode a big canyon, even in granite, with small local processes. It can form nicely in a few million years. Why not? Canyons weren't formed just in the Pleistocene."

Inside the canyon Greene found that these local processes were in fact very active. Detailed seismic and sonar studies of the canyon revealed the tracks of turbidity currents, or undersea landslides. Piles of sand and gravel would build up in the upper reaches of the canyon to some critical point of instability and then cut loose, moving down the canyon in a flume of rock and sand that eroded the canyon even further.

There was no shortage of debris being dumped into the head of the canyon either. In places, the shoreline was eroding by as much as eight feet per year. Carried along by currents near shore, an estimated three hundred thousand cubic yards of sediment was being dumped into the canyon every year—the equivalent of a dump-truck-size load every fifteen minutes. Elsewhere offshore Greene found slump blocks, places where large pieces of the seafloor and canyon wall had broken loose and slid downhill, widening and deepening the canyon. Some of these blocks were twenty miles across, suggesting the possibility of major earthquakes accompanied by tidal waves.

The canyon was not young and static, formed by processes that came to an end when the area sank below sea level—it was old and constantly changing, formed by processes that are still taking place today.

■ ■ ■

While the ancient buried canyon solved some questions, it raised others—namely, how had it been buried and then dug out again? Here, too, there were answers to be had in the data Greene collected.

The same seismic surveys that detected slumps and turbidity currents also found that the area offshore was laced with faults. In places, they seemed to direct the course of the canyon, bending it into a broad S-curve at its seaward end. Elsewhere, both on land and offshore, they seemed to divide the area's crustal rocks into blocks. Layers of rocks on these blocks suggested that not only had they moved up and down but they had also done so at different times, like the pistons in an engine. Such up-and-down movement could account for the ancient submarine canyon being filled in and then dug out. In fact, there was evidence to suggest that the ancient head of Monterey Canyon had been excavated and filled in at least twice in the past.

The driving force for all this up-and-down motion is, of course, the northward drift of the Pacific Plate. This, too, was recorded on the seafloor.

North of Monterey Canyon there is evidence to suggest that this buried canyon may have been a "father of canyons," responsible not just for the present-day canyon that slices through Monterey Bay, but a string of other canyons found on the nearby seafloor as well. The northward drift of the Pacific Plate is not entirely taken up by movements along the San Andreas Fault. In places some of that motion is taken up by movements along side faults running parallel to the San Andreas. Here on the central California coast, several of those side faults run through the Monterey Bay, slicing pieces of the coast and the seafloor into thin slivers of rock that are being slowly driven northward. North of Monterey Bay the continental shelf is notched by a series of truncated submarine canyons: Pioneer, Ascension, and No-Name—canyons that seem to have no heads or tails and whose origins were even more puzzling than that of Monterey Canyon. It seemed, as Greene saw it, as if the ancient head of Monterey Canyon had repeatedly cut canyons in the continental shelf as slivers of it drifted by, only to have them slide northward and out of reach.

The canyon here, scientists now believe, is not just an isolated product of erosion and random faulting, but as much a part of plate tectonics as the mountains bordering the coast and the nearby San Andreas Fault. Greene downplays his own role in creating this new picture of the canyon. "I had the advantage of looking at the canyon in three dimensions. The poor geologist onshore sees things only in one dimension, and then only when it breaks the surface, so he is a bit limited. I had more at my fingertips."

Over the past twenty years continued research has refined this image of an ancient and mobile canyon, but by no means have all of the problems concerning its history and the processes at work inside it been worked out. As for his own ideas of movement and its relationship to the canyon, Greene is confident but cautious. While he links the history of the canyon to the theory of plate tectonics, he is careful to label his own ideas as hypotheses. He considers the distinction important:

"A hypothesis is something you come up with based on a proliferation of good hard facts. A theory is something that develops on more than just data. I think my ideas on the origin of Monterey Canyon are a hypothesis. They certainly can be challenged and they're not to the point where you can call them a theory. They're not conclusive yet. Of course nothing is ever conclusive in geology."

As for the ultimate origin of the canyon, that may be unknowable: a fault or freshwater seep, for example, may have created a zone of weakness in the rock that led to the start of the canyon. There is no point, however, in looking for signs of that origin here in Monterey Bay. The San Andreas Fault is several miles inland, and like the rocks of the continental shelf farther offshore, even the head of Monterey Canyon has drifted northward along the coast. It did not start here. Like so many other things on the Pacific Coast, it is just passing through.

•••

By four o'clock *Alvin* was on its way back to the surface and out on deck preparations were under way for the sub's return. In most parts of the world, a trip to the deep sea also means a trip far from shore. Here, however, Monterey Canyon cuts so far into Monterey Bay that we were only a few miles from shore. Smaller ships had brought a steady stream of daily visitors over the past week. That day the ship was crowded with reporters and photographers from newspapers and television stations in the San Francisco and Monterey Bay areas. Local scientists working on the ship had also taken the opportunity to bring their families out to see the ship and submarine firsthand.

After successfully finding the *Titanic* in 1984, *Alvin* is a celebrity in its own right. The research submarine has explored oceans and seas

around the world. It has discovered new forms of life living on volcanic vents on the seafloor and explored the rift zone that created the
Atlantic Ocean. Off the coast of Spain it helped retrieve a missing
atomic bomb. In more than three thousand dives there has been no
loss of life.

The tiny sub can dive to more than twelve thousand feet in the
sea and withstand pressures of more than a thousand pounds per
square inch. Almost every part of the submarine—from its mechanical
arms to the fiberglass cowling that surrounds the seven-foot-diameter
titanium sphere that lies at its center—can be jettisoned to free the sub
if it becomes entangled on the bottom. In the event of emergency it
carries enough oxygen to support its three-person crew for several
days.

In many respects, however, the most dangerous point in the dive
is at the surface, where the tiny sub can be battered by waves. If the
sub has been sent out, it must be retrieved, regardless of the weather.
Rough seas can be even more dangerous to the crew members on
board the *Atlantis* working to bring the sub in.

Today, however, the sea was relatively calm. On the bridge, crew
members scanned the surface waiting for the sub to rise. A quarter-
mile away, the chase boat—a small launch with two divers on board—
rose and fell in the swell.

When the bright orange conning tower of the sub broke the surface, both the crew and the ship came alive. Moving ahead of the
Atlantis the chase boat streaked through the water at full throttle, skip-
ping over the swells toward the wallowing submarine. On board the
Atlantis the engines went from full stop to full-speed ahead, sending a
plume of black smoke pouring out the stack. By the time we pulled up
alongside, divers from the chase boat were already in the water, secur-
ing a tow line and talking with the crew inside the sub via a plug-in
telephone jack.

From the stern a large cradle-mounted crane hoisted the sub out
of the water and placed it gently on deck. When the hatch opened and
the crew climbed out, there was a burst of spontaneous applause and
hoots from those on deck. The crew seemed to have an air of electric
excitement about them, like astronauts returning from a space mission. They had seen a part of the earth today that had never been seen
by human eyes. It had been a good dive. The sub's sample baskets were
filled with a collection of rocks and clams pried from the floor of the
canyon.

• • •

After sunset, with the *Alvin* out of the water and the day's visitors back
onshore, the *Atlantis* was gearing for a full night's work. There was a

gentle roll and swell to the ship as we headed out toward the next day's dive site at full speed to deploy a series of sonar beacons, or pingers, to be used by the submarine to track its location on the bottom. In *Alvin's* on-deck "hangar," engineers and technicians were already at work recharging its batteries and making small repairs and changes in equipment for the next morning's dive. Research time is precious and we would spend the night not anchored over the dive site but sailing back and forth over the fan of sediment that spreads out across the seafloor from the mouth of the canyon like a river delta, timing our trip to arrive back at the dive site by sunrise.

After dinner the scientists and crew members from the *Alvin* gathered in the ship's library for a debriefing session. The day's dive had been located near the spot where Carmel Canyon, a small steep submarine canyon that starts a few feet from shore at Monastery Beach near Carmel, merges with the larger and deeper Monterey Canyon. In addition to the sub's pilot, two biologists had been on board: Chris Harrold, from the Monterey Bay Aquarium, and Mike Foster, from the Moss Landing Marine Laboratory.

The dive was successful but not without its hair-raising moments. Carmel Canyon is narrow, and on the way down the sub had started bouncing off the canyon walls like a pinball. "It was very disorienting," Harrold explained with an uncomfortable laugh. Farther down they saw sand streaming over ledges, a potentially dangerous situation because a full-scale landslide or turbidity current could have buried the sub on the bottom.

On the floor of the canyon signs of the world above were everywhere. In places, branches and tree trunks littered the bottom. In Carmel Canyon they found other surprises: hundreds of golf balls lost, no doubt, from the oceanside holes at nearby Pebble Beach. At more than ten thousand feet deep in the ocean, it lent new meaning to the term "water hazard."

The deep water was rich with a life of its own as well. Sea lilies, sponges, and anemones carpeted the canyon walls, while the floor of the canyon was covered with a thick layer of silt. In places, brittle-armed starfish carpeted the bottom arm-in-arm. Organic debris—the remains of plankton, jellyfish, and other animals in the water above—floated through the water like snow.

• ■ ■

Tiny floating marine plants are the basis of the marine food chain, and that productivity is concentrated in the sunlit upper layers of the ocean. Sunlight penetrates no deeper than three hundred feet and often less if the water is not perfectly clear. Life below this upper sunlit layer is dominated by a host of scavengers and predators who depend on the organic debris raining down from above. Survival gets increasingly more difficult with depth. Not only does pressure become increasingly greater, but the food and organic debris is picked over by those living above and becomes increasingly harder to find.

In light of these basic biological realities, the unexpected richness and diversity of life on the canyon floor is surprising. Biologists have begun to suspect that just as the canyon connects the shallow waters of the bay to the deep sea, it may also connect the deep sea to the productivity of the ocean above. In and among the animals and sediments on the seafloor, Harrold and Foster found decayed pieces of kelp, a type of algae that grows only in the uppermost layers of the ocean. It was what they had been looking for.

A fast-growing treelike form of algae, kelp often dominates the shallow water near shore in central and southern California. While plankton are still the base of the marine food chain, in places kelp make up as much as half of the total biomass. Anchored to the bottom by roots called holdfasts, they reach up toward the sunlight like trees. Clustered together in groups, beds of kelp weave and sway in the surf and currents near shore like an aquatic forest.

Like a coral reef, the kelp forest is a locus of life in the open ocean. It supports an elaborate array of fish, shellfish, and other marine animals. Unlike the reef, however, the kelp forest lives only from year to year. Uprooted by winter storms, it all but disappears. Some is tossed onto the beach in tangled mats, but somewhere between 60 and 90 percent of the kelp is unaccounted for—and biologists have no clear idea where it goes. Here near Monterey Canyon some of it seems to make its way into the deep sea. Here, too, it is a source of life—food for scavengers and opportunists in the deep sea.

The appearance of kelp on the seafloor was promising, but much more work is needed before any definite understanding of its importance is gained. It does, however, suggest that Monterey Canyon, which connects the shallow waters of Monterey Bay to the deep ocean, may work in the opposite direction as well: connecting the deep ocean to the more productive layers of the upper ocean. The canyon is a two-way street.

• • •

The world outside was silver and blue as the *Ventana* floated at the surface, its underwater camera alternately in and out of the water, offering glimpses of the sky overhead and the mirrorlike underside of the sea. At times its wide-angle lens offered a fish's eye view of crew members on the deck of the mother ship, the *Point Lobos*, making a last check of equipment before the dive began.

When everything was ready we turned the electric motors on full and headed straight for the bottom, leaving a cloud of bubbles behind. The world outside faded from turquoise to indigo to black as the *Ventana* dived toward the canyon floor.

The ocean below was not barren but full of life. Near the surface we surprised fish: a solitary anchovy that flitted in front of our camera for nearly a minute; later a starry-eyed flounder that scurried off in alarm.

In the soft, blue half-light of the deep water just below the surface were floating sacklike animals: tunicates, cyanophores, and tinaphores, soft-bodied relatives of the jellyfish. Elsewhere, larvaceans floated by, surrounded by a basketball-size net of mucous used like a spider's web to collect food. Delicate and sheer, it rippled in the currents kicked up by our sub like a curtain blown by the breeze.

As the deep-sea twilight faded into darkness we turned on the lights, attracting a large skate that came by to regard us eyeball to eyeball. Sunlight does not reach into the deep sea, but more than half the animals living here are bioluminescent—capable, like fireflies, of generating their own light. Beyond the reach of the sub's light the dark water was lit by small points of blue and yellow light from a collection of fish and tiny floating animals that shimmered and danced like a galaxy of stars.

Clouds of sediment billowed over the *Ventana* as we reached the bottom. Rising up a few feet, we began skimming over the floor of the canyon, nearly one thousand feet below the surface. The bottom was covered with a thick layer of silt and sediment that seemed to be tinted with reds and greens. Cut shark-egg cases litter the bottom, shaped like small rectangular boxes with four needlelike spines projecting out from the corners. In places, the canyon seemed to be carpeted with animals: sea cucumbers and brittle starfish. Flounders quietly buried in the sand erupted in an explosive burst of speed, leaving a cloud of silt behind. Elsewhere, rattail fish weaved their way across the bottom, leaving an S-shaped path in the sand behind them like a sidewinder rattlesnake weaving across the desert.

The images were so intimate and immediate that it was hard to remember that we were actually not on the bottom but topside in the *Point Lobos,* guiding the *Ventana* through the upper reaches of Monterey Canyon by remote control. Pictures from the remotely operated vehicle (ROV) are fed to the surface via the fifteen-hundred-foot cable that connects it to the ship on the surface. There is no need to worry about air supplies or battery levels. The crew is not cramped in a deep-diving submarine but sitting topside seeing the world below on a bank of television screens. The same cable that carries pictures and data to the surface also carries power and instructions down to the ROV.

The *Ventana* is the latest tool for deep sea research created by MBARI, the Monterey Bay Research Institute. Started in 1987 by David Packard, cofounder of the Hewlett Packard computer company, MBARI is unique in that its focus is not global but local. While it has managed to attract a staff of world-class scientists, the institute's research is focused on Monterey Bay and, at the moment, on Monterey Canyon in particular. Its goal, according to institute director Dick Barber, is to function as an observatory: to examine a small stretch of the sea so closely that its scientists are able to see patterns and relationships in the ocean as a whole that other researchers miss.

The institute is also unique in that it is interested in acting as a catalyst for research among the collection of marine laboratories and institutes ringing the bay. Today's cruise was no exception to that goal. On board was a collection of scientists from around the bay: Bruce Robeson (from MBARI), Chuck Baxter (from Stanford's Hopkins Marine Station), and Gary Greene (from the U.S. Geological Survey).

• • •

After an appreciative sigh aimed at the tapestry of deep-sea life floating across the screen in front of him, Robeson leaned back in his swivel-mounted captain's chair and smiled. "I hate to sound like a company guy, but this thing is just great. We're turning deep-sea biology into an everyday thing. Out here everything runs contrary to what we thought we knew. Of course no one had ever looked before."

Looking is something Robeson places great stock in, and in the deep sea it is something that is relatively hard to come by. While biologists on land can walk through a forest or spend hours in a field studying plants and animals, biologists in the deep sea have to get most of their information secondhand: from fish and other animals caught in nets or dredged from the seafloor. In Robeson's view that perspective is incomplete. But it is only in the past few decades that biologists have been able to regularly study the deep sea firsthand through vehicles like the *Ventana* and *Alvin*.

"I started working in the area with trawls that we dragged behind ships," Robeson explained. "And over the years I have come to believe that there are only certain kinds of things you can learn that way. You tow a midwater trawl at two knots and it's lumbering through the water making all kinds of noise and squeaking and creaking with a big pressure wave in front of it so it's glowing. It's a wonder we caught anything at all. It's embarrassing really. We had been catching the stupid ones and sleeping ones. I know now from trying to emulate a net that fish laugh. I'm sure they do. They swim backward out of the net. They dance in and out on their tails."

At the University of California at Santa Barbara Robeson was a pioneer in the use of manned submersibles for observing deep-sea life. Borrowing the same small submarines used by oil companies to inspect and work on offshore oil platforms, Robeson explored the deep-sea basins of southern California's Continental Borderland. Once in deep water and able to observe things with his own eyes, it became readily apparent to Robeson that there was more to deep-sea life than the fish they had been capturing in nets. Soft-bodied animals and the webs of mucous left by animals like larvaceans seemed to be everywhere. Getting others to appreciate what he was seeing was hard.

"When I started diving with submersibles," he said, "I would come back excited and tell people, 'You guys aren't going to believe all this mucous down there!' I had some video footage to show people and I would jump up and down and people would say, [here he drops his voice to indicate scholarly displeasure] 'Yeah. . . Right. . . Been down too long, eh? You're a fish guy. What do you know about invertebrates?'"

When Bruce Robeson joined MBARI one of the first things he wanted to look at was the soft-bodied animals. What he found seemed to confirm ideas that had been taking shape since he began working with small research submarines in southern California. From the ROV's cameras it became apparent that these soft-bodied animals—the tunicates, cyanophores, and tinaphores, as well as the feeding webs of animals like larvaceans—were everywhere. There was no doubt that they had a tremendous effect on the structure of the marine environment. They attracted swarms of fish and tiny floating marine animals.

It soon became apparent to Robeson that these floating congregations of life in the ocean were also linked to the rich life found on the floor of the canyon. On some of the *Ventana*'s first trips into the bay, Robeson and his coworkers spent entire days following larvaceans around and studying their behavior. No one had ever been able to do it before and no one had ever been able to identify them. They found that the larvaceans were tiny peanut-size animals that lived inside mucous-feeding webs, "houses" that they could construct with considerable ease, like a spider spinning a web. Some of those houses were the size of medicine balls. When the webs became clogged with debris, they were abandoned and left to sink to the bottom and the animal began to work on a new one. When they used the *Ventana* to follow one of these sinking houses toward the bottom, they found that they were sinking by as much as 800 feet per day—fast enough, Robeson reasoned, to carry organic debris to the bottom before it was picked over by other scavengers living in the middle reaches of the ocean.

A few days later Robeson was on the floor of Monterey Canyon in *Alvin*, almost directly under the site where he had been watching larvacean houses fall through the water. His suspicions had been correct.

"Here were all these things on the bottom hanging from worm tubes and flapping in the breeze," Robeson said. "I was really jazzed. Partly because of seeing the top end and bottom end of the same site at the same time. It really hammered home the idea that these things made it to the bottom and that they were an important source of carbon to the benthic community—a fairly big elevator that's taking organic matter to the bottom. In video they're covered with phytoplankton cells and shrimp shit. All sorts of things which fall down and stick to it. It's not just the house that's important on the bottom, but what's stuck to it."

• • •

On the bottom of the canyon, signs of that link to the more productive waters above were everywhere. Sea cucumbers and anemones sprouted from the seafloor. Snails and starfish inched their way across patches of sand and silt. The sharp surfaces of rocks were softened and dusted with silt. In places, the canyon walls were covered with worm tubes and sea lilies.

"I wish those guys wouldn't do that to my rocks," said Greene, who was hoping for a glimpse of the rock layers underneath.

"A rock is a rock. Some are big and some are small," said Baxter with a chuckle.

Time passed quickly. The world outside was new and intriguing. Our purposes were no different from those of sailors who explored the coast above nearly three hundred years ago. We were here to observe and record. Until our arrival, this spot in the canyon was unseen and unexplored. We wandered over the canyon floor following stray fish and stopping here and there to examine an interesting starfish or snail. Like the crew on a plane we were connected by headphones, and the comments flew back and forth:

"Let's go explore that crevice on the right for a minute."

"Can you follow that sablefish for a while?"

"Zero in on that starfish!"

"What is it?"

"I don't know."

Higher up in the canyon we came across stones and pebbles rounded by waves in the surf zone and large pools of sand. Gliding over a series of sand bars, we came across a field of octopuses, hundreds of them, spread out across the bottom in windrows. Above them the rocks were cluttered with starfish. Two brightly colored ones—one bright red, the other purple with white spots—inched their way across the seafloor.

"Anyone want to put bets on this race?" Baxter asked.

"Look at all those colors and patterns and they live in a black world where no one can see them!" said Robeson.

"That's not true," Baxter said. "They were just waiting for us to come by."

• • •

The tide was low, almost six feet below normal. South of Monterey, amid the rocks offshore, the Great Tide Pool was fully exposed. A large, almost circular pool of water, it opens into the ocean only along its western edge. Two curving arms of rock curl around its side. From the shore it looks almost like a miniature version of Monterey Bay itself. It has not changed much since John Steinbeck described it in his novel *Cannery Row*. It is still "ferocious with life," filled with a dense and diverse array of plants and animals, an image of the productivity and diversity of the adjoining bay.

Out on the seaward edge of the Great Tide Pool, the pockets and pools of water were a mass of improbable colors. Pink and white layers of coralline algae coated the surfaces of rocks. Green and brown feathery branches of marine plants like encladia and nail brush reached inward from the edges of pools like a fringe. Inside the pools themselves were blue and green anemone spread open like flowers, the largest being four and five inches across. Orange and purple starfish were anchored to the rocks, while scarlet patches of velvety sponge added bright points of color.

I had walked out to this edge of the Great Tide Pool with Chuck Baxter from Stanford's Hopkins Marine Station to learn about the plants and animals here and their links to the bay and canyon that lies just offshore. For the past sixteen years Baxter has been teaching a fifteen-unit course at Hopkins, a demanding quarter-long exposure to marine biology and scientific research. As Baxter explains, "It is not so much a course in marine biology as one in general-purpose problem solving. It's a course in how to run your life." That assessment is one a number of students take quite literally. Baxter is the pied piper of marine biology, and not a few promising students have ended up pursuing a career in the field after taking his class.

The tide pools bordering Monterey Bay are an invaluable resource for teaching marine biology. Not only are the plants and animals readily visible, they are also extremely concentrated. While the offshore rocks provide an anchor in the sea, the environment is rugged, filled with resources and pressures that change in the space of a few feet. Near the high-tide line plants and animals must be able to withstand hours, even days, out of water. A few yards farther down the shore pools of water are left by the tide, but there are stresses here as well. Once cut off from the ocean, the water temperature inside these shallow pools can rise to dangerous levels for the animals clustered within them. When the tide is out they are also easy prey for land-based predators like raccoons and sea gulls. When the tide returns they must be able to withstand the pounding of breaking waves. This boundary between land and sea, however, is remarkably rich, and the farther one heads out in this world between tides the richer the marine life. Near the low-tide line the competition for food and space can be as keen as that on a coral reef. From an ecological standpoint, the differences are so sharp that it is almost like walking through a succession of desert, prairie, forest, and wetland in the space of a few feet.

For biologists this concentration of life is a useful window on the world, but the work is often tedious. "In studying marine ecology you have to do a lot of things you'd be embarrassed to tell your friends about—like counting the number of barnacles on a rock. But it's important because small animals are good for understanding larger ones. It's easier to work with barnacles than elephants," Baxter said. Many of biology's basic theories about the interactions between predator and prey were formed by studies done here in the intertidal zone.

In places along the shore the role that environment and competition plays in structuring the marine environment is evident to even the untrained eye. Alongshore the high-tide area is marked by a bleached-white line of barnacles that marks the rocks like the ring around a well-used sink. Farther offshore there are other markers as well. The outermost rocks of Point Piños are thickly coated with mussels whose blue-black shells glisten in the late afternoon sun. Exposed by the low tide, they are cropped below in an almost perfectly straight line. That lower line is not due to any weakness in the mussels but is a sign of how far the pilaster starfish that prey upon them can safely crawl up the rocks when the tide is out. Unlike the tightly closed mussels, they can withstand only short periods of time out of water.

For years biologists had assumed that these ecological laws of supply and demand were solely responsible for the "structure" of the marine environment. More recently, however, they have begun to realize that local populations are also strongly tied to current patterns and nutrient supplies in the ocean at large. "It's an exciting time to be in marine biology," Baxter said as we picked our way through pools whose rocks were slick with seaweed and barnacles. "You have to be as much an oceanographer as a biologist to understand population dynamics," he said.

Among the outermost rocks the richness of the bay and the canyon offshore is reflected in the tide pool. The floor of the Great Tide Pool is not smooth, solid rock but covered with large melon-size boulders of granite. Sitting half in, half-out of the water, each one is a maze of small habitats and environments and covered with a bewildering variety of plants and animals. Porcelain crabs scurried through the shifting crevices between rocks, snapping their claws.

Moving a rock to see what lay beneath it, we uncovered an eel that flipped and skittered away through the wet stones in search of a more private resting place. Its gills are stronger than those of fish and it can survive out of water for short periods of time by keeping them wet. Fish suffocate out of water not because of a lack of oxygen but because their delicate gills, which are adapted to breathing the oxygen dissolved in water, collapse when they are taken out of the water.

Turning the rock over in his hand, Baxter reveals a new world within the tide pool: tiny clear-bodied tunicates, relatives of the larger animals we saw floating offshore from the ROV, cling to the bottom of the rock along with brightly colored sponges. Others hold iridescent green flatworms and tiny gold-colored proliferating anemones, each one surrounded by a flock of younger animals clustered around its base that will eventually move off on their own. We placed each rock carefully back the same way we found it. The animals are so carefully attuned to their small niche in the world that putting a rock back wrong-side up would kill most of the animals living on it.

In the small pools between rocks we found jet-black turban snails, the tip of their shell a bright silver. Baxter picked one up and placed it in the palm of his hand. The snail quickly retreated inside its shell, closing off the open end with its operculum, a tiny hatch that looks almost like a sliver of mica or isinglass. No inch of space is wasted in the tide pool. The tegula, only two inches in diameter, has three tiny slipper snails attached to the back of its shell.

Even empty shells are not overlooked. While most snails inched their way across the bottom, others seemed to be possessed with frenetic energy, running wildly back and forth on the bottom. Inside these fast-moving shells were not snails but tiny hermit crabs, which use old shells to protect their own vulnerable bodies. When I picked one up and held it in my hand it quickly retreated inside its borrowed shell. After a few minutes it came back out, waving its stalked eyes and tiny claws as if it were looking for a fight. Its body and claws were brightly colored with red and neon blue.

Other pools were filled with blades of eel grass. Small snails clung to the underside of grass blades. On a rock near the edge of the pool Baxter pointed out a small abalone, its back vented by a curving row of breathing holes. Crawling among the grass was a Hermissenda nudibranch, or sea clown. Its body was a golden orange, and thin electric-blue stripes ran down its sides. White-tipped stinging rods called rhinophores covered its back like a fringed shawl. Nudibranchs are relatives of snails that have evolved to live without shells, protected by a

bad taste and rows of stinging rhinophores. Sometimes called the butterflies of the sea, their bright color helps ward off would-be predators.
In and among the grass we found other nudibranchs in an array of colors: neon pink, white and yellow, another a collection of blues, blacks,
and yellows. They are voracious predators but capable of delicate acro-
batics underwater. On a calm day they can crawl upside down across
the thin and delicate interface between air and water.

As the tide began to rise, the blades of eel grass billowed in the
soft push and pull of the incoming water. The sharp smell of seaweed
and salt water hung in the air.

By the time we reached shore it was almost sunset. The tide was
already pounding the outermost rocks of the tide pool with four-foot
waves, and what were shallow pools only a few minutes earlier became
two and three feet deep. The fog was beginning to roll in, setting off
the booming call of the foghorn at Point Piños. Cracks and crevices
higher up on the shore were filled with broken bits of shell and dry
strands of kelp—a sign of the rich life just a few feet from shore.

"I haven't been able to get out and do that for a while," said Baxter as we sat on the tailgate of my pickup peeling off our hip waders.
"Every time I get out, I realize how much I miss it."

The diversity and density of things, I remarked, was almost overwhelming, and Baxter readily agreed: "Oh it's a real zoo out there," he
said. "We probably saw representatives of nine or ten different phyla
today. In the intertidal zone you can see diversity in a few feet that you
would have to travel through several thousand feet in elevation to see
on land. It's like walking from here to the Sierras."

•   •   •

Two weeks later I went back to the Great Tide Pool with a friend from
out of town, a wildlife photographer from Montana. Earlier in the
week the first winter storm blew through, a three-day storm whose
high waves had cut five feet from the beach at Carmel and left it littered with tangles of uprooted kelp.

The Great Tide Pool was almost bare. Rocks and boulders had
been rolled and thrown by the high waves. In more than an hour of
slippery and precarious walking we saw almost nothing, only a few
barnacles and starfish. It gave me an appreciation of how fragile the
life was that I had seen only a few days earlier. With time, I knew, it
would all be back—working its way up from the rich waters of the bay
to cover the rocks with a new layer of life.

•   •   •

The mid-June sun was bright, but its warmth was offset by the twenty-
knot wind blowing in from offshore. Inland it was already hot and dry,

but along the coast strong seasonal winds pull cold water to the surface, making the summer months the coldest of the year.

Here at Point Lobos a few miles south of Monterey Bay you could almost see the wind. It caught the tops of waves, blowing clouds of spray through the air that fell like rain on the rocks near shore. Farther back the twisted, windblown shapes of cypress trees offered a permanent record of the wind's passing. Offshore, beds of kelp swayed in the sea swell, while large rocks and sea stacks were covered with hundreds of nesting seabirds. The ocean was deep blue and covered with purple streamers of fog.

The rocks here are unusual: a collection of buff- and tan-colored sandstones cut by channels of multicolored boulders and stones that have been carved into rounded, irregular shapes by the waves. For years, the point, with its striking rocks and twisted trees, has captured the attention of photographers like Ansel Adams and Edward Weston. Today its scenery attracts tourists by the thousands who gather at overlooks and walk down the pathways that lead out among the rocks.

For Greene, however, the point is more than just admirable scenery. It is one of the few places on land where you glimpse the canyon that lies just offshore—not Monterey Canyon itself, but the boulders and stones that once tumbled down the sides of an ancient submarine canyon.

To Greene the sandstones of the point with their channeled deposits of boulders and stones look almost identical to layers of rock and sandstone he has seen in Monterey Canyon from the window of a research submarine. They look, he says, like a turbidity current—an undersea landslide of pebbles, boulders, and sand, like something one might expect to find in the upper reaches of a submarine canyon. In geology it is axiomatic that the present is the key to the past. At times the converse is also true. In these sixty-five-million-year-old rocks at Point Lobos you can see events taking place in Monterey Canyon today.

As we walked out across the point, Greene pointed out the features hidden in the rocks. Layers of pebbles and stones seemed to slice through the tan-colored sandstones of the point in channels, like wet concrete tumbling down a chute. The boulders and stones were jumbled and chaotic, as if they had been caught in midslide by a high-speed photograph. It was not just the large stones that carried such a strong suggestion of motion, but the thin layers of sand and silt within the rock as well. In places, those thin layers looked folded and squeezed, like toothpaste forced out of a tube. "You get the idea," Greene had said, "that this whole area is just moving downhill."

These jumbled marine conglomerates are underlain by a solid layer of granite. On the backside of the point you can actually see the contact between them. Like the conglomerate, that granite is also distinctive: a pale, almost white-gray granular rock, its regular fabric studded with rectangular twinned crystals of plagioclase three and four inches across shaped like the facing pages of an open book. Sixty million years ago these rocks were all part of a submarine canyon. Since then they have been not only lifted out of the water but moved northward as well.

The jumbled collection of boulders and stones that lie on Point Lobos are a collection of granites and cherts in a variety of purples and grays. Some are almost green. They are not like those of nearby hills but are more typical of rocks found in the high mountains around Big Sur farther south. Like Monterey Canyon, these rocks may have been part of a canyon that was once farther south.

Here near Monterey Bay the San Andreas Fault is more than thirty miles inland. It is still active and represents the boundary between the stable North American Plate and the mobile Pacific Plate. Not all of that plate's motion, however, is taken up by the fault. To the east and west there are smaller faults running parallel to the San Andreas that are also the sites of earthquakes and motion. Offshore, these faults slice the rocks into small pieces that are being carried slowly northward.

As for the San Andreas, it moves closer to the ocean as it heads north, finally reaching it near San Francisco, where it has given rise to bays and inlets along the coast. These movements of rocks and canyons on the seafloor near Monterey are linked to those farther north. They are a sign not only of movements in the past but a sign of movements and earthquakes to come.

SAN FRANCISCO

A SEARCH FOR SOLID GROUND

GARY GREENE'S HOUSE IN THE SANTA CRUZ MOUNTAINS IS LESS THAN five miles from the main trace of the San Andreas Fault. In five million years or so it will not be here some sixty miles south of San Francisco but a few miles north of the city, where Point Reyes now juts out into the Pacific. With offices at the U.S. Geological Survey's Pacific Marine Geology branch in Palo Alto and the Moss Landing ·Marine Laboratory on the edge of Monterey Bay, Greene can head east or west when going to work, downhill toward either San Francisco Bay or the open Pacific. Although its present location near the crest of the mountains puts it roughly halfway between the Pacific and San Francisco Bay, his backyard gives a view not of open water but of ridge after ridge of high, forested peaks—products of the drifting Pacific Plate.

On October 17, 1989, Greene had spent the day at his office in Palo Alto. In the afternoon he left work early to avoid the evening rush hour on Highway 17, the main route through the mountains between San Jose and Santa Cruz, and arrived home a few minutes before five. It was the last trip he or anyone else would make on Highway 17 for several weeks.

At four minutes past five Greene was standing half in, half out of the doorway between his kitchen and deck when, at a distance several miles beneath his feet, the Pacific Plate suddenly slipped past the North American Plate. One minute he was talking over the day's events with his wife, the next he was being thrown from side to side in the doorway—first his right shoulder then his left striking the door frame with a force that nearly knocked him off his feet.

After a lifetime of mapping faults on the floor of the nearby ocean and in the surrounding mountains, he watched with a combination of amazement and fear as wave after wave of tremors rolled through the house. In the living room where his wife Lynn was sitting, the brick fireplace trembled and then exploded outward as if it had been wired with a demolition charge. Picture windows tried to bend and flex with the waves of energy rolling through the house and then broke into shards. The force of the shaking popped nails halfway out of the walls and then lifted the house into the air and threw it sideways, bending the bolts that secured it to the foundation.

Outside, the water in their pool sloshed back and forth until the bank of dirt that held up its south wall slid away, ripping the pump and filter unit from the pool and releasing a flood of water. Near the driveway, Gary's caboose, purchased from the Southern Pacific Railroad Company and trucked up into the mountains as a visible sign of his love affair with trains, broke free from the welds that secured it to a short length of track and rolled a few inches farther down the line.

In less than ten seconds it was all over. Both Greene and his wife were unhurt. Worried about the risk of fire, they quickly turned off the electricity and gas and then surveyed the damage. Their house was in shambles but still standing. With aftershocks likely to follow, they decided they would be safer outside than in.

Opening the door of their walk-in kitchen pantry to pull a cache of emergency supplies from a bottom shelf, they found their supplies buried waist deep in cans and boxes of food that had been thrown from the shelves. It took several minutes to dig them out. Before the next quake, he thought, they would have to find a safer place to store their emergency supplies. Carrying flashlights and sleeping bags, they headed up to a grassy clearing near the main road where their neighbors were beginning to gather.

From the clearing, the view was apocalyptic. It had been a dry year and the quake had struck during the peak of the fire season. A half a dozen miles away they could see brush fires burning in the mountains. Clouds of smoke rose straight up into the still air while planes circled overhead dropping fire retardant on the flames below. As night fell, aftershocks rocked the ground, adding to the air of general unease as they watched the news on a small battery-powered TV and speculated on the damage in the cities and towns below.

• • •

It took only a few seconds for the quake in the Santa Cruz Mountains to make itself felt throughout the Monterey and San Francisco Bay areas. Waves of energy radiated outward from the break on the San Andreas Fault like the ripples created by a stone thrown into a still pond. The amount of energy released, scientists would later calculate, was roughly equal to the explosion of a thermonuclear bomb.

Near the intersection of Highway 17 and Summit Road on the crest of the mountains, the shaking was so intense that it snapped the tops off tall trees and left cracks in the ground more than two feet wide and two thousand feet long. Homes in the rural area looked like they had been picked up and actively shaken. Farther back in the mountains huge blocks of ground, some the size of several city blocks, broke free and slid downhill during the quake. Elsewhere, landslides and mudslides sealed off roads and highways with mountains of rock and rubble that would take several weeks to clear.

Ten miles away in Santa Cruz on the northern end of Monterey Bay, six people died as dozens of buildings and homes in the seaside town were all but destroyed by the quake. Hardest hit was the Pacific Garden Mall, a downtown shopping district of tightly packed stores and office buildings—most of them more than fifty years old. Taller buildings collapsed on top of smaller ones as the shaking pulled floors and walls apart.

The quake left its mark on the coast as well. At the city's small yacht harbor, shaking from the quake created a rhythmic pattern of small waves that rocked the ships and sailboats in their berths. North of town, boulders and stones fell from the high coastal cliffs, closing Highway 1. To the south, in the flat fields and crop lands bordering Monterey Bay, the shaking broke levees and irrigation pipes. In places, jets of sand shot up through the ground like miniature volcanoes. At Moss Landing near the center of the bay the Moss Landing Marine Laboratory broke in half as the soft beach sand shifted underneath it. Five miles inland at Watsonville, the center of the region's farming and canning industry, thousands were left homeless as houses and apartment buildings were thrown off their foundations by the quake.

• • •

On the other side of the mountains the earthquake was making itself felt as well. Tremors rolled through San Jose and the suburbs that ring the southern end of San Francisco Bay in a visible wave, rippling down city streets and sidewalks and rattling homes and office buildings. Destruction here was not so complete, but as the force of the quake moved through the area it seemed to search out weaknesses, leaving a scattered trail of damage behind—toppling chimneys and cracking sidewalks and streets in some areas while neighboring blocks were left all but untouched.

At Stanford University, halfway between San Jose and San Francisco, the quake threw more than 750,000 books from the shelves of campus libraries. Most buildings and homes on the sprawling campus survived intact, but some were so badly damaged that they would eventually have to be completely rebuilt. Strangely enough, one of the most heavily damaged buildings on campus was Geology Corner on the main quadrangle—proof, or so it seemed, that knowledge of a problem does not always lead to its solution.

Twenty miles north at Candlestick Park on the edge of San Francisco Bay, the third game of the World Series between the San Francisco Giants and the Oakland A's was about to get under way. The Giants were down two games to none. The day before a particularly savvy out-of-town reporter had written that "only an earthquake could save the Giants now." With cameras trained on announcers wrapping up their pregame remarks, the tremors rolled through the stadium. The Loma Prieta earthquake, as it would later be called, became the first major earthquake to occur live on television.

When the shaking stopped, the crowd in the stadium let out a boisterous cheer—a mixture of mindless exultation and outright fear. One fan held up an impromptu sign: "That was nothing. Wait 'til the Giants bat!" Others had heart attacks. While locals assured out-of-town visitors that it was just a routine shaking, the news from radios and portable TVs told quite a different story: the quake was anything but routine. The area had been heavily damaged.

By the time the tremors reached San Francisco, they had traveled more than sixty miles from their point of origin in the Santa Cruz Mountains. Although they carried only a fraction of the energy that struck the towns of Santa Cruz and Watsonville, they were still capable of causing substantial damage.

•　•　•

In the downtown financial district tall office buildings swayed by as much as eight feet as the waves of the quake passed by and then settled back into place. Most of these tall buildings, a collection of forty- and fifty-story skyscrapers, were less than ten years old and designed with earthquakes in mind, but the lower, older buildings scattered around the city center were harder hit. Ornamental facades and window ledges cracked and broke, showering the sidewalks below with a rain of deadly debris. In the South of Market district, a brick warehouse collapsed onto the street, killing five people as they sat in their cars. Up at Union Square the mirrored glass windows of the I. Magnin department store shattered, sending a shower of glass onto the crowded sidewalks below.

Damage in the city was linked to something most residents had never considered—the type of rocks and soils under city streets and sidewalks. Out toward the Golden Gate Bridge the quake rolled through Pacific Heights with barely a tremor—the rocks below that well-to-do neighborhood are solid granite. Less than a half-mile away on the edge of the bay, however, the Marina district turned as fluid as water when the quake hit. Buildings and streets heaved up and down like ships at sea in a heavy swell. A stately neighborhood of stuccoed apartment buildings and townhouses, it had been built on top of what had once been a coastal lagoon.

After the 1906 San Francisco earthquake, that lagoon had been filled in with rubble and debris from the quake to create a park for the 1912 Pan Pacific Exposition. The city had hosted the exposition in part to demonstrate its recovery from the quake. Later, homes and apartment buildings were built on top of the exposition grounds—the start of what later became known as the Marina district. For eighty-three years the area had remained quiet, but when the shaking started again jets of sand and mud shot out of the unstable ground like toothpaste squeezed out of a tube, bringing flecks of tar paper and bits of charred redwood timbers to the surface—a poignant reminder of the past.

Homes and apartment buildings in the neighborhood had been built with garages and carports that took up most of the ground floor, what engineers would later describe as a "soft first story." Lacking the support of four solid walls, most fared poorly during the quake. Corner buildings collapsed like overloaded card tables, while those between them were left cracked and twisted.

Damage was also heavy underground, where the quake ruptured gas lines and water mains. The fires started almost as soon as the shaking had stopped. Leaking natural gas fed the flames, while the ruptured water mains meant that there was no water in the city hydrants to fight the fires. The situation was almost identical to the disastrous fires that had all but burned the city to the ground after the deadly 1906 earthquake. This time, however, the fire department had had the foresight to purchase a portable hosing system, and the fires in the Marina district were eventually fought with water pumped from the bay. With firemen in short supply in the confusion that followed the quake, neighborhood residents helped man the hoses.

Out in the middle of San Francisco Bay, shaking caused a section of the upper level of the Bay Bridge to collapse onto the deck below, severing the main link between San Francisco and Oakland. Miraculously, no one was killed in the initial collapse, but two people later died when their car fell through the trapdoorlike gap in the upper deck in a panicked attempt to drive back across the bridge.

The deadliest effects of the quake, however, were felt not in Santa Cruz or San Francisco but even farther away from its center—in Oakland on the eastern side of San Francisco Bay. On the edge of the city, a mile long bilevel section of the Nimitz Freeway known as the Cypress Structure collapsed, killing forty-two people. Survivors recall watching the tremors roll down the freeway like waves: the upper deck rising up into the air and then collapsing onto the lanes below, its fall punctuated by explosions and fires. Cars underneath were flattened right down to their axles. For most, death was instantaneous. Others were trapped in the rubble for hours.

•   •   •

By nightfall, broader details of the damage caused by the quake were beginning to emerge, few of them good. More than one million homes were without power. The phone system was still working, but it was swamped with incoming calls—more than twenty-seven million in the first twenty-four hours—as those outside the area tried to reach friends and relatives. As a result, emergency communications inside the city were all but paralyzed. Traffic was surprisingly light in the emergency rooms of area hospitals, an encouraging sign that suggested that injuries and deaths from the quake might be lighter than expected. Television coverage, however, left one with indelible images: the broken Bay Bridge, a curtain of fire in the Marina district, and the remains of bodies and cars being lifted from the rubble of the Cypress Structure.

News organizations scrambled for facts in a world of rumor and worry. More concerned with getting it first than getting it right, morning newspapers would lead with the worst: more than two hundred dead in the collapse of the Cypress Structure. Mercifully, the totals were less than a third that number, but the truth was still bad enough: sixty-two dead and more than 3,500 injured. Twelve thousand people were left homeless. Eighteen thousand homes and more than 2,500 businesses were damaged. Total damage to the area was estimated at more than $6 billion. It was one of the most expensive natural disasters in U.S. history. The earthquake had lasted just six seconds.

•   •   •

The first thing Allan Lindh noticed was the wind. As a seismologist with the U.S. Geological Survey in Menlo Park he had spent a lot of time thinking about earthquakes in the Bay Area, but the wind was something he had never considered. When the earthquake struck, he was watching his son's soccer game in Palo Alto. As the ground shook beneath them the tall trees bordering the field began to sway back and forth, their moving branches kicking up a windstorm that sent clouds of dust and leaves swirling through the air. "It was a great place to watch an earthquake," Lindh later remarked. "There were all these big trees around and things were flying around. And I knew where my kid was and I was perfectly safe."

He had no idea how large the quake was or where it was located, but as soon as the shaking had stopped both he and his son headed back to his office at the USGS. For Lindh it was the start of several weeks of little or no sleep.

When they arrived at the USGS others were already at work trying to bring the computers back to life and pinpoint the quake's location. Power to the building was cut off, and while backup power supplies had kept some computers running, the shaking had been strong enough to cause others to crash, or shut down. While his son helped set up the emergency generators and rewire the building, Lindh went to work with the others on getting the computers back on line and sifting through the incoming data. In less than thirty minutes they had pinpointed the quake's location: the break had occurred on the San Andreas Fault sixty miles south of San Francisco and some ten miles north of Santa Cruz on the flanks of a 3,700-foot-high peak in the Santa Cruz Mountains known as Loma Prieta. The quake's hypocenter, the exact three-dimensional location of the initial fault break, was more than eleven miles underground.

Seismology laboratories around the world were also busy analyzing the quake, and over the next few hours a detailed picture of the direction of movement along the fault and the magnitude of the quake began to emerge. By nightfall, scientists would put the quake's magnitude at 7.1 on the Richter scale. It was the first major earthquake on the San Andreas Fault in northern California since the 1906 San Francisco quake. For Lindh and his co-workers, it was the quake they had been expecting.

■ ■ ■

In 1981 Allan Lindh and fellow USGS seismologist Bill Ellsworth published a paper predicting that the most likely site of the next major earthquake in northern California was along the San Andreas Fault in the Santa Cruz Mountains. While that portion of the fault had broken in the 1906 San Francisco earthquake along with segments of the fault on both the San Francisco and Point Reyes peninsulas, the patterns of past quakes in the area seemed to suggest that it broke both independently and more frequently than those other segments—giving rise to a major earthquake of its own every seventy-five years or so. From there the mathematics were simple and straightforward. By 1981 the San Andreas Fault in the Santa Cruz Mountains was ripe for another major earthquake. The next quake, Lindh and Ellsworth had predicted, would be of a magnitude of 6.5 or 7.0 and it would hit the area sometime before 1996.

For the next seven years nothing happened. But then in June 1988 a magnitude 5.0 earthquake rocked the Bay Area. It was much larger than any of the small tremors or quakes that had hit the area over the past few years, but its location was even more troubling to

Lindh and Ellsworth. The quake was centered near Lake Elsman just south of San Jose—at the northern end of the fault segment they had picked as the most likely source of the next major earthquake. Prodded by Lindh and other scientists who were concerned that it might be the foreshock of an even larger and more destructive quake, the state issued an "earthquake advisory," warning area residents that a major quake was likely over the next five days.

That same year, the USGS's own working group on earthquake probabilities had identified the same fault segment as the most likely source of the next major earthquake in northern California, reaffirming Lindh's earlier predictions. They had put the chance of a major quake in the Santa Cruz Mountains at 30 percent over the next thirty years, a number that suddenly looked conservative.

The warning period, however, came and went with no sign of a major quake. For the next fourteen months everything was quiet. Then in August 1989 another magnitude 5.0 earthquake struck the area. This time the shaking was strong enough to cause visible damage to homes and office buildings in the South Bay. In Los Gatos one man was killed by a falling wall.

The quake was centered in almost exactly the same spot as the 1988 quake, and for Lindh the situation was even more troubling. "You don't have to know quantum mechanics to know that two of something you're worried about are more worrisome than one," he said. The state issued another five-day earthquake advisory and this time it was taken to heart. Hundreds of workers and schoolchildren stayed home, stocking up on supplies of bottled water and flashlights. Others made use of the warning time to bolt water heaters and bookshelves to floors and walls—preparations that would later be put to good use. Once again the warning period came and went with no sign of a major quake. The Loma Prieta earthquake was just six weeks away.

• • •

In spite of all the warnings and predictions, when the expected quake finally struck on October 17, it took most area residents by surprise. Although the path of the San Andreas Fault through the Bay Area is one of the most intensively monitored fault zones in the world, there was no sign of creep or movement in the days and hours that immediately preceded the quake. Even though the quake had been predicted and preceded by a pair of warnings, collective memory is a short-term thing—and for the next several weeks the perception that the quake had caught scientists off guard would dominate the news.

Outside of a small handful of scientists, few were familiar with the predictions Lindh and others had been making about the high probability of a major quake in the Santa Cruz Mountains. Most local

residents, as well as a surprisingly large number of geologists and seismologists who were not intensively studying the area, had come to
think of the next major quake in the Bay Area as a repeat of the 1906
earthquake and had focused their attention on segments of the fault
that passed through the San Francisco Peninsula, believing that others
to the north or south were of little concern.

Scientists were also puzzled by the nature of movement along the
fault. Since the advent of plate tectonics in the early 1960s, the press
and general public had come to believe that movement along the San
Andreas was almost exclusively horizontal. In the Santa Cruz Mountains, however, movements along the fault during the Loma Prieta
quake had been more vertical than horizontal. In the main shock area,
rocks on the western side of the fault had moved some four feet northward and six feet upward. Like the quake itself, this too was pre-
dictable, but it would be understood only several days after the fact.

The San Andreas Fault makes a small bend as it passes through
the Santa Cruz Mountains, and the plates do not slide smoothly past
one another here but collide head-on. The force of this collision is
pushing rocks along the length of the bend up into the air just as it
does in the Transverse Ranges farther south near Los Angeles. "In ret-
rospect we were somewhat naive," Lindh said. "There are mountains
all over the place down there. How do you think they got there?"

• • •

When the Loma Prieta earthquake struck, Kerry Sieh was working in
his office at the California Institute of Technology in Pasadena. A few
doors away at the university's seismological laboratory, the seismographs skittered and danced as the waves of the tremor passed by, but
for Sieh and others in southern California who did not happen to be
staring at the needle of a seismograph, the quake was all but imperceptible.

A few minutes later a reporter from KWFB radio in Los Angeles
called to ask Sieh what he could tell them about the earthquake in San
Francisco. "What quake?" Sieh asked. It was the first he had heard of
the quake. After assuring them that he knew nothing about it, he suggested they call the seismology laboratory for more details.

As a professor of geology and geophysics at Caltech and one of
the fathers of modern earthquake geology, Sieh had grown accustomed
to calls from the press. They came after almost every small tremor in
California. Privately he thought to himself, "Oh boy. Another 4 or 4.5
earthquake and somebody is going to spend another fifteen minutes of
my life trying to get a sound bite."

Just as he was about to head out the door to go to the seismology
lab and check up on the quake himself, a student stopped by his office

with a hand-held portable TV. When they turned it on the first thing they saw on its tiny two-inch screen was the collapsed section of the Bay Bridge. As reports of damage up and down the bay began to trickle in, Sieh knew beyond any shadow of a doubt that this was a major quake. And, like Allan Lindh in Palo Alto, he knew almost immediately where it was centered—somewhere along the San Andreas Fault in the Santa Cruz Mountains, the fault segment that he, like Lindh and the others, had picked as the most likely site of the next major earthquake in northern California.

• • •

In 1975 Kerry Sieh was on the verge of dropping out of graduate school at Stanford when Dick Jahns, the dean of the university's School of Earth Sciences, took an interest in his ideas for dating prehistoric earthquakes. As an undergraduate Sieh had studied the San Jacinto Fault in southern California with a grant from the National Science Foundation in the hope of finding evidence of past earthquakes, but he had been unable to find anyone on the faculty at Stanford who would allow him to continue that work in graduate school—until Jahns gave his blessing to the project. Privately, the dean believed the project was doomed to failure. But he also believed, as Sieh later found out, "that an individual's initiative should be given a chance to bloom."

Instead of working on the San Jacinto Fault, however, Jahns suggested that he work on the San Andreas, the "master actor," as he termed it, in the drama of plate tectonics. And, although the fault ran right along the edge of the Stanford campus, Jahns urged him not to focus on the San Andreas Fault in northern California, where the heavy rainfall and thick vegetation make work difficult, but in southern California, where its features were clearer and better preserved.

One of the first things to do, Jahns believed, was to walk along the San Andreas and get a feel for its size and features. Sieh started walking and eventually covered more than one hundred miles of the fault in southern California, looking for a site to start his research. East of Los Angeles on the edge of the Mojave Desert in the banks of a small, seasonal stream known as Pallette Creek he found what he would later come to regard as the Rosetta Stone—a thirty-foot sequence of sediments and peats less than two thousand years old that contained evidence of more than a dozen major earthquakes.

For decades, geologists and seismologists had been limited to studying earthquakes as they happened or by sifting through the all-too-often anecdotal accounts of earthquakes from the recent past. Sieh's work, however, provided a way of studying prehistoric earthquakes. Layers of peat and silt alongside Pallette Creek had been repeatedly ruptured by earthquakes. Orderly bands of silt and peat would be broken or displaced by faulting and then buried by new layers of sediment. By dating the uppermost layers of sediment broken by these buried faults, Sieh could determine the timing of past earthquakes, while the size of the offsets gave a rough idea of magnitude. What he found was that great earthquakes on this segment of the San Andreas Fault seemed to occur every 120 years or so. It was the first time anyone could say with any certainty whether earthquakes on the San Andreas Fault were likely to occur every five years or every five hundred.

Although the San Andreas Fault is a continuous feature, scientists in the late seventies were already beginning to understand that different segments of the fault moved at different rates. Some areas produced dozens of small quakes each year, while others remained locked for decades, releasing their stored-up energy in sudden destructive quakes. Sieh's work provided a way of studying these patterns, and over the next few years others would begin using his techniques up and down the length of the fault to make predictions about the timing and frequency of large earthquakes in California.

■ ■ ■

Less than four hours after hearing about the Loma Prieta earthquake, Sieh and a group of graduate students from Caltech were on their way north with maps and surveying equipment to study the surface breaks left by the quake. After more than a decade of digging through fine layers of peat and silt looking for the subtle signs of past earthquakes, studying the cracks and fissures left by a major quake was the chance of a lifetime.

There was a full moon that night, and by the time they reached Salinas one hundred miles south of San Francisco it was already high in the sky. Everything around them seemed to be lit by a cool, blue light. The streets, stores, and homes around them were completely dark. The quake had left the area without power. It was about that time that they also realized that they had stopped only once to buy gas on the drive up from Los Angeles and that buying gasoline was suddenly a problem. With no electricity gas pumps were as dead as the street lights and they had little more than a quarter of a tank left in their trucks. Phone lines were out of order. There was no way to get messages in or out. Just before dawn they were finally able to buy gas, filling up their trucks at the Salsipuedes Fire Station in the Santa Cruz Mountains and promising to send a check later to pay for it all.

With a quake as large as the Loma Prieta earthquake, the trace of the fault should have been easy to find. But as they worked their way northward on back roads through the mountains following the known trace of the fault, there was no sign of the quake at all. The ground was almost untouched. There were no visible offsets or fault scarps; no sign of sheared fences or displaced roads and tunnels. By the time they reached the northern end of the epicentral area near the intersection of Highway 17 and Summit Road, Sieh was thoroughly astonished. Aside from a pattern of ground cracks near the junction of the main highway, the quake had not even broken the surface.

On Highway 17 he ran into Pat Williams, a former student who had gone on to the University of California at Berkeley. Williams was studying slumps, or small landslides of soil and rock alongside the highway, his every move being followed by a camera crew from CBS. Sieh, who had been able to avoid the press while traveling, agreed to do an on-camera interview for television if they would allow him to use their cellular telephone to call the Seismology Lab at Caltech and tell them that there had been no surface faulting. It was all a profound disappointment. With no surface faulting, there was not much he and his group could do to help analyze the quake. Less than twenty-four hours after he had left, Sieh was on his way back to Pasadena.

• • •

Over the next several days geologists from the U.S. Geological Survey and area universities scrambled over the rugged mountains and confirmed what Sieh's early-morning trip had already suggested: although the quake had damaged homes and buildings more than fifty miles away, it had not even broken the surface along the fault. The nature of movement and the size of the offset along the fault were all hidden underground. For geologists who were interested in other aspects of the quake, however, there were still plenty of features to study. For the next several weeks they went to work mapping the ground cracks and landslides left by the quake and studying the damage to roads, homes, and other man-made structures.

Seismologists were also busy after the quake. In addition to the data picked up by the array of strain meters, creep meters, and seismographs strung out along the fault, there was also a steady stream of information coming in from the aftershocks that followed the main shock—as many as seventy per hour in the first twenty-four hours. Scientists did not always wait passively for the data to come rattling in, but repositioned monitors and instruments to take advantage of these smaller tremors to learn more about side faults and the behavior of different types of soils and rocks to shaking.

While the main shock had come without any kind of short-term warning, these aftershocks were expected and closely monitored. Seismologists set up an early-warning system of sorts for rescue workers digging through the unstable rubble of the Cypress Structure in Oakland—monitoring the aftershocks emanating from the quake's center with instruments in the Santa Cruz Mountains and then sending a warning to Oakland by radio to give them a few seconds of warning before the tremors arrived.

For rescue workers those first few days after the quake were critical. Experiences from the deadly quakes in Mexico City and Armenia had shown that the first forty-eight hours were "golden"—the time before shock and dehydration take their toll and buried survivors are most likely to be found alive.

Other needs were no less pressing. More than twelve thousand people had been left homeless by the quake. They were in need of not just shelter but food and water as well. In the five months that preceded the quake there had been just two inches of rain. In the week that followed more than ten inches fell, making rescue work and repair work even more difficult and shelter even more essential. From Santa Cruz to San Francisco there were literally thousands of homes, buildings, bridges, and overpasses that needed to be inspected for damage. Others were without power, water, or gas, and in some hard-hit areas restoring them would take days.

An uneasy sense of calm prevailed over the area for the remainder of the week as businesses closed and thousands of workers stayed home. On Monday the city would be going back to work, but with the Bay Bridge out of commission and key sections of freeway in both San Francisco and Oakland indefinitely closed, getting there was something most people did not want to think about. In the meantime a flotilla of ferries was pieced together to shuttle workers back and forth across the bay when work resumed. By Friday the newspapers were predicting gridlock.

• • •

After two years of traveling up and down the West Coast and talking to geologists about the mobile Pacific Plate, I had no doubt that earthquakes were inevitable in California. But like most everyone else I had no idea that a major earthquake would strike northern California in October 1989.

On the day of the quake I was watching sea lions at Cape Arago near Coos Bay on the central Oregon coast. It had been a bright, sunny day with no sign of clouds or trouble. The research for my book, or so I had thought, was nearly finished. I was driving back to my hotel at about sunset with National Public Radio's news program "All Things Considered" fading in and out on my car radio when I heard the news: a major quake had just struck the San Francisco Bay Area and a portion of the Bay Bridge had collapsed. I pulled over to the side of the road and for the next half-hour I sat and listened, wondering how friends and relatives in the area were doing and thinking that the research for my book was now far from finished.

My first thought was to head south immediately, but after thinking it over for a few hours that night, I decided that I would simply be underfoot as reporters from around the world flocked to the area. It would also take scientists several weeks or even months to sort out what had happened. With interviews and field trips scheduled for the rest of the week in Oregon, I decided to stay put.

By the weekend I finally headed south, driving down the coast and taking an extra day to drive the back roads out around the mountainous shoulder of Cape Mendocino near the California–Oregon border where the San Andreas Fault makes an abrupt westward bend and disappears. By Sunday morning I finally reached San Francisco.

• • •

There were few signs of damage from the Marin Headlands on the far side of the Golden Gate Bridge. The view of the city was framed by the bridge's red cables. No fires burned in the Marina district. The Bay Bridge seemed whole and complete. Peregrine falcons soared in the updrafts along the steep coastal cliffs.

Traffic was surprisingly light on the bridge even though the tolls had been suspended in the aftermath of the quake. I parked near Golden Gate Park and took the metro downtown. I wanted to walk around the city and see things on foot without having to worry about traffic or parking. Signs of damage were slight in the neighborhoods near the park—cracked windows and walls here and there that would have been easy to overlook if one had not known that the area had been hit by a major quake just six days earlier. Near Van Ness the metro went underground, heading toward the Embarcadero Center and the heart of the downtown financial district. The end of the line.

Up at street level in the Embarcadero the signs of the quake were everywhere. Streets and sidewalks were cracked and tilted—like the pieces of a broken plate that have been imperfectly glued back together. The cable cars that normally carry tourists up California Street and Chinatown were strangely quiet, shut down in deference to the quake. The tall, modern office buildings that towered over the narrow city streets seemed to have escaped untouched, their smooth exteriors concealing the jumbled offices and conference rooms inside. Notices and inspection certificates pasted on doors and street-level windows gave a clearer picture of the conditions inside. Some were safe for occupancy with no restrictions on use; others were still without power or natural gas.

The signs of damage on the lower, older buildings scattered among the high rises were plainly visible. Walls and windows were peppered with cracks. Broken bricks and windows littered the sidewalks. Up at the corner of California and Kearny an eight-story office building listed precariously to port, surrounded by police barricades. Tenants wheeled desks and filing cabinets into waiting trucks and trailers, while others carried potted plants and framed posters to cars parked nearby.

Elsewhere, preparations were already under way for the coming workday. At the foot of Market Street, workers were busy surrounding the cracked facade of the Southern Pacific Building with scaffolding and covered walkways to protect pedestrians from falling debris. Building inspectors with hard hats and clipboards walked from building to building, stepping inside to look for signs of damage. On top of the Ferry Building's campanile-like tower, workers tried to straighten with a block and tackle the flagpole that had been knocked askew during the quake. From street level the workers looked almost like a reenactment of the marines raising the flag at Iwo Jima.

The damage downtown was not random but closely tied to the rocks below—or, to be more explicit, the lack of them. Since the early 1850s, more than one-third of San Francisco Bay's original area had been filled in with mud, rubble, and sand to make room for freeways, factories, airports, and even homes. While that filled land feels deceptively solid underfoot, during an earthquake it becomes as soft and unstable as Jell-O. Most of the downtown business district, as well as the heavily damaged Marina district on the other side of town, had been built on top of loose bay fill.

•••

Up at Union Square the streets and park were nearly empty, overlooked by the boarded-up windows of the I. Magnin department store. A few blocks away in Chinatown red and white banners were strung across Grant Street like holiday decorations, but down at street level one's eyes were drawn not so much to the banners as to the police cordons and barricades that sealed off buildings and warned pedestrians away. The old storefront buildings and tenement apartment houses of the neighborhood were still standing, but many had been seriously damaged.

In North Beach the restaurants and coffee shops were almost empty. Traffic was so light that you could walk out into the middle of Broadway and stop in the center of the street. Looking toward downtown you could see the Transamerica Building and behind it the clean, curving lines of the Bay Bridge cutting across the skyline. It looked almost normal—until one realized that there was no traffic at all on the bridge.

Climbing to the top of Russian Hill you could look out over hills of tightly packed homes and apartment buildings and see almost no sign of the quake. Sailboats jockeyed for position out in San Francisco Bay, circling Alcatraz Island and passing under the Bay Bridge. In a small neighborhood park at the top of Vallejo Street a chicken picked through the grass. In the fires that followed the 1906 earthquake, this neighborhood had burned to the ground. This time it had escaped almost untouched.

• • •

Thirty blocks away in the Marina district, however, the force of the quake was plain to see. Barricades manned by police sealed off streets leading into the neighborhood, while bordering streets and sidewalks were filled with thousands of people. Looking past the barricades and down the empty streets you could see rows of apartment buildings and homes knocked askew by the quake. Some leaned against each other at precarious angles while others were propped up with makeshift shores.

Lombard Street looked like a vast flea market as homeless residents huddled on street corners amid piles of salvaged clothing and furniture—a chaotic collection of stereos, tables, chairs, potted plants, and paintings. Others loaded dressers and plastic trash bags crammed full of clothes into waiting cars and rented trucks. After tying a mattress onto the back of a fully loaded pickup truck, two men let out a victorious whoop for anyone within earshot: "Pacific Heights here we come!" Although it was just ten blocks away on top of the hills that rose above the Marina district, Pacific Heights had ridden out the quake with barely a tremor.

On television the aftermath of the disaster had seemed fast paced, but in person everything seemed to move with painful slowness. Entry into the neighborhood was by permit only. Tight knots of people gathered around policemen manning the barricades, some showing permits, others pleading for another chance to go in to salvage more of their possessions. For some, there had been no chance to save anything. At the corner of Northpoint and Fillmore a bulldozer was already at work on the remains of an apartment building that had collapsed during the quake. The piles of rubble pushed about by the machine were filled with glimpses of the lives that had once been lived inside: a splintered bookcase and broken couch; a red shirt and a pair of jeans floating in a sea of broken plaster and plumbing.

Down at Marina Green on the edge of the bay more than a thousand people were walking up and down the park, stopping here and there to study the huge cracks in the ground or the twisted shapes of the million-dollar homes overlooking the edge of the park and bay. I had never known that so many people could be so quiet. On a normal day the air would have been filled with small talk about families and work or an upcoming vacation. That day, however, almost no one talked. It was as if the sudden, destructive power of the quake was something best contemplated in silence.

That night I was on my way to Los Angeles to catch a mid-week flight to Baja California. When I checked into my hotel near the Los Angeles Airport, the clerk behind the desk was so busy reading the emergency pages at the front of the phone book that she didn't notice me step in.

"Looking for something?" I asked.

"No," she said. "I was just reading about what to do in case of an earthquake. It seemed like a good idea."

•   •   •

During the 1906 San Francisco earthquake the San Andreas Fault broke in a clean line all the way from San Juan Bautista 100 miles south of San Francisco to Point Arena on the northern California coast, a distance of more than 350 miles. Registering an estimated 8.3 on the Richter scale, it was more than ten times the size of the 1989 Loma Prieta quake. It was felt all the way from Los Angeles to Coos Bay on the central Oregon coast, a distance of more than 750 miles and as far inland as central Nevada. On the Point Reyes Peninsula where the quake was centered, the fault slipped by more than fifteen feet.

The quake struck at seventeen minutes past five in the morning, and in the Bay Area its force threw sleepers from their beds and stopped pendulum clocks. Drawers and cupboards were thrown open and their contents scattered across rooms and floors. Those who happened to be up and about when the tremors hit were knocked off their feet. Outside, the shaking was so intense that trees swayed from side to side, with their tops all but touching the ground. Riders on horseback were thrown to the ground. Here and there along the fault fissures three and four feet wide tore the ground. Roads and fencelines crossing the fault were torn in half as land on the western side of the fault surged northward.

In San Francisco hundreds of buildings and homes collapsed. Toppled stoves and lanterns touched off fires that started almost as soon as the shaking had stopped. When firemen arrived they found that the quake had also ruptured the city's water mains. With no water in the city hydrants, helpless fire crews fought the fires with dynamite, blowing up whole city blocks in a futile effort to create firebreaks to check the spread of the flames. The fires burned out of control for three days. When it was all over more than a third of the city had been burned to the ground.

While the 1989 Loma Prieta earthquake played to a live TV audience, news was heavily censored in the chaos that followed the 1906 quake. Looters were strung up on city streets, while those who lit fires to keep warm in defiance of police orders were shot on sight. Pictures of the devastation were so disturbing that city officials confiscated photographers' photographic plates in a futile effort to cover up the impact of the quake. Months later the army would put the official death toll at 498, but the real number was far higher. Historians studying city documents today believe that the quake and the fire that followed it claimed more than three thousand lives.

It was not the first earthquake to strike San Francisco, but at the time almost nothing was known about the causes of earthquakes or the location of faults in the area. Less than a week after the quake the California Earthquake Investigation Commission was set up. Funded by the Carnegie Foundation, it brought geologists from around the United States to California to study the effects of the quake and the nature of movements along what would eventually become known as the San Andreas Fault.

Although they lacked the sophisticated surveying techniques and monitoring devices of science today, geologists in 1906 did such an exceptional job of mapping the fault and the features left by the quake that their reports and field notes are still an invaluable source of information. They were also quick to grasp the fault's larger reach. It was not, they speculated, merely a local feature but part of a larger system of faults and breaks that extended southward all the way to the Mexican border and northward for several hundred miles toward Oregon.

But while geologists of the early 1900s could perceive the larger scope of the fault, the nature of movement along it was harder to understand. While rocks in places like the Olema Valley in the Point Reyes Peninsula had shot past one another by as much as fifteen feet during the quake, they assumed that movements along the fault were actually vertical. Although they had carefully noted that rocks on the western side of the fault were granitic and those on the eastern side were marine, they attributed the difference to uplift and erosion, not continental drift. The theory of plate tectonics was still more than fifty years away. They had no trouble imagining that the continents had moved up and down (after all, the presence of mountain ranges left no doubt that rocks could move upward), but horizontal movements were all but unthinkable.

• • •

That perception of the San Andreas Fault was not seriously questioned until the early 1950s, when two geologists from Richfield Petroleum in Los Angeles, Tom Diblee and Mason Hill, began studying the patterns of rocks alongside the fault. As a field geologist for Richfield, Diblee had probably mapped more of California than anyone else; and as Diblee's supervisor, Hill had probably spent more time than anyone else looking at Diblee's maps. Together they had a broader perspective on California geology than almost anyone before or since, and what they saw was inescapable: you could follow a band of rock from central California to the San Andreas Fault only to find the same band of rock on the opposite side of the fault a hundred miles or more farther north—as if it had been displaced. Movements along the fault, they suspected, were not vertical but horizontal. Land on the western side of the fault had been driven northward, not by a few feet but by more than one hundred miles.

Diblee and Hill decided to present their revolutionary ideas at a 1952 meeting of the American Association of Petroleum Geologists in Los Angeles. They were not, Hill recalls, hooted down but received polite applause. "I was surprised how well it was received," Hill told me one morning at his home in the Whittier Hills near Los Angeles. "But later I found out people just thought we were crazy."

It would take more than fifteen years to prove Hill and Diblee's ideas to the rest of the scientific community. Final acceptance, however, would not come from more work on land but from the discovery of midocean ridges and patterns of magnetic anomalies on the seafloor—evidence that would open the door to plate tectonics and the understanding that the continents had drifted across the face of the earth. In this new framework the San Andreas Fault was no longer just a mysterious source of earthquakes but a boundary between two mobile plates of the earth's crust.

Hill is not bitter that it took so long for his ideas to be accepted. "There is no such thing as truth in science. You come up with ideas based on what you've seen. If you want to upset the conventional theories there's a lot of opposition. People don't like to see their ideas change. That isn't bad. It forces you to marshal your evidence."

Ironically enough, by 1989 Hill and Diblee's idea of northward movement along the fault had become so well accepted that scientists were puzzled by the fact that fault movements during the Loma Prieta earthquake had been more vertical than horizontal. It was a sign of how much the conventional wisdom had changed.

• • •

Six weeks after the Loma Prieta earthquake more than three thousand earth scientists from around the world came to San Francisco for the annual fall meeting of the American Geophysical Union (AGU). In an average year, talks and conferences at the meeting cover everything from atmospheric chemistry to deep-sea mineral deposits, but in 1989 the Loma Prieta earthquake occupied center stage. For those who had been studying the quake it was a chance to meet and compare data.

The October quake had caught most scientists flat-footed. Although it had occurred along one of the most intensively monitored fault segments in the world, it had come without any sign of a short-term warning. Not only did the direction of movement along the fault during the quake—more vertical than horizontal—seem to contradict conventional wisdom, but the damage it caused was unexpectedly widespread. Although it had been centered in the Santa Cruz Mountains, some of the most heavily damaged areas were in San Francisco and Oakland more than seventy miles away.

By December, however, the puzzling features of the Loma Prieta earthquake were beginning to make sense. While there was still no sign of a short-term warning, after having had time to look over the data scientists could see that their long-term predictions had been right on the money.

Damage from the quake had been predicted as well. Maps of seismic hazards in the Bay Area made in the 1970s were almost interchangeable with those of the damage caused by the Loma Prieta earthquake. Santa Cruz's Pacific Garden Mall, the Marina district in San Francisco, and the site of the Cypress Structure in Oakland had all been clearly identified as areas of known seismic risk almost twenty years earlier.

Not only was damage limited to areas of known risk, it also typically occurred in structures of known risk as well. From Watsonville to San Francisco the most heavily damaged buildings were those that had been built out of unreinforced masonry or had not been bolted to their foundations. State officials, for example, had known for years that the Cypress Structure on the Nimitz Freeway would fare poorly during a major quake, but with tax dollars in short supply, the money had been spent on other projects. All of these hazards had been known for more than a decade. "The quake," in the words of one keynote speaker at the AGU meeting, "was a reminder of what we already knew." But not everyone was so reassured.

• • •

A month and a half after the Loma Prieta earthquake, Kerry Sieh was still puzzled that the quake had not reached the surface. "At the time, I was quoted as saying that I was flabbergasted that there wasn't any surface offset," he said. "Of course, then people started calling, saying, 'Gee, were you really flabbergasted?' Then I felt that maybe I should have really understood what was going on and that there shouldn't have been any surface faulting. But no. The fact is that I was really, utterly astonished that for a magnitude 7 earthquake there had been no surface faulting."

The fact that large earthquakes like Loma Prieta could occur without rupturing the ground was disturbing because Sieh's work on the timing and frequency of prehistoric earthquakes was predicated on the idea that major quakes left cracks and breaks behind that could be detected and dated by careful study. Far from invalidating his work, however, the absence of surface faulting with the Loma Prieta quake suggested that some large, prehistoric earthquakes may be undetectable—and that studies like Sieh's may have actually underestimated the frequency of damaging quakes in some areas.

That possibility is particularly troubling in Los Angeles, where Sieh lives and works, because the area is cut by a number of "blind thrusts"—buried faults whose movements, like those seen on the San Andreas during the Loma Prieta earthquake, are nearly vertical and seldom break the surface. They are not part of the San Andreas but part of another added source of risk, which Sieh refers to as the "rogue elephant" of earthquake prediction in southern California.

For the past two to three million years the Pacific Plate has been moving not only northward but slightly eastward as well. As mentioned earlier, geologists see evidence for this in the mountains that fringe the West Coast all the way from Mexico to Alaska as well as in faraway places like the Hawaiian Islands, where kinks and bends in the trend of the island chain record changes of direction in the Pacific Plate. In southern California, satellite data suggest that this eastward movement is actively compressing the area by as much as one centimeter per year. From time to time this buildup of strain is released by sudden upward slips and earthquakes along the thrust faults that cut through the region.

These buried faults not only played a critical role in the formation of the region's phenomenally productive oil fields, but they underlie most of its prominent topographic features as well. The Santa Monica Mountains, the Hollywood Hills, and the Palos Verdes Peninsula are all the products of thrust faults. And while those buried faults are much smaller than the San Andreas, they are still capable of producing damaging earthquakes. Both the 1971 San Fernando earthquake and the 1985 Whittier Narrows earthquake were caused by movements along buried thrust faults.

For Sieh the Loma Prieta earthquake also showed how much damage a major quake on the San Andreas Fault in southern California could cause in Los Angeles. "That's one thing this earthquake helped me understand," he said. "Just because we're fifty-five kilometers away from the fault doesn't mean there isn't the potential for tens of billions of dollars of damage from a large earthquake on the San Andreas Fault." Although the fault does not run through the center of the city as it does in the Bay Area farther north, its path through southern California down the backside of the San Gabriel Mountains is twenty miles closer to downtown Los Angeles than the site of the Loma Prieta earthquake was to Oakland and San Francisco.

• • •

There has been no major earthquake on the San Andreas Fault in southern California since 1857. Since that time Los Angeles has grown from a small village into one of the largest cities in the world. But because of that long period of quiet, many southern Californians have come to view earthquakes as more of a problem for San Francisco than for Los Angeles. But from the Salton Sea near the Mexican border to Cajon Pass, the San Andreas Fault in southern California is ripe for another major quake. Instead of relieving pressure, the cluster of large quakes that rocked the deserts of southern California in 1992 at Landers, Big Bear, Joshua Tree, and Mojave, seems to have made the possibility of a major quake on the San Andreas Fault in southern California even more likely. All of these quakes, including the massive 7.5 quake that struck the Landers area, were located on side faults bordering the San Andreas.

As for the San Andreas, Sieh and other scientists typically divide its two-hundred mile reach in southern California into as many as three segments. The probabilities of a magnitude 7.0 or larger earthquake on those three segments range from 20 to 40 percent. When all of these segments are considered together, the probability of a major quake occurring somewhere on the San Andreas Fault in southern California is thought to be 60 percent over the next thirty years—more than twice as likely as the Loma Prieta earthquake was in northern California.

"My own hunch," Sieh said, "is that the next great earthquake in southern California will be a 7.5 near Palm Springs or Indio, with rupturing all the way to the Salton Sea, a two-hundred-kilometer length." There is also the possibility of a magnitude 8.0, he added, with the fault breaking all the way from the Salton Sea to Cajon Pass, a distance of more than three hundred kilometers.

With the probability of a major quake running in at greater than 50 percent in both northern and southern California, some critics have accused scientists of trying to cover all the bases. In Sieh's view, however, those high numbers are nothing more, unfortunately, than an accurate picture of the risks involved. It was not, he added, just a coincidence that the Loma Prieta earthquake was located on the fault segment that had been identified as the most likely source of the next major earthquake in northern California. The probabilities scientists have been publishing for the last five to ten years have been based on solid science. And while it may seem as if most of the state is covered by one prediction or another, scientists actually believe that well over half of the state—as well as more than a third of the San Andreas Fault, the final segment that runs from Point Reyes to Cape Mendocino—has little or no chance of a major earthquake over the next thirty years. The problem lies not with the predictions of scientists, but with the fact that the state's two largest cities have been built on top of one of the most active fault zones in the world.

As for predicting the day and hour when that next quake will strike, Sieh doubted that would happen anytime soon. "We can't understand earthquakes for the same reason psychiatrists can't understand schizophrenia. There are simply too many variables." In the meantime there is no doubt that the Loma Prieta quake has thrust earthquakes and earthquake prediction into the public eye.

"It's become clear to me that in spite of our desire to predict earthquakes, what really drives earthquake science is the occurrence of earthquakes," he said. "We learned a phenomenal amount about earthquakes in the Los Angeles Basin after the San Fernando earthquake and we're going to learn a lot about faulting up here because of the Loma Prieta earthquake." But while those lessons may take several years to learn, the biggest lesson of all for California, in Sieh's opinion, was evident almost as soon as the shaking had stopped: "Big earthquakes happen. Big earthquakes happen in populated areas."

• • •

During the 1989 AGU meeting in San Francisco, Allan Lindh was wired to the San Andreas Fault. By coincidence the meeting happened to fall during one of Lindh's weeks of being on call. A beeper on his belt linked him to computers at the U.S. Geological Survey in Menlo Park tracking earthquakes in the state. On a typical day there are more than one hundred small earthquakes in California. The beeper goes off for every 3.5 or larger earthquake in the state, sending Lindh or whoever is on call down to the office to decide whether state officials should be notified. As a matter of habit they notify the state of every earthquake larger than a magnitude of 4.0 and the press of anything larger than a 3.0 in the Bay Area—"basically," Lindh said, "anything that's large enough to be felt and frighten people."

Earthquakes large enough to trigger the beeper happen two or three times a month on average and sometimes more frequently. They are as likely to strike at three in the morning as at three in the afternoon, and at times Lindh's wife has been known to remark that being married to a seismologist in the Bay Area is like being married to an obstetrician, without the Mercedes-Benz. It comes with the territory. "We drive old pickup trucks, but we have to get up in the middle of the night, too," he explained.

Since the Loma Prieta earthquake, sleep has been in short supply. As the survey's leading guru on earthquake prediction and the author of the report that most closely predicted the Loma Prieta quake, Lindh has been inundated with requests for talks and interviews both on camera and off. Visually, his role as a geological prophet is enhanced by a full-length beard of blond hair that reaches more than half-way to his waist. He is not a man one loses sight of easily in a crowd and since the quake Lindh has grown as accustomed to crowds and interviews as a seasoned politician. What were once considered the finer points of seismology are now front-page news, and after a small quake the press no longer waits for scientists at the USGS to call them.

"If there is a 3.5 at 6:00 A.M. you know damn well somebody's going to want an interview on the morning news telling them what that means," he said. "And if you tell them that it doesn't mean any-thing, that's not an acceptable answer anymore. The question is: Is it an aftershock? And if the answer is no, the question is: Is it a fore-shock? They want to know what the history of earthquakes is on that segment and what the probabilities are."

But while the debate has grown more sophisticated, there is still a wide gap between public expectations and scientific capabilities. While the public wants predictions in terms of days and hours, Lindh and other scientists talk about earthquakes in terms of thirty-year proba-bilities. And while critics are far from satisfied, Lindh believes that the era of earthquake prediction has already arrived.

"When you get to the point where you're observing the patterns in real time and doing your best to understand them and people are ask-ing you pointed questions—that's earthquake prediction. As for the crowd that wants you to really predict, you look them in the eye and say, 'We're going to further refine the probabilities.'"

"My proudest moment was when Jimmy Breslin [the New York newspaper columnist] came into my office after the quake to talk to me," he added. "He wanted to discuss probabilities. He was skeptical at first. But after a while he decided that it was like handicapping horses. And in northern California we had decided that the Santa Cruz Mountain segment of the fault was likely to go next, and in fact it came in first in the race. After that, Breslin decided that our predictions weren't too bad."

■   ■   ■

In the 1960s and 1970s earth scientists went through what Lindh refers to as the "alchemical period" of earthquake prediction. "We weren't quite using magic formulas and secret symbols, but there was this approach to earthquake prediction which consisted of measuring something and looking for changes and then trying to decode the signals. It was as if there was going to be something magically coded within them that would tell us the secret of when the next earthquake would be. You didn't even have to understand what you were measuring because you didn't want to get too caught up in preconceptions which might blind you to the truth."

By the late 1970s, however, it was clear that there was no magic formula for predicting earthquakes. Scientists, Lindh said, "switched to plan B, which should have been plan A all along," Lindh said. "That plan was that you treat earthquakes as physical and geological problems and that you try to understand the problem physically and geologically. And you do good geology and good geophysics and you try to understand things from the bottom up in the hope that your understanding will someday get you to the point where you can predict earthquakes."

Instead of looking for magic symbols and formulas, geologists and geophysicists began trying to learn about the timing and frequency of prehistoric earthquakes through a variety of detailed new techniques. In addition to the trenching studies like those pioneered by Kerry Sieh in the mid-1970s, two of the most important approaches centered around using geologic evidence to determine slip rates to calculate how fast different segments of the fault were moving, and using seismological evidence to identify seismic gaps—areas that were loaded and ready for a major earthquake.

• • •

Of these two key techniques, the calculations and importance of slip rates along the fault are probably the easiest to understand. If, for example, you know that a certain fault segment slipped by fifty centimeters during the last major earthquake and that rocks alongside the fault are moving at roughly a centimeter per year, you could predict that an earthquake was likely to strike the area every fifty years.

It was these kinds of calculations that Allan Lindh used to make his first prediction about the probability of a major earthquake in the Santa Cruz Mountains. During the 1906 San Francisco earthquake the fault slipped by about five feet or roughly 150 centimeters. With the faulting moving at about two centimeters per year in the area, Lindh reasoned that this fault segment was due for an earthquake in about seventy-five years—anytime after the seventy-fifth anniversary of the

San Francisco earthquake in 1981, the same year he published his first prediction of the likelihood of a major quake in the Santa Cruz Mountains.

The calculations are so simple that a grade school student could do them, but the difficulty lies not in the equations themselves but in deciding what numbers to put into them. While the speed of the Pacific Plate is easy to track in the Gulf of California along the East Pacific Rise, that steady six centimeters per year of movement in Mexico is split between literally hundreds of side faults in northern California. Determining the rate of movement along a given segment of the fault or a side fault can involve everything from years of careful fieldwork measuring creep and movement to the split-second calculations of earth-orbiting satellites tracking the drift of mountains and continents like the movements of ships at sea.

While Lindh's 1981 calculations had an uncanny accuracy, he believes the results were somewhat deceptive. Although such calculations can provide a good ballpark figure of timing and frequency, no one should think they are foolproof. "We weren't that smart," he said. "That was a very, very lucky guess. Usually when you make a guess like that you're wrong."

Slip rates, however, were not the only piece of information that suggested that the Santa Cruz Mountains were ripe for a major quake. Several years of plotting small tremors and quakes along the fault in northern California had also shown that the area was a "seismic gap." And by the late 1980s it was this combination of evidence—seismic gaps and slip rates—that again led Lindh and his fellow scientists to predict that the Santa Cruz Mountains segment of the fault was the most likely sight of the next major earthquake in northern California.

• • •

The idea behind the so-called seismic gap theory of earthquake prediction is complex, but it is much easier to understand if you think of a fault as a giant spring being slowly compressed and loaded with energy. That energy can be released only by movements that are inevitably accompanied by earthquakes. Scientists are unsure why, but different segments of the fault give rise to earthquakes at different rates. Some steadily release their built-up energy in hundreds of small quakes each year, while others store their energy up for decades, releasing it in a single giant earthquake.

For seismologists these zones of inactivity are easy to spot. By plotting the location and depth of earthquakes along the fault on a graph, patterns of activity quickly become apparent. While some parts of the graph are soon filled in by hundreds of small quakes in a decade or two, others are left almost blank—a "gap" of seismic activity in the midst of a sea of small earthquakes.

Prior to the Loma Prieta earthquake there was a seismic gap between San Juan Bautista one hundred miles south of San Francisco all the way to the Portola Valley halfway up the San Francisco Peninsula on the edge of the foothills west of Menlo Park. But although the Loma Prieta earthquake filled that gap, the Loma Prieta Bay Area's problems are far from finished. A troubling set of both geologic and seismological data suggest that the San Francisco Bay Area is poised for yet another giant quake.

• • •

The 1906 San Francisco earthquake released so much energy that the San Andreas Fault was quiet and earthquake-free for nearly forty years. By the late 1940s, however, small tremors and quakes began to rock the area again, suggesting that strain was building up on the fault. Since then the frequency of small earthquakes has grown exponentially.

While some Bay Area residents took comfort in the small quakes that rocked the area with increasing frequency in the 1970s and 1980s—assuming they were releasing pressure on the fault—the amount of energy released by those small 3.0- and 4.0-magnitude quakes was insignificant in comparison to the amount of energy released by a major quake like the ones in 1989 and 1906. In fact, they did little to relieve the strain that was building up on the fault and were merely a sign of growing instability underground.

The Richter scale is the internationally recognized scale of earthquake magnitude, measuring the amplitude of ground waves created by a quake at a set distance from its center. Those waves can vary greatly in size—anywhere from a few thousandths of an inch to several feet—and as a result the Richter scale is logarithmic: each whole-number increase equals a tenfold increase in size. A magnitude 5.0 earthquake, for example, produces ground waves that are ten times larger than those of a magnitude 4.0. A 7.0 earthquake, in turn, is one hundred times larger than a 5.0.

While the differences in wave size are great, the differences in the amount of energy released are even greater—and multiply even faster with increasing magnitude. A magnitude 7.0 earthquake, for example, releases more than nine hundred times as much energy as a 5.0 quake and more than thirty thousand times as much energy as a 4.0 quake. While the 1988 and 1989 Lake Elsman earthquakes near San Jose were strong enough to damage homes and be readily felt throughout the Bay Area, it would have taken nearly a thousand of these relatively large quakes to release the amount of energy given off in the Loma Prieta earthquake.

While the Loma Prieta quake filled the seismic gap that had existed in the Santa Cruz Mountains record, it did nothing to relieve the strain that has been building up along the fault on the San Francisco Peninsula. The next major seismic gap on the San Andreas Fault runs from a small cluster of quakes near Portola Valley to the San Francisco city limits. After passing through the Santa Cruz Mountains, the San Andreas Fault runs right down the middle of the San Francisco Peninsula along the eastern edge of the Coast Ranges—right through the middle of the suburbs and cities that crowd the western shore of San Francisco Bay.

Not only is this seismic gap in the midst of the San Francisco Peninsula troubling in its own right, but the patterns of past quakes in the area also paint a disturbing picture. Over the past 150 years, damaging earthquakes have tended to strike the Bay Area in pairs—both on the San Andreas and on major side faults like the Hayward, which, in a mirror image of the San Andreas, runs right down the middle of the East Bay.

••••

In 1865 a magnitude 6.5 earthquake centered in the Santa Cruz Mountains nearly killed Mark Twain, who was a young reporter in San Francisco. Later he would write about the quake in his book *Roughing It*. Three years later, that quake was followed by an even larger quake, a magnitude 7.0, on the Hayward Fault. Thirty years earlier, a similar pairing had taken place. An 1836 earthquake on the Hayward Fault that registered an estimated 7.0 on the Richter scale was followed two years later by a 7.0 on the San Francisco Peninsula segment of the San Andreas Fault.

Both pairs of those early quakes destroyed buildings and claimed lives, but their effect today would be almost unimaginable. San Francisco is no longer a frontier boomtown of a few thousand but a sprawling metropolis of more than five million. Both the San Andreas Fault and the Hayward Fault run right through the middle of some of the most heavily populated areas in the region. Figures from the Federal Emergency Management Agency predict that a magnitude 7.5 earthquake on the Hayward Fault could kill 4,500 people, injure 135,000, and cause more than $40 billion worth of damage.

With both the San Andreas Fault and the Hayward Fault to worry about—as well as a host of smaller faults such as the Calaveras, San Gregorio, Healdsburg-Rogers Creek, Stanford, and Concord—scientists today put the probability of another major quake in the San Francisco area at 50 percent over the next thirty years. In all likelihood, according to a U.S. Geological Survey report, another major earthquake will strike the Bay Area during the lifetime of most residents who lived through the Loma Prieta quake.

• • •

The high probability of another major quake has put scientists like Lindh on the spot to predict earthquakes more closely. But while the public and politicians would like predictions in terms of days and hours, Lindh warned that that may never be truly possible. "We don't know enough and we don't know if the earth knows enough to predict exactly when earthquakes will occur."

The fear of earthquakes is palpable in California and Lindh is exposed to it in odd ways. "Lately I've gotten a number of phone calls based on rumors," he said. "For example, that there's going to be an 8.0 in Los Angeles. That's a favorite one. Scientists believe it's coming and won't tell people. Honest to God. We couldn't keep a secret if we tried." The problem, he explained, "is that people are afraid scientists will know things that might kill them and not tell them."

"What we're left with is a contradiction—at what point is what you know sufficiently serious that you can tell people about it? I personally believe that the only way out of the contradiction is to involve people in the whole process and basically, like a weatherman, tell them everything you know. And if what you tell them ends with the statement, 'And I don't know what this means,' then good. At least you won't leave them with the nagging fear that you know something you're not telling them."

Instead of focusing on learning exactly when the next earthquake will strike, Lindh believes that public attention should be focused on upgrading homes, buildings, and highways in areas that are prone to earthquake damage. While scientists' understanding of just when and where earthquakes will strike is still developing, they have had remarkable success in predicting where those quakes will cause damage. The patterns of damage seen in the Loma Prieta earthquake, for example, had been predicted and carefully documented for more than a decade. But while that damage captured the public attention, it was also highly concentrated, and Lindh is afraid that it may have left some area residents with the false impression that the buildings and structures that survived are earthquake proof.

"Outside of the most sophisticated engineering circles, I'm not sure that people really appreciate how far away this earthquake was," he said. "If you're going to have a magnitude 7.0 earthquake in an urban area it's hard to imagine how you could design a safer one than this one. You couldn't put a 7.0 anywhere else in the Bay Area and do so little damage." While no buildings or freeway overpasses built to modern earthquake safety standards collapsed, none were located closer than ten miles to the quake's center.

"As a measure of how far we have to go is the fact, I'm told, that half the homes in the Bay Area are not bolted to their foundations," Lindh said. It's not only their family's life, but the equity in their house that's going to send their kids to school and take care of mom and pop in their old age. For the sake of a few hundred dollars in most cases they are literally preparing to throw that away. We've already told you that there's a 50 percent chance of a major quake in the next thirty years. If that's not going to inspire you to bolt your house to the foundation, would it do any more good to tell you that there was some increased probability in the next three years?

"It's my hope that faced with concrete evidence of what an earthquake like this means, people in the Bay Area will get on with treating that earthquake as if it is a certainty that will occur. I honestly don't know what seismologists can tell people beyond 50 percent in the next thirty years. I think maybe in terms of the message that we're really responsible for getting across, we're going to have to repeat it until we're sick of it. It's not only that this earthquake did not help—this earthquake did not reduce the probability of a big earthquake in the Bay Area. The next one is just going to be so much worse."

POINT REYES

THE MOBILE LANDSCAPE

PAST STINSON BEACH, THE FOG THAT HAD SHROUDED THE GOLDEN GATE Bridge returned, covering the coast in a soft, gray mist. As we headed north toward Point Reyes along the edge of Bolinas Lagoon, the white shapes of egrets feeding on the tidal flats near shore appeared through the fog like bright points of light.

Since early morning I had been following the path of the San Andreas Fault with Jim Ingle. Inland from Monterey the San Andreas Fault cuts through the Salinas Valley and then climbs into the Santa Cruz Mountains, running right down the middle of the San Francisco Peninsula. After heading out to sea just south of San Francisco, the fault comes back onshore here, slicing inland through the lagoon to the Olema Valley on the edge of the Point Reyes Peninsula. From Bolinas Lagoon to Tomales Bay the fault cuts through the landscape like a knife, all but severing the peninsula from the mainland. Farther south the path of the fault is marked by fractured and folded layers of rock. Its features are not subtle here but so large-scale that they are almost easier to see on a map than on the ground. Farther south the moving Pacific Plate has created desert seas and deep basins filled with oil. Here in Marin County just north of San Francisco that same movement has created a wedge-shaped peninsula cut by bays and lagoons—one of the richest and most varied stretches of coastline in California.

Its richness, however, does not come without a price. In intervals that vary from a few decades to a century or more, the ground here is rocked by giant earthquakes—sudden large slips along the San Andreas Fault that are felt as far away as Los Angeles and central Oregon. During the 1906 San Francisco earthquake the fault here slipped by more than fifteen feet.

As we headed up into the Olema Valley, Ingle sipped on the remains of a cup of coffee picked up in Stinson Beach. Mountains more than one thousand feet high rose up to either side. In places, the valley was so narrow that it seemed like we were traveling at the bottom of a canyon. The valley floor, however, was not flat as logic and erosion declared it should be but broken by low ridges fifty and one hundred feet high that weaved back and forth across it like the strands of a rope. They cut across streams and drainages, knocking the landscape off kilter. Streams in the valley less than a half-mile apart run in opposite directions. Although virtually no rain falls here for six months of the year, there are no less than forty-seven ponds in the fourteen miles between Bolinas Lagoon and Tomales Bay. They sit incongruously on the tops of small ridges and halfway up hillsides—the product of water seeping up along the fault.

The ridges, Ingle explained, are what geologists call shutter ridges. They were formed by movement along the fault. "Imagine you and a friend standing on opposite sides of a rug—one pulling left, the other pulling right. In between you would create a series of wrinkles and ridges. That is what is happening here. Only the pulling is being done by the fault."

At the northern end of the valley near the edge of Tomales Bay we drove right across the top of the fault. Although the San Andreas has not moved here since 1906, the ground is still unstable. Its trace across the highway was marked by an asphalt patch, as if a construction crew had torn up the pavement to put in a culvert or pipe. Ingle slowed the van to a crawl as we passed over. When we reached the other side he said: "We've left the continent behind. We're now traveling on the Pacific Plate."

The land looked no different on the other side of the fault, but the peninsula is a displaced island of rock. Rocks on the eastern side of the fault are almost entirely marine—a variety of deep-water shales known as cherts and submarine lavas known as basalts that are collectively known as the Franciscan Formation. Rocks on the peninsula, however, are part of the Salinian Block, a thin slice of the continent that has been driven northward by the Pacific Plate. Its core, like that of Baja California farther south, is built out of solid granite. Similar juxtapositions of rock occur up and down the length of the San Andreas Fault, but they seem more accentuated here because the peninsula is all but surrounded by water.

Away from the high coastal cliffs, however, these differences are hard to see. The rocks are hidden from view by a thick forest of pine, live oak, and madrone. As the road climbed over the foothills toward the outer reaches of the peninsula, however, Ingle pointed out weathered exposures of granite and pale cream-colored layers of Monterey shale in road cuts and stream banks. They looked almost identical to rocks I had seen farther south near Monterey Bay and Los Angeles.

• • •

Out toward the coast the forest gave way to a rolling, windswept coastal prairie. Gusts of wind rocked the van as we headed out toward Point Reyes on the southern edge of the peninsula. Dairies and ranches appeared through blowing streamers of fog, their herds of cattle huddled on the lee sides of barns and hills. Meadows and hillsides were green with winter rains and dotted with the first small, yellow flowers of spring.

In 1579, forty-one years before the Pilgrims reached Plymouth Rock, Sir Francis Drake sailed behind the protective arm of Point Reyes and into the bay that now bears his name. The land was not uninhabited, but Drake and his men found the natives peaceful and resourceful. They spent six weeks here repairing their ship and replenishing supplies. The rolling hills and high white cliffs that bordered the bay—a product of the soft and easily eroded marine rocks that cover the peninsula's granite core—reminded him of the chalk downs of southeast England. He called the land New Albion.

Out at Point Reyes the wind was so strong it nearly blew us off our feet. The air was cold and wet, soaking the windbreak of cypress trees that bordered the path leading out to the tip with a fine mist that fell like rain as we passed by.

On a clear day, standing at the tip of Point Reyes is like standing in the crow's nest of a ship at sea. Its high cliffs rise more than four hundred feet above the ocean. The view offshore seems almost limitless. That morning, however, the point was closed in by a swirling layer of fog. Steps led down the cliff face to a lighthouse perched on a ledge below. The low moan of its foghorn bounced off the rocks, mingling with the sounds of waves breaking on the rocks below. The point reaches more than ten miles out into the Pacific. In the winter and spring, gray whales pass by on their yearly migrations between the warm coastal lagoons of Baja California and the Bering Sea. Their path takes them only a few hundred yards from shore. We spent time looking for them, scanning the sea for their spouts, but the fog closed us in.

The whales, however, are not the only facet of this landscape which is moving. The headland is made not out of soft marine rocks like the low, rolling landscape behind us but of hard granite that rises out of the sea in a solid wall. Here near the top it is covered by a layer of conglomerates, buff-colored sandstone laced with a jumbled collection of red and green stones. They are, Ingle says, not unlike the conglomerates that make up Point Lobos more than one hundred miles to the south on the edge of Monterey Bay. The resemblance becomes even more striking when one begins to notice the peculiar texture of the granite below—light gray and filled with large, twinned crystals of plagioclase three and four inches across. Aside from Point Reyes, these distinctive layers of rock are found nowhere else on the coast except Point Lobos.

Geologists, Ingle explained, believe that the rocks of Point Lobos and Point Reyes not only resemble one another but were in fact once joined together. Over the past thirty million years they have been slowly separated by the steady drift of the Pacific Plate. Like Baja California and Los Angeles, the Point Reyes Peninsula is slowly heading north. If you looked hard enough at these displaced rocks, you could see not only the movements of points and peninsulas but the earthquakes and tremors that have rocked this mobile edge of the continent—quakes that have shaken not only this all-but-uninhabited peninsula but also the ground where San Francisco, San Jose, and Santa Cruz now stand. In time you would develop a sense that earthquakes in this mobile landscape are as regular and inevitable as the migrations of whales.

■ ■ ■

It was a warm, still afternoon in early June, and the only signs of movement or life in the Olema Valley were the patterns of ridges and hills that ran across the valley floor—the products of movement and creep along the San Andreas Fault. Grassy hills on the far side of the valley were already a dry, golden brown, but closer at hand the grass was still tinged with green, as if the soft, spring rains were not all that far away.

"On this side of the fault we have lots of water, but on the other side it's dry. Our springs run all year long," said Clare Scott, who lives here in the Olema Valley. "Drive a truck out into that meadow in the spring," he added, gesturing with his hand, "and you will sink right down to your hubcaps." Like most everything else here on the Point Reyes Peninsula, the presence of water is linked to the San Andreas Fault. It seeps up through the ground along the plane of the fault. Scott was not surprised with all of this but described it in a calm, matter-of-fact voice. "If you take a general look at topography there's obvious evidence of fault movement here in terms of stream diversions, isolated little hilltops, and things like that," he said. "They're all fault phenomena. There's fault phenomenon halfway up the hillsides here."

Unlike most Bay Area residents, Scott was interested in the history of faults and earthquakes in northern California long before the 1989 Loma Prieta earthquake. A swami with the Vedanta Society, a religious order founded in India in the late 1800s, he lives almost directly on top of the San Andreas Fault at the society's two-thousand-acre retreat in the Olema Valley. Although the retreat is dedicated to religious meditation, for Bay Area geologists it is a favorite place to come and meditate about the nature of earthquakes along the San Andreas Fault. The retreat sits almost squarely on top of the epicenter of the 1906 San Francisco quake.

You did not need to understand plate tectonics or the finer points of earthquake geology to know that a major quake would eventually strike the San Francisco Bay Area. All you had to do was come to the Point Reyes Peninsula and look down the narrow knifelike reach of the Olema Valley and see its peculiar patterns of ridges and hills to see how the moving Pacific Plate has shaped this unstable edge of the continent. Here on the Point Reyes Peninsula almost every facet of the landscape has been shaped by the fault. Its steep hills and shallow bays—even its rich mix of plants and animals—have been shaped by the fault.

• • •

Over the years, Scott has seen geologists come and go. While his duties, strictly speaking, are centered around matters more theological than geological, he has spent no small amount of time listening to ideas about faults and earthquakes in the area and walking over the retreat grounds with visiting geologists looking for signs of the fault. With their help he has also collected copies of almost every scientific paper written about the San Andreas Fault on the Point Reyes Peninsula as well as a series of aerial photographs and maps. Living on top of the fault helps foster a certain degree of attention to geologic details.

The title of *swami* conjures up certain images, but Scott looked more like seasoned farmer or field hand as we walked down the dirt road that wandered through the fields toward the high peaks of Inverness Ridge on the western edge of the retreat. He was not dressed in a turban or robe but in a pair of blue-and-white herringbone overalls and well-worn high-top work boots. A wide-brimmed straw hat covered a full head of gray hair, its loose weave casting a checkered pattern of shade across his face like the shadows created by sunlight shining through a screen door.

Near the main house he had pointed out small details—the welt in the base of a redwood tree, for example, that had been all but split in half by the 1906 earthquake. The tree sits almost right on top of the main trace on the San Andreas Fault. A few yards away a tool shed leaned at a precarious angle. Like the tree, it too straddled the fault. While its rear wall rested on the mobile Pacific Plate, the rest of it was anchored to the more stable ground of the continent. Although it was built after the 1906 quake and there has been no fault movements here since then, the ground is settling at different rates on the two sides of the fault. The shed's rear wall has sunk almost a foot. Inside, a crack on its concrete floor marked the trace of the San Andreas.

Out here in the hills, however, Scott's eyes were drawn to larger features: fault scarps cut into the sides of hills and ridges like terraces; the notch or break in a ridge that marked a sudden sideways jump in the path of the San Andreas Fault.

Taking off cross country we headed straight up a forested hillside. "Now we're getting to the point where things get weird," he said as we neared the crest. At the top was an inexplicable craterlike cusp more than twenty feet deep with an enormous bay tree nearly four feet in diameter growing right in the center of it. Like many of the features here, geologists have no clear idea how this cusp was formed, but there is little doubt it is related to the fault. Off to one side a ravine led down from the cusp in a straight line toward a small lake that sits on a known side fault of the San Andreas.

Although the fault has not moved here since 1906, there is no doubt in Scott's mind that it will jump again. The details of fault scarps, cracks, and cusps are all part of a larger picture in his mind that has to do with the movements of continents and ocean basins. "It used to be that when I went to school they considered geological movements on a scale of a foot in ten thousand years. That's been wiped out. The present rate of fault movement is thought to be two or three centimeters per year," he said. "When I went to school, if I had said that, they would have laughed me out of the room. The interesting thing though, is that while the plates are moving we know the fault is locked here. So the next movement is going to be a jump like the last one. That's when you have the fun."

• • •

If you followed the crack in the floor of the Vedanta Society's tool shed southward, in less than a half-mile you would come to Tina Niemi's trench site on the San Andreas Fault. Some people dig for gold. Others drill for oil. Niemi, a graduate student in geology at Stanford University, is looking for the San Andreas Fault.

Although the Point Reyes Peninsula was extensively studied after the 1906 San Francisco earthquake, firsthand knowledge of major quakes along this stretch of the San Andreas Fault essentially begins and ends with that giant quake. While conventional wisdom has it that major earthquakes are not likely here in the immediate future because the fault slipped so far in 1906, it would be nice to find some solid evidence to support that idea. Like Kerry Sieh, who was able to learn something about the timing and frequency of major earthquakes in southern California by trench studies, Niemi is hoping that the buried layers of soil and silt here in the Olema Valley will tell her about the timing and frequency of past earthquakes.

It was the first week of October and the grass in the surrounding fields was brittle and dry. By midmorning the sun was almost unbearably hot, but there in the shade of the tall trees the air was cool and scented with the sharp smells of eucalyptus and dry, fine dust: earthquake weather, according to popular superstition in northern California, although there is nothing to suggest that earthquakes are any more likely during these dry fall days than any other time of year.

The Loma Prieta earthquake was still more than a year away, but the importance of Niemi's work had not been lost on Bay Area geologists and seismologists. As the 1906 San Francisco quake proved, fault movements here can have a devastating impact on San Francisco. After Niemi had been at it for little more than two weeks, more than twenty geologists and seismologists from the U.S. Geological Survey and the California Division of Mining and Geology had come by to see what her trenches had uncovered. After a day of entertaining visitors, it was time to get back to work.

Niemi stood at the head of the trench with her arms folded, thinking, not saying a word. A few feet away Danny Manning from Point Reyes Custom Excavators sat on his backhoe with his engine idling. After digging foundations and culverts, looking for the San Andreas Fault was something new. Manning looked from Niemi to the trench and then back again with his hands poised over the controls waiting for instructions. A few feet away, Eric, a black Angus bull, regarded us with large dark eyes, his four feet planted solidly on the ground. After a while he let out a plaintive bellow aimed at the heifers in a neighboring field, separated from his amorous attentions by a double row of barbwire fencing. The San Andreas Fault was somewhere under our feet.

After the 1906 San Francisco quake, Grove Karl Gilbert, one of the leading geologists in America and the head of the California Earthquake Commission, came to the Point Reyes Peninsula to study the effects of the quake. Here in the Olema Valley, roads and fence lines that crossed the fault had been knocked askew, pulled ten and fifteen feet apart by the sudden surge of movement along the fault. The quake opened up giant cracks and fissures in the ground—some of them three and four feet across. Elsewhere the trace of the fault was marked by a giant molelike track across the ground.

Gilbert's field notes and photographs clearly show the fault passing through Niemi's trench site, which is located in a gap between two low ridges running across the valley floor. After eighty-two years, however, the trace of that break—to say nothing of those left by the earthquakes that preceded it—has proved all but impossible to find.

"Take it down another meter," Niemi said.

There was a moment of silence as Manning eyed Niemi and shifted uncomfortably in his seat. Finally his face broke into a wide grin. "Fine. What's a meter?"

"About three feet!" Niemi said with an incredulous laugh.

"Hey, you've got to remember," Manning said with an equally amused laugh, "I'm just a country boy."

Deepening the trench took only seconds. The new cut revealed layers of fine sand, what looked like the remains of an animal burrow and the roots of old trees—suggesting that this area was once a stream or creek bank. Niemi looked hopefully up and down the trench, but there was still no sign of the fault.

"It gets kind of embarrassing after a while, you know," Niemi said with a groan. Here we are, I thought to myself, sitting on top of one of the largest and most active fault zones in the world and we can't even find the thing. It was something I had never considered.

The San Andreas Fault is more than seven hundred miles long, the active boundary between two large plates of the earth's crust. Movements along it have not only created this narrow valley and the bay and lagoon that lie to either end of it, but have moved the whole Point Reyes Peninsula northward as well. Yet while earthquakes along it can be felt for hundreds of miles, in places the fault itself might be no thicker than this book. Pin-pointing its exact location requires equal amounts of guesswork and luck.

Even with no clear sign of the fault, there are still valuable details to learn about the layering and patterns of the soil below. While Manning took the backhoe and cleared the reeds from a nearby pond Niemi planned to study the following week, we propped up the trench walls with hydraulic shores and climbed inside the fifteen-foot-deep trench to study its details more closely. As I looked up from under the brim of my hard hat to the narrow slit of sky overhead it occurred to me that this would not be the best place to be in the event of an earthquake.

• • •

In a few hundred thousand years the layers of soft sand and soil that now cover the floor of the Olema Valley will become hard rock. Niemi, however, is not willing to wait that long. She would like to know what they have to say right now—or at least some time in the next year.

Over the past several thousand years successive layers of sand and silt have been deposited on the floor of the Olema Valley. For the most part that slow and steady accumulation has been uneventful, but from time to time it has been interrupted by earthquakes that have cracked the ground, shifting the delicate layers of soil below and opening deep fissures that have in turn been filled in with new layers of sediment. By finding these buried cracks and fissures Niemi hopes to determine the timing and frequency of prehistoric earthquakes on this critical segment of the San Andreas Fault.

Niemi's work is an effort to bring geology right up to the present. While other graduate students measure their field area in square miles, Niemi measures hers in square inches—marking off the trench walls with a gridwork of nails, an aid to mapping the details of fine layers of soils and pebbles. Hard-rock geologists are constantly looking for a fresh rock face whose features have not been altered by air and water. Niemi is looking for clean dirt, soil whose delicate features and layers have not been smeared and disfigured by the backhoe's shovel.

With whisk brooms and hand trowels we started scraping a thin layer of dirt and loose pebbles off the freshly dug trench walls—"cleaning the trench," as Niemi called it, a dubious term at best. In less than ten minutes we were covered from head to foot with dirt and mud.

As we scraped the walls clean, however, small features began to emerge. At first they were almost imperceptible—subtle changes in texture or color that marked different layers of soil. Some, Niemi says, were derived from the deep-water marine rocks of the Franciscan Formation that lie on the eastern side of the fault. Others had been eroded from the granitic rocks of the Salinian Block. Details on a scale of millimeters and centimeters were described with the same terms that other geologists typically use for features occurring on a scale of yards or miles. Small ripples in a layer of sand became folds. Fine cracks became faults. Mapping the details of a single trench can take days.

Digging through the dirt is nothing new to Niemi. As an undergraduate at Wooster College in Ohio she did a double major in geology and archaeology, spending her summers working on archaeological digs in Israel, Greece, and Jordan. While the search here is for soil, artifacts would be a welcome find. "All of this would be a lot easier if we were working in the Middle East," she said, pausing to push her hard hat back on her head. "We would have pottery shards and old coins to help date things." Here age is determined by carbon-14 dating of bits of charcoal and wood buried in the soil. Until the dates come back from the laboratory there is no way of knowing for sure if one inch represents ten years or ten thousand.

To a traditional geologist these layers of soil are overburden, something to be stripped away to get at the more solid facts of the hard rocks below. Niemi, however, could care less about the rocks below. They are old stuff. Rigid. Unimaginative. "We call that under-burden," she said with a laugh. If you want to learn about recent earth-quakes, the soil is where it's at.

• • •

After lunch Niemi's thesis adviser, Tim Hall, and his wife, Lisa Rowles, also a geologist, arrive to lend a hand with the new trenches. Hall is not at all surprised that the fault has been so hard to find. "You never know what you're going to find until you start digging," he said philo-sophically. In addition to teaching classes at both Stanford and Foothill Community College, Hall also works as a geologic consultant, and he has spent a lot of time in trenches like these looking for signs of faults and earthquakes under the proposed sites of office buildings and schools as well as at controversial projects like the Diablo Canyon Nuclear Power Plant in southern California.

Tim Hall is the acknowledged expert on the San Andreas Fault in Marin County and has studied the fault here for the better part of a decade. He can recite the details of field notes by famous geologists who have worked in the area—like Grove Karl Gilbert, who came here after the 1906 earthquake—the way some people quote batting aver-ages in baseball or chapter and verse from scripture. For several years he has also brought his introductory classes in geology here to see the fault firsthand. Less than a half-mile away is the Earthquake Trail, an interpretative trail Hall set up to teach visitors about the San Andreas Fault. Like Clare Scott, Hall's interest in earthquakes is more than aca-demic: his home in Woodside on the San Francisco Peninsula is less than a half-mile from the main trace of the San Andreas Fault.

For the better part of an hour Hall and Niemi looked over Niemi's field notes, searching for clues about the probable position of the fault. By midafternoon they had a new site picked out. Manning wheeled his backhoe into position and started digging. For the first few minutes everything looked promising. Suddenly the shovel hit a perched aquifer—a buried, water-rich layer of sand—and the walls began to slowly cave in. There had been no rain for more than five months, but in a matter of seconds the floor of the trench was more than a foot deep in water.

It was not the end of the day's work, but merely a delay. Working in the waterlogged trench full time would require a pump to keep its floor dry, but Niemi was eager to press ahead. To temporarily drain off the water, she had Manning dig a sump, or deep hole, off to one end.

Clare Scott came by just as Manning was finishing the sump.

When he started digging the trench again, every eye was following the backhoe's shovel. "Just a minute!" Niemi shouted. "That's it!"

At first glance it didn't seem like much—a wedge of yellow sand cutting into a layer of red sand. It thrust upward, as if it had been punched into the sand above. We popped in the shores and climbed inside for a closer look, running our hands over the fault—the boundary between the North American and Pacific Plates. The thin line that separated those two layers of sand represented the trace of an earthquake that once broke the ground here.

• • •

It was nearly sunset by the time the trench was finished. We spent what was left of the daylight covering its top with sheets of plywood and a makeshift barbwire fence to keep wandering cows and clerics from falling in. It had been a promising day.

As we worked, Hall talked about the fault. Like most geologists studying the San Andreas, he has no doubt that the ground here will move again. "Just because we're here doesn't mean that geologic processes are going to stop," he said. The Bay Area's five million residents have little or no say in the matter. Like Lindh and Sieh, he doubted that scientists would ever be able to predict the exact day and hour of the next major quake. The San Andreas Fault is like an intricate puzzle, and scientists may never completely understand all its pieces. While trenching studies are invaluable, they only scratch the surface. In the meantime, Hall says, both scientists and the general pubic have learned a great deal about faults and earthquakes over the past twenty years.

"I think people have become much more sophisticated about their geologic environment over the past couple of decades," Hall said. "When I first started teaching there were folks who thought a fault like the San Andreas ought to look like something out of a Hollywood movie—a great crack in the ground that everything was going to fall into. When they actually saw the plate boundary and how, in some sense, there didn't appear to be a great difference across it, they wondered, 'How significant is this feature?' Of course it's not what's on the surface that is really important. It's the processes going on down below our direct observation at a depth of several kilometers that have implications for our safety."

• • •

The hillsides out at Chimney Rock were so steep that it was almost impossible to stand. A few feet below us they turned into cliffs that dropped straight down to the beach more than four hundred feet below. Farther inland we had been protected from the wind by a low ridge, but out here the wind rose up under our jackets and blew us along like kites. We walked along the steep slopes in a crouch, pausing now and then to look over the edge of the cliffs at the rocks below.

Offshore, the Pacific was slate gray, its surface alive with white-caps and windblown spray. Closer to shore the surf was almost tan-colored—laden with silt and sand kicked up by the high storm waves as they approached the beach. Behind the protective arm of the point, Drakes Bay was smooth and still, its normally empty waters filled with a flotilla of more than fifty small fishing boats waiting out the wind.

It was the first week of May and I was walking with Bob Stewart, the Marin County naturalist, out near the tip of the Point Reyes Penin-sula at Chimney Rock. The Point Reyes Lighthouse was less than a half-mile away. A little more than a year earlier, I had been here with Jim Ingle to look at the granites and conglomerates that make up the high cliffs of the Point Reyes headland. Stewart, however, was inter-ested not so much in the rocks as in the plants that grow on top of them.

Summer was only a few weeks away, but the grass out here in the coastal prairie that covers the outer reaches of the peninsula was still full and green. In places it was almost knee-high and studded with wildflowers: a collection of purples, yellows, reds, and whites. At first glance they looked no different from flowers that had been blooming on the mainland, but here and there Stewart's well-trained eyes picked up subtle differences and variations that set them apart. In a steep-sided swale that shielded us from the wind he crouched on his hands and knees, comparing the leaves and flowers of nearby plants with those in the three-inch-thick hardcover guide to the flora of Marin County that he had carried out in his backpack. In the space of a few feet we found examples of Point Reyes lupine, Point Reyes paintbrush, and Point Reyes blennospurna—flowering plants that are found, as their name suggests, only on the peninsula. "The history of this land is not here, but farther south," Stewart said.

• • •

On the Point Reyes Peninsula the San Andreas Fault has juxtaposed not only different rocks but different plants as well. All together there are more than sixty different species of plants here that are not found on the nearby coast of the mainland. Most of them are small shrubs and grasses, but not all of the differences between plants on the two sides of the fault are so subtle. Redwoods, for example, thrive on the eastern side of the fault at nearby spots like Muir Woods and Mount Tamalpais State Park, but they are not found on the peninsula in places where the rainfall, terrain, and temperature are similar. It is thought that the trees fare poorly in the peninsula's well-drained soils, which are not derived from the deep-sea rocks of the Franciscan Formation like those on the mainland but from the granitic rocks and Monterey Shales of the Salinian Block.

The San Francisco Bay Area and its six million people are less than thirty miles away, but the Point Reyes Peninsula is an island of open space preserved by a patchwork of federal, state, and county parkland. Elk and mountain lion still roam its outermost reaches. More than four hundred species of birds have been sighted here, making it one of the richest bird habitats in North America. While more than thirty thousand acres of parkland have been preserved, it is not the size of the area that makes it so diverse but the variety of habitats it contains. In the space of a few square miles the peninsula encompasses shallow bays and inlets, mountain forests, and rolling coastal prairies.

Not only does the peninsula sit on the edge of two vast plates of the earth's crust, but its particular location in space and time puts it on the edge of an important ecological boundary as well. Located halfway between Baja California and Alaska, the peninsula straddles both the northern limit of the coastal scrub community that begins in Baja California and the southern boundary of the northern coniferous forest that runs all the way to Alaska. The balance between these two worlds is so tenuous here that you can actually see it in the steep-sided canyons that run toward the sea on the peninsula's southern edge. South-facing slopes are covered with manzanita, scrub oak, and sage and look almost like hillsides near Tijuana or Ensenada—while north-facing slopes are covered with thick stands of Douglas fir, huckleberry, and fern that are all but indistinguishable from the cool, coastal forests of Washington and British Columbia.

Like the small, rare plants which are unique to the peninsula, however, these differences are easy to overlook. They are hidden in the pattern and fabric of the landscape. Even Stewart, who has been working in the area as a naturalist and bird researcher for more than twenty years, finds that its diversity still takes him by surprise:

"Maybe it's just some little special spot riding on the Pacific Plate," Stewart said. "You can't go anywhere up or down the coast and find this rich variety of habitats. There isn't the concentration of life that there is here. It's so special. It's so unique. It really is a wonderful island of habitat. Even the configuration of the land, the way it sticks out into the ocean, plays a role."

• • •

Out on the edge of the high coastal cliffs the peninsula seemed more like an island than part of the mainland. Gusts of wind carried the barks of sea lions and the cries of gulls up from the sea several hundred feet below. Steep cliffs and offshore rocks were alive with nesting seabirds—penguin-like murres and black, swan-like cormorants nesting on the rocks. For some, like the murres, Point Reyes is their one point of contact with land.

The point is a gathering place for marine mammals as well. Elephant seals and sea lions haul out on the rocky ledges and offshore beaches. Harbor seals by the hundreds bask in the sun on isolated beaches. In the winter and spring gray whales pass by on their way to and from Alaska. Biologists are unsure just how these large whales find their way along the coast, but it is thought they use bottom contours to chart their progress, navigating from point to point along the coast.

•••

While the peninsula's role as a gathering point for marine animals like seabirds and seals is striking, perhaps the most elegant example of its richness and diversity are the migrating warblers that are drawn to the peninsula each fall. "They are the most beautiful of all land birds," Stewart said. "They are nature's little jewels."

Brightly colored birds, smaller than sparrows, warblers fly up and down the continent in yearly migrations that mirror those of the whales offshore. They spend their summers in the boreal forest of northern Canada and Alaska and their winters in Central and South America. Although they weigh only a few ounces, some, like the blackpoll warbler, make the trip nonstop—going from Canada to Brazil in as little as seventy-two hours and flying at altitudes as high as twenty thousand feet. Others make the trip in short hops, stopping along the way to feed and rest.

These migrations, however, are not foolproof, and every year a certain percentage of these migrating warblers lose their way. Point Reyes seems to draw these lost birds like a magnet. In the fall, the peninsula is one of the few places on the West Coast where you can see magnolia warblers, blackpoll warblers, parula warblers, and black-throated green warblers in the same day, Stewart says. All of these birds are vagrants, birds not typically found here in California. In fact, most are generally found only in the eastern United States. The reason they are drawn here, researchers believe, is an intriguing combination of bird behavior, weather, and the peninsula's curving reach into the sea. Most of the vagrant birds that arrive here are less than a year old and making their first trip south. It is thought that they make a "mirror-image mistake," confusing right with left and following a course that eventually takes them out over the open Pacific instead of diagonally across the continent toward the Atlantic. For the birds, tired and lost at sea, the peninsula's projecting arm and its flashing lighthouse might easily be their first sight of land after several days at sea.

For the most part, migrating warblers travel only at night, using the earth's magnetic field (birds are much more sensitive to magnetic forces than humans) and the night stars to navigate. Clouds, however, apparently confuse their sense of direction, and the greatest number of vagrants are seen after a high overcast that obscures the night sky.

Not only are these lost birds almost magically drawn to Point Reyes, but they are drawn to specific places on it. Forest birds, they flock not to the open grasslands of the point but to the scattered stands of cypress trees planted as windbreaks by early ranchers and farmers more than one hundred years ago. The birds come and go, and every day is different, Stewart says. On a good fall weekend there can be as many as three hundred bird watchers out on the peninsula looking for warblers and other rare birds. Like its unique plants and the mix of coastal scrub and northern forest that fill its canyons, these rare birds are another subtle detail that reveals the richness and diversity of the Point Reyes Peninsula.

"It seems to me that as I live my life and wonder about why I'm doing these things, you know, birds are very hard to see," Stewart said. "And somehow getting people to focus on something that they have a hard time seeing, it gets them a little deeper into the natural world. I think what it does in a general way is that it gets people more in tune with the wonder and beauty of nature and the incredible complexity of it. As a naturalist I find it romantic that people come here to see both warblers and whales, some of the largest and smallest animals on earth."

• • •

Tom Moore measured off a small area of tidal flat with a square frame of plastic tubing. On his hands and knees he began digging up black, wet handfuls of mud and throwing them into a sieve. Each scoop sent a sour, sulfury smell wafting through the air. "Biology is not glamorous," he said. "I've done this before in San Francisco Bay. There has to be a better way."

Behind the crisp and clean conclusions of almost any study in marine biology are hours of less-than-tidy fieldwork like this. Moore, a biologist with the California Fish and Game Department who has been monitoring the clam populations in Tomales Bay, has no illusions about his work. After serving in Vietnam, he went to Humboldt State University in northern California and earned a master's degree in wildlife biology. His first job out of graduate school was at the San Onofre Nuclear Power Plant in southern California, studying the effects of the plant's water intakes on marine life. A large part of the job, as Moore described it, consisted of "sitting inside a hot, rank dumpster sorting through piles of rotten kelp and dead fish." Plant workers making their rounds would come by and watch Moore and the others at work for a few minutes and them mumble sympathetically, "Are you guys prisoners?"—assuming that they were convicts from the nearby state penitentiary who had been sent over for a few days of hard labor.

Today's work, however, was not without its fringe benefits. We were sitting not inside an overheated dumpster but out in the middle of Tomales Bay, the long, narrow reach of water that lies between the northern edge of the Point Reyes Peninsula and the mainland.

We had left for work a few minutes after sunrise from Lawson's Landing on the northern end of the bay, heading south at full throttle, cutting long, graceful arcs through the water as we followed the narrow winding channel into the bay. The tide was falling and alongshore the tidal flats that fringed the bay were slowing rising up out of the water. Their wet, smooth surfaces were alive with birds. Herons and egrets stalked through the shallows searching for small crabs and fish. Flocks of seagulls burst into the air as we passed by. Rounding a corner at full speed we surprised a herd of more than thirty harbor seals hauled out on a sandbar. Lying side by side with their silver fur drying in the early morning sun they looked almost like driftwood washed up by the tide.

We beached the boat on a tidal flat in the middle of the bay to start work. I could feel the pull of the tide around my feet as we waded ashore and could see the current break in tiny waves alongshore. Beds of eel grass swayed and danced in the moving water. Here and there among its bright green blades were the broken shells of clams and crabs.

■ ■ ■

Although I had been living on the edge of the bay for more than eight weeks, I felt like I was seeing it for the first time. Onshore it had seemed like a narrow arm of the sea surrounded by mountains and high coastal hills. On the water, however, one's perception was dominated not by the surrounding peaks and hills but by the flat open reach of the water and the gently sloping surfaces of the tidal flats. The bay seemed to go on forever. Even points on land looked new and different. Nearby hills and ridges were reduced in size, as if the broad reach of the water and the tidal flats had somehow robbed them of height.

At low tide the mudflats seemed barren and lifeless. But as I poured a bucket of water into the mud-filled sieve while Moore rocked it back and forth in his hands, a collection of small treasures emerged: the shell of a moon snail, a curling tube of sand shaped by a small bay worm, and more than a dozen tiny horseneck and gaper clams each no larger than a fingernail. In a few months those tiny clams would be larger than a fist and buried several feet deep in the rich bay mud, filtering their food from the bay with tubelike siphons that reach up into the water above.

Although Tomales Bay is just twelve miles long, it contains the largest clam fishery in California. While nearby San Francisco Bay is more than five times the size of Tomales Bay, its productivity has been tainted by pollution and sewage. More than one-third of its original area has been filled in to make room for the cities and towns that now crowd its shores. Here on the western edge of Marin County, however, cows still outnumber people. For the time being Tomales Bay is one of the most pristine and productive bays on the West Coast.

• • •

Tomales Bay is an estuary, a place where the salt water of the sea meets the fresh water of creeks and streams. Tides here rise and fall by as much as six feet per day—leaving broad patches of the bay alternately in and out of the water. These shallow reaches are neither wholly part of the land nor part of the sea. Like the mix of plants and soils found on the neighboring Point Reyes Peninsula, this constantly changing world of marsh and tidal flat is unusually productive—an integral part of the larger marine world that lies just offshore. Loaded with nutrients from the decay of plants in the fringing marshland and supplied with others by currents in the neighboring Pacific, the waters of Tomales Bay are more than three times as productive per acre as a comparable area of the open ocean.

The bay supports not only large populations of clams and shore birds but a host of other animals as well. Its shallow protected waters serve as a nursery for oceangoing fish like halibut, flounder, rockfish, and herring, which depend on the bay to raise their young.

In January and February schools of herring fill the bay, returning from the open sea to lay their eggs in the beds of eel grass that fringe the tidal flats and shoreline. A single school of herring can be several miles long and contain literally millions of fish. For a few weeks the bay is filled with an explosion of life as seagulls, small fish, and other predators feast on both the herring and their newly laid eggs. A single female can lay as many as forty thousand eggs, but less than one in ten thousand make it to adulthood. Even though their chance of survival is low, it is still several times higher inside the bay's protected waters than it would be in the ocean offshore.

On the southern side of the peninsula Bolinas Lagoon plays a similarly important role for marine life in the ocean offshore. More than any other facet of the landscape, these shallow bodies of water are responsible for the phenomenal productivity and diversity of the Point Reyes Peninsula.

• • •

Bays and estuaries are relatively rare on the West Coast. While the sinking East Coast is fringed with a maze of bays and inlets, the rising West Coast fronts the Pacific with a wall of high mountains and cliffs. From Cabo San Lucas to Seattle there are only a handful of bays and harbors worthy of the name. Here in northern California, however, the same mobile Pacific Plate that has shaped this mountainous coastline has created bays and lagoons. Like most everything else on the Point Reyes Peninsula, both Tomales Bay and Bolinas Lagoon are products of the San Andreas Fault.

As it angles across California, the path of the San Andreas Fault is often marked by a shallow trench or trough. Constant movement over the past thirty million years has made the ground over the fault unstable and prone to sinking and settling. Farther south on the San Francisco Peninsula this natural trough has been dammed to create lakes and reservoirs: Lake Elsman and Lexington Reservoir near San Jose and Crystal Springs Reservoir and San Andreas Lake near San Francisco. Here on the Point Reyes Peninsula that trough has been filled by the sea. The San Andreas Fault runs right down the middle of Tomales Bay and Bolinas Lagoon.

The processes that created these two protected bodies of water have not stopped. During the 1906 San Francisco earthquake the eastern side of Bolinas Lagoon sank by more than a foot, drowning large mats of marsh plants like pickleweed. The dead plants turned bright orange, puzzling scientists and local residents for months until they figured out that the suddenly deeper water had killed the plants. Farther north in Tomales Bay the effects of the shaking were preserved in series of ripples and waves in the soft bay muds—some of them several feet high—that lasted for months. Movement along the fault bent piers and jetties like wire, a visible sign of slip. Without the continual disturbances of quakes and tremors, these two shallow bodies of water would have long since been filled in by sands and silts eroded from the surrounding hills.

Such movements have of course played a similar role in shaping the San Francisco area farther south. Even San Francisco Bay, bordered on the west by the San Andreas and on the east by the Hayward Fault, has been shaped by the steady drift of the Pacific Plate. Without it, the Bay Area would be as flat and featureless as Florida. Earthquakes are simply a cost of doing business. They are as much a part of the landscape as the hills and bays.

• • •

It was the last Sunday in April, the greenest day of the year. The grass in the rolling coastal prairies out at Tomales Point was full and tall, topped with plumes of seeds that waved and danced in the wind like a vast field of wheat. Purple stands of wild irises lay in the damp swales between the rolling hills. Hillsides were colored with the knee-high yellow and white blooms of lupine.

A thick morning fog that lasted until well into the afternoon had kept the weekend crowds away. Coming out, the point had been covered in a milk-white sea of fog, but by late afternoon the air was sharp and clear. Out in the Pacific I could see the mountainous shapes of Farallon Island twenty miles offshore.

Tomales Point lies at the northern tip of the wedge-shaped Point Reyes Peninsula. It does not project out into the ocean like Point Reyes itself but runs parallel to the coast like a spear point, shielding Tomales Bay from the open ocean. Looking south along the long, straight line of Point Reyes Beach you could see the flash of the lighthouse more than fifteen miles away at the peninsula's outermost tip.

Halfway to the end of the narrow point I sat on a rock, which lifted me just above the top of the grass, to watch the last of the sun. The water offshore was almost copper-colored in the late afternoon sun. Red-tailed hawks soared in the updrafts from the high coastal cliffs, tethered in the sky as still and motionless as kites. Behind my back a herd of tule elk grazed on the side of a steep ravine that led down to the edge of Tomales Bay. Through my binoculars I watched a harrier skim over the ground, and I caught the movements of a bobcat weaving through the tall grass. After a few minutes he climbed up on top of a pile of rocks and then sat facing out toward sea.

It was high tide down at San Francisco Bay, and for the next hour the water offshore filled with passing ships: container ships and freighters bound for Tokyo and Taiwan and small fishing boats heading north to try their luck along the coast. I was less than thirty miles from San Francisco, but it felt much farther away. The city itself was hidden from view by successive ridges of mountains—the high peaks of Inverness Ridge and Mount Tamalpais. The windblown point seemed almost untouched and unchanged, a world apart from the cities and suburbs that encircle San Francisco Bay.

One hundred years ago the freighters and cargo ships offshore would have been bound for the Point Reyes Peninsula. From the 1850s until the early 1900s, Point Reyes was a point of supply for San Francisco. Its forests were clear-cut for firewood and building timbers. Dairies and cattle ranches sprang up on the cleared land, sending milk and meat to San Francisco via a fleet of schooners and small freighters that docked in Tomales Bay, Bolinas Lagoon, and Drakes Estero. Almost nothing was overlooked. Clams and oysters were gathered in nearby bays and lagoons. Seabird eggs were collected by the hundreds of thousands from offshore rocks and islands to be sold in city markets.

By the 1920s the land and its wildlife were all but played out. San Francisco found other sources of supply in Washington and Oregon. All but severed from the mainland by the fault, the peninsula was left untouched and almost forgotten for more than thirty years. When the peninsula was threatened by redevelopment in the 1960s, this time in the form of beach-front vacation homes and marinas, a dedicated collection of local and national conservation groups began to push for its preservation.

Today it is a near wilderness of forested peaks and coastal prairies cut by bays and lagoons. There is almost no sign of past abuses. The traces of slaughterhouses, piers, and lumber mills have nearly disappeared. Once-bare slopes are now covered with trees. Pockets of wild coastal prairie still thrive on the outer reaches of the coast. Fish and shellfish have come to fill the bays and lagoons again. Seabirds flock to the offshore rocks and islands. In places it is all but indistinguishable from the land that greeted Sir Francis Drake more than four hundred years ago.

But while the abuses of the late 1800s proved how quickly it could all be destroyed, its recovery and rejuvenation today is a sign of nature's power and endurance. Like earthquakes, this ability of the land to heal itself and cover the past is a subtle reminder that the natural world is much larger and more powerful than ourselves and that our own time and place in the landscape is tenuous and by no means guaranteed.

After running down the middle of Tomales Bay the San Andreas Fault heads out to sea again. On its way north it skips along the coast, shaping and cutting projecting points of land: Bodega Head, Point Arena, and Fort Ross. Near Cape Mendocino, however, the fault disappears and a spreading ridge appears offshore. Unlike the East Pacific Rise in the Gulf of California, however, this spreading ridge is not pushing the continent apart but sliding rocks underneath it. Closer to shore is a subduction zone—a place where the mobile oceanic crust is being thrust under the edge of the continent. It runs all the way to Alaska. At depth this subducting crust melts, transformed into magmas that rise to the surface to form volcanoes—Mount St. Helens, Mount Shasta, Mount Rainier, and more than a dozen others—a ring of fire on the edge of the Pacific that runs all the way to British Columbia.

Outward of this spreading ridge, however, the Pacific Plate is still heading north toward the Bering Sea and the Aleutian Islands. North of Cape Mendocino one passes into a different world.

CAPE MENDOCINO

ROUGH WATER AND RUGGED MOUNTAINS

T HE TWO-LANE ROAD THAT LEADS OUT TO THE TIP OF CAPE MENDOCINO is
barely one and a half lanes wide. Its route through the high peaks
that rise up behind the small town of Ferndale seems to have more
to do with tectonics than traffic—twisted and bent into a series of
improbable curves as it clings to the side of the mountains.

Cape Mendocino reaches farther out into the Pacific than any
other point in the continental United States outside of Alaska. Its outermost edge is seventy miles farther west than San Francisco. Unlike
Point Reyes, it is not a narrow wedge of rock but a mountainous shoulder of land. In places its peaks are growing even faster than those of
the Transverse Ranges of southern California, shooting upward by as
much as three inches per year. Like the creep on the San Andreas
Fault, however, those movements are not smooth and regular but more
likely to come in sudden jumps of several feet, accompanied by powerful earthquakes.

But while the rocks are active here, they are also hard to see.
Traveling north along the California coast is a journey into steadily
increasing rainfall and a steadily increasing cover of trees. North of
San Francisco the coast is cut not by dry canyons and small seasonal
streams but by wide, wild rivers like the Russian, Eel, Klamath, and
Smith. Here near the Oregon border more than sixty inches of rain fall
each year. Hillsides and stream banks are covered not with dry grass
and coastal scrub but with giant trees—groves of giant redwoods, hem-

locks, and Douglas firs with a flowering understory of dogwood and rhododendron.

It is only near the coast that the trees give way, held in check by the harsh salt air and the ever-present wind. The outermost edge of Cape Mendocino is a windblown wilderness of grass and coastal scrub. Steep grass-covered hillsides tumble down to the Pacific. Scattered flocks of sheep and herds of cattle graze in the rolling coastal meadows. Roadside fenceposts are drunk on their feet—they lean and tilt at precarious angles, not from a lack of care or maintenance but from the constant slumps and landslides triggered by the unstable ground below. Rocks and stones hidden by the forest farther inland rise out of the grass here like the remains of some natural Stonehenge. Offshore the coastline is dotted by rocks as well—rounded mounds of stone shaped like giant haystacks the size of small islands. Down at the water's edge the thin, rocky beach is bordered by a low cliff barely six feet tall and a narrow coastal plain. It seems to border the sea like a stair step. You can almost see how the landscape here was lifted right up out of the sea.

From the top of the steep grassy slopes bordering the beach you can look out over the Pacific and see how the waves approach the coast. They roll in from the northwest in a regular rhythm, their evenly spaced crests only a few seconds apart. Gusts of wind blow across their tops, sending streamers of spray and foam above the water. As they break against the sides of offshore rocks and islands they send walls of blue and white water twenty and thirty feet into the air.

Cape Mendocino reaches so far out to sea that it faces the wind head on. Average wind speeds in the ocean offshore are among the highest in North America, and they not infrequently top fifty miles per hour. The cape's high peaks funnel and shape the wind like a mountain pass on land, creating blasts of wind that make the offshore waters here some of the most dangerous and unpredictable in the United States outside of Alaska. In the words of one local fisherman: "It can be a beautiful day in town, but you go out and turn the corner and *Bam!* all hell breaks loose ten miles from shore." But while the waters offshore are risky, they are also rewarding. The same wind that makes the water rough also drives currents that pull cold, nutrient-rich water to the surface, turning the ocean offshore into a vast feedlot for fish like salmon and albacore and making the area off Cape Mendocino one of the richest fishing grounds in North America.

Like the neighboring sea, the cape itself is an area of intense activity. It sits right on the edge of a triple junction—a place where three moving plates of the earth's crust come together. North of Cape Mendocino the San Andreas Fault disappears. Farther offshore a rift zone appears on the floor of the sea—the Gorda Ridge. Unlike the East Pacific Rise in Baja California, this spreading ridge is not pushing the continent apart but pushing a small fragment of the earth's crust, known as the Gorda Plate, toward it. Somewhere near the cape, and possibly right underneath it, the Gorda Plate, Pacific Plate, and North American Plate all come together. As a result the land here is being squeezed in all directions, both by the northward drift of the Pacific Plate and the eastward thrust of the Gorda Plate.

Alongshore the San Andreas Fault is replaced by a subduction zone. Known as the Cascadia Subduction Zone, it is nearly one thousand miles long and reaches all the way to the northern tip of Vancouver Island in British Columbia. Offshore, faulting has separated the spreading ridge into segments—the Gorda, San Juan de Fuca, and Explorer Ridges. Rocks driven toward the coast by these spreading ridges are not running into the continent, but underneath it to fuel volcanoes farther inland.

Offshore on the western side of the spreading ridge the Pacific Plate is still heading north. What lies inland of it is a sideshow on the edge of the continent, an ongoing demonstration of the geologic forces that have been building up the western edge of North America for the last two hundred million years.

One hundred eighty million years ago the entire western coast of North America was bordered by a subduction zone like the one now found off the coast of northern California and the Pacific Northwest. The Atlantic was little more than a narrow rift between North and South America and the coasts of Europe and Africa. As it began to open, however, North America began to slide westward into the Pacific, scraping up rocks from the floor of the sea to build up the edge of the continent. Not only did rocks collide with the continent, they also slid underneath it, giving rise to molten bodies of rock that rose to the surface in the form of volcanoes and ranges of high granite peaks like the Sierra Nevada of California and the Sawtooth Mountains of southern Idaho.

While the continent slid westward the rocks it collided with were not those of the Pacific Plate, but those of a pair of ancient and now all-but-vanished pieces of the earth's crust known as the Kula and Farallon plates. While the Farallon plate was moving eastward and disappearing underneath the western edge of the continent, the Kula plate was heading northward towards Alaska and the Arctic where it was thrust down into the earth.

Thirty million years ago, however, this orderly picture of development came to an end when the western edge of the continent ran over the East Pacific Rise. The collision was not head-on but angular. Geologists believe that the original point of impact was probably located somewhere near what is now Los Angeles, or possibly as far south as the Vizcaino Peninsula in Baja California.

After that collision, movements along the coast were no longer uniform. Where the rise met the continent it created a triple junction much like the one that sits near Cape Mendocino today. To the north of it subduction was still taking place, but to the south the continent was no longer bordered by a plate that was moving east but by one that was moving north—the present-day Pacific Plate.

• • •

It is a low gray day at Trinidad, California, a small town on the coast twenty miles north of Cape Mendocino. It was the first week of June, but the wind was cold and damp, blowing out of the northwest at a steady twenty knots. A few miles from shore upwelling was in full swing. It was just a few minutes past eight and down at the town pier fishermen were already returning from their early morning trips, trundling down the docks with gunny sacks laden with twenty- and thirty-pound salmon.

Gary Carver walked out to the far end of the pier among the returning fishermen looking for a light for his morning cigarette. His faded blue-jean jacket was buttoned right up to his collar, leaving only a glimpse of his green flannel shirt and the red bandana knotted around his neck. Were it not for a temperamental boat engine, Carver would have been out fishing himself this morning. From time to time he has considered fishing full-time—not as a hobby but as a profession. As a field geologist, Carver has spent a great deal of his life outdoors. He has worked in both South America and Alaska where he did reconnaissance geology for the route of the Alaskan pipeline. For a time he even considered homesteading in British Columbia. At present, however, he is a professor of geology at Humboldt State University in nearby Arcata, California. This particular corner of northwest California is his area of expertise.

At the end of the pier we turned around and looked back toward the high cliffs ringing the small harbor. "Geology here has a different style," he said as we looked over the rocks. "It's more like Alaska than California. In most of California you can walk from one plate to the other. You can stand with your left foot on the Pacific Plate and your right foot on the North American Plate. Here, however, the plates are stacked up on top of each other. The crust is not eight kilometers thick, but sixteen." Rocks are not being pushed along the edge of the continent here but underneath it.

• • •

In 1973 Carver committed what his colleagues thought to be academic suicide. Fed up with the more settled life of the East, he quit his tenure-track teaching job at the State University of New York in Binghamton to teach at what was then the still relatively unheard-of Humboldt State University. "Of course at the time no one knew there was an unstudied triple junction nearby," he said. With the junction of the Pacific Plate, the North American Plate, and the Gorda Plate only a few miles away, the movements of rocks here to the north of Cape Mendocino are as active as they are confusing. For Carver the area's complex geology has been an academic gold mine—and one that he has had largely to himself. "There aren't many people poking around up here," he said. "And those that are haven't been at it for very long."

While most geologists spend their time rethinking and reworking the geology of areas that have been studied for decades, Carver has spent the last several years discovering and naming the area's major faults and folds. At times, he says, working here is almost like working at sea. Unlike the deserts of Baja California where you can follow a single layer of rock for miles, here you can travel for miles between outcrops—small exposures of rock in stream banks and road cuts, islands of rock in a vast, rugged ocean of trees. In Carver's words, geology here is "largely speculative," pieced together from small, widely spaced pieces of evidence. "You have only the vaguest idea of structural features."

It is only here along the coast that the rocks are in plain view and, as we look back at the cliffs, Carver points out the details. The uppermost layers are a gray siltstone, laced with the shells of clams and sand dollars that are almost indistinguishable from those found in the ocean offshore today. They are, in fact, some five hundred thousand years old. Part of a collection of rocks known as the Falor Formation, they were deposited in shallow water not far from shore. Today they are more than one hundred feet above the beach.

The layer below is still different—a thick pile of dark, blue-gray sandstone. You can see it offshore as well, rising above the waves in the form of rocks and sea stacks the size of small islands. Unlike the rocks above them, however, they are not from shallow water but were once deep in the sea. Soft and easily eroded, they have been worn and weathered into rounded shapes. In places they seem to have been notched by the waves, suggesting that they too were recently lifted out of the sea. They are, Carver says, "blue goo," part of a package of rocks known as the Franciscan Formation that stretches all the way to southern California. "It makes up the world up here," he said. "When it rains they just run."

• • •

The Franciscan Formation contains not only soft sandstones but also sharply folded and fractured layers of chert and basalt like those that anchor the northern end of the Golden Gate Bridge in Marin County. East of the San Andreas Fault this seemingly disparate group of rocks makes up the bedrock of coastal California. A collection of shales, lavas, and sandstones from the deep sea, they have been faulted and folded so many times that their history is often as unintelligible as they themselves are unstable.

Until the advent of plate tectonics the origins of these twisted and unstable rocks were enigmatic. They made no sense in the conventional scheme of things. Their complex patterns of fractures and folds could not be readily explained by the orderly rise and fall of mountains or the advance and retreat of glaciers. Developing theories to explain their origin was something of a cottage industry among California geologists.

Today geologists believe that these rocks were once part of a wedge of sediments that collected along the edge of the subduction zone like the shavings of wood that build up in front of the blade of a wood plane passing over a board. Scraped off the descending slab of oceanic crust, they have been repeatedly folded and shattered as the plates slide past one another.

Sandstones like those found here in the cliffs at Trinidad suggest that those rocks were at one time actually thrust down into the subduction zone. Their color here is the product of altered clays, but in places these same Franciscan rocks are laden with exotic minerals like lawsonite and glaucophane that give them an even stronger blueish tint. Formed only under conditions of high pressure and low temperature, the appearance of these minerals suggests to geologists that parts of the Franciscan were subducted—thrust down into the earth and then quickly uplifted before the heat of burial could take effect.

The only difference between these Franciscan rocks north of Cape Mendocino and those found farther south is age. While Franciscan-type rocks in southern California are more than 150,000,000 years old, these are less than 15,000,000 years old. Here in northern California geologic forces that once shaped the coastline farther south are still taking place. The wildly varying ages of Franciscan-type rocks are due to nothing less than the migration of the triple junction that now lies off of Cape Mendocino.

It is not just continents and ocean basins that have drifted across the face of the earth, but fault zones, spreading ridges, and subduction zones. At one time the triple junction that now sits near Cape Mendocino was located as far south as southern, or possibly even Baja California. As it moved northward, it slowly pinched out the subduction zone that had once bordered the continent, driving the rocks of the Franciscan Formation up onto the continent. In its wake the San Andreas Fault appeared, a link between the spreading ridge that runs down the center of the Gulf of California and those that now lie off the coast of the Pacific Northwest.

That migration of the triple junction helps explain the wide age differences in the rocks of the Franciscan Formation. While those farther south were pushed up onto the edge of the continent twenty and thirty million years ago, those farther north were driven up out of the sea only a few million years ago. Here near Cape Mendocino the rocks from an ancient subduction zone sit right on the edge of an active one.

• • •

Spring comes early in northern California, beginning in December with the soft winter rains that send green shoots of grass up through the dry, barren ground. By March the hillsides are in full bloom. Once-dry slopes are covered with a green blanket of grass and flecked with the bright orange blossoms of California poppies and the deep purple of lupines and wild irises. It all fades quickly. By April the grass has already started to turn, its green blades topped with golden plumes of seeds. By May the hills are dry again, burnt to a bright, golden white. The ground looks sparse and barren.

Offshore in the ocean, however, the bloom is just beginning. Prevailing northwest winds, part of the same persistent high-pressure zone that steers rain away from the area during the summer and fall, begin blowing again, setting up a pattern of currents in the ocean offshore that pull cold, nutrient-rich water to the surface. As it flows to the surface this rich water triggers an explosion of growth that lasts as long as the wind blows—sometimes all the way into September. In a matter of days the ocean offshore is transformed into a rich feeding ground for literally millions of fish. Alongshore you can feel the sudden arrival of this cold water in the thick summer fogs that blanket the coast from late afternoon until midmorning—a product of the interaction of warm air and cold water.

Biologists call these areas upwelling zones. Although they cover less than 1 percent of the ocean's total area, they account for more than 50 percent of the world's fish catch. They are some of the most biologically productive areas on earth. Off the western coast of the United States the prime upwelling zone stretches roughly from Monterey Bay to central Oregon, with the coldest, richest water of all centered around Cape Mendocino like a bull's-eye.

Plants are the basis of the food chain in the ocean just as they are on land. Here, however, the dominant plants are not grasses and trees but tiny floating ones known as phytoplankton that are almost micro-scopic in size. On the floor of the sea their shells and remains fall on the bottom like rain, building up thick piles of debris that are later turned into deep-sea shales or possibly even oil. At the surface they are food for a host of microscopic predators like copepods and krill as well as fish. In theory they can grow almost anywhere in the upper sunlit layers of the ocean, but in reality they are often held in check by the limited availability of key nutrients like carbon and nitrogen. Although the ocean contains the largest pool of fixed carbon and nitrogen in the world, most of that pool is locked up in water that is several hundred meters below the surface—far below the reach of the sunlight the phy-toplankton and other marine plants need to survive.

Outside of upwelling zones that water seldom makes it to the sur-face. As a result, most stretches of the surface ocean are little more than biological deserts. The limited supply of nutrients is quickly used up by the sparse populations of plants and animals already living there. Where upwelling occurs, however, the supply of nutrients is almost unlimited. And where these currents of rich water reach the surface, nutrient levels increase a hundredfold in the space of a few days or hours. The result is an explosion of growth that starts with the phytoplankton and cascades up the food chain through a host of graz-ers and predators to larger animals like salmon, seals, and seabirds. Like the plankton, their lives are also keyed to this seasonal surge of growth.

· · ·

Upwelling was first discovered in the sixteenth century by Spanish sailors traveling off the coast of Peru who were surprised by the pools of unusually cold water that lay just offshore. Although they quickly learned that these pools of cold, rich water made for good fishing, it took observers and scientists almost four hundred years to understand the natural forces that created them. The first real breakthrough did not come until the 1930s in work being done off the coast of California by the oceanographer Otto Sverdrup.

What Sverdrup found was that while prevailing winds blew southward almost parallel to shore along the California coast, the net direction of water movement below the surface was actually perpen-dicular to shore. What causes this difference in motion is nothing less than the earth's rotation, something known as the coriolis effect.

At the surface, water moves in the same direction as the wind. The layers below it move as well, of course, but as they do they are gradually deflected farther and farther away from that original direc-tion of movement at the surface as the earth turns beneath them—like a deck of cards being fanned out across a table. The effects of all this can be visualized by trying to draw a straight line on a spinning piece of paper—while the pencil moves in a straight line, it leaves a spiral-shaped path on the paper. Add up all the subtle changes in direction that occur in these stacked up "layers" of moving water and you end up with a net direction of movement that is at right angles to the wind—away from shore along the California coast. Something has to move in to take the place of all that displaced water, and that some-thing, Sverdrup discovered, is the deeper, richer water below.

While it may take some effort to see how these forces all come together, there is nothing subtle about their effect in the ocean. "You don't have to be a precision analytical chemist to work in the coastal ocean," Dick Barber, the director of the Monterey Bay Research Insti-tute in Pacific Grove California, told me one afternoon in his office above Cannery Row overlooking the semicircular sweep of Monterey Bay. "When the bloom comes on I can see it from my office window. The water gets cloudy. You can look over the side of your boat and see what's going on."

• • •

The only problem with this rich system is that it periodically breaks down. Every four to eight years, and sometimes more frequently, these rich currents of water fail to arrive. Coastal winds die down. The ocean becomes stagnant. For plants and animals whose lives are keyed to these seasonal surges of productivity these breakdowns are catas-trophic—wiping out whole fisheries in some cases, as well as the local populations of sea lions, seabirds, and seals that feed on them. Those effects, in turn, are often projected around the world in unexpected ways.

After upwelling currents failed to arrive off the coast of South America in 1972, for example, the Peruvian anchovy fishery collapsed. Up until that time it had been the world's largest source of protein meal. When it momentarily vanished, buyers began looking for other sources of supply. Soybeans were a viable substitute, and in the space of a few months their price shot up from $100 per ton to more than $300 per ton. In South America millions of acres of rain forest were cleared to make way for growing the suddenly profitable beans. In the United States thousands of acres of ecologically invaluable wetlands were filled in and destroyed for the same reason.

These breakdowns in coastal upwelling are known as El Niño events. The name, Spanish for "the Christ child," was originally used to refer to the normal relaxation of upwelling currents that occurred every year off the coast of Chile and Peru shortly after Christmas. (Upwelling, like the seasons, is reversed in South America, with its spring and summer corresponding to our fall and winter.) In time, however, the term came to refer to the irregularly spaced breakdowns of this rich system.

Scientists had originally blamed El Niño events on local wind and weather changes, theorizing that they were caused when prevailing seasonal winds failed to materialize. In fact, while El Niño events were often accompanied by flat, windless seas, what appeared to be causes were actually effects: it was the El Niño event itself that changed local winds and weather. The forces behind these phenomena, scientists were soon to discover, were truly global in nature, linked to the distribution of water in the ocean and regional changes in sea level.

"It's important to remember that while we're on the western edge of a continent, we're on the eastern side of an ocean basin and that affects things enormously," Barber explained. The surface of the sea is not level but has its own subtle topography of hills and valleys. Prevailing winds cause water to pile up on the western sides of ocean basins. Sea level, for example, is actually some sixty centimeters higher in the Philippines than it is off the western coast of North America. That influences upwelling tremendously because it means that cold water is closer to the surface off the Pacific coasts of North and South America than it is in the South Pacific or the Far East.

"It's a trick," he explained. "We tend to think of deep water as being cool and nutrient rich, but on the eastern edge of an ocean basin that water actually cuts through to the surface because the basin is so tilted. If you're working off Hawaii and you bring up water that is twelve degrees centigrade and has thirty micromoles of nitrogen per liter, that's deep water. Off of California we get that kind of water at the surface. Upwelling is essentially sucking like a straw. But the length of that straw is not very deep. Certainly no deeper than two hundred meters." That also explains why upwelling areas are found off the Pacific Coast of North America but not off the Atlantic. In the Atlantic, that cold, rich water lies far beneath the surface.

The same tilt that makes upwelling possible, however, also makes its breakdown inevitable. On intervals that typically run from four to eight years, and sometimes more frequently, the ocean levels out as water piled up in the western Pacific flows back across the ocean like water pouring out of a broken dam. The force of its shifting weight is so great that the earth's rotation actually slows down by a few fractions of a second. As it moves across the Pacific its effects are felt all the way from Alaska to Chile as sea level rises and water temperatures at the surface soar by ten degrees or more. Layers of cold, nutrient-rich water that were once near the surface are suddenly several hundred meters below it—far beyond the reach of the wind-driven currents that normally pull it to the surface.

Piecing all this together—the links between global changes in sea level and the breakdown of coastal upwelling—required a new look at the earth, one that was possible only from space, from the eyes of earth-orbiting satellites that were able to track global changes in sea level and sea temperature. Before that perspective was available, scientists had assumed that El Niño events, like upwelling, were entirely local—linked to changes in local winds rather than the sea itself.

"There were a lot of intellectual traps," Barber said. "The wind drives upwelling and brings cold water to the surface and all of a sudden the water gets warm and you say, 'Gee . . . Upwelling stopped.' But if you put out momentum sensors instead of temperature sensors you can see that upwelling continues during most El Niño events." The only problem is that it is bringing only warm, barren water to the surface. To understand how it all works, "you have to have a pretty big-picture perspective to hang things on," he added. "When you don't have it you construct a logical picture without it. Like Kafka's dog, who figured out everything. Except the fact that he was a dog."

• • •

Over the past twenty years scientists have come to realize that El Niño events affect not just marine life but world climate as well. As this wave of warm water spreads out across the globe, it changes the heat content of the tropical ocean. As a result, thunderstorms in the tropics become more intense, changing the course of the jet stream and projecting climate anomalies around the world.

An El Niño event during the winter of 1977, for example, created a devastating drought in California. Elsewhere in the country related anomalies in the jet stream sent winter temperatures in the eastern United States plunging down to record levels for weeks on end. The Ohio River froze solid; the Mississippi was blocked with ice as far south as St. Louis. With barge traffic on those two major rivers literally frozen in place, there was no way to bring needed shipments of coal and oil into the region. Schools and factories in the Midwest were forced to shut down for several weeks to save precious supplies of fuel.

Six years later another El Niño event during the winter of 1982–1983 soaked the deserts of Peru and Chile, creating lakes that lasted for several years, while Australia was hit by a crippling drought. California, for a change, had plenty of rain, but with sea level some thirty centimeters higher than normal, the coast was pounded by the unusually heavy winter storms that rolled through the area. South of Monterey Bay along the Big Sur coast the storms triggered landslides that closed the Coast Highway for more than a year.

• • •

While scientists today have a much clearer picture of just how far reaching the effects of an El Niño event can be and the global forces behind them, what triggers them and why they occur in such irregularly spaced intervals—anywhere from two to ten years on average—is still not known. It is thought that unusually high winds in the South Pacific may play a role by pushing water levels there up to some critical point of instability. Others have suggested that these events may be linked to changes in the snow cover in Asia or the monsoon season in India. "We have a lot of ideas, but we certainly don't have the stuff nailed down," Barber said. "What is clear, however, is that it is one big connected system."

While the trigger for these global changes in sea level and weather remains elusive, over the past decades scientists' understanding of El Niño events has started to come full circle. While the events were once considered an unmitigated disaster, researchers are finally beginning to appreciate that El Niños, like forest fires, are also in some way essential for the long-term survival of the marine plants and animals they seem to destroy.

Marine animals, like clams, mussels, and fish, reproduce by releasing free-floating larvae that drift with the prevailing currents. While upwelling areas provide a rich source of food and nutrients for the plants and animals living within them, the net circulation of water is offshore—"a one-way conveyor belt away from the food," in Barber's words. "This is the great biologic irony," he added. "The same upwelling that makes an area rich also makes it hard to recruit a new population because the larvae are blown offshore." El Niño events, as disastrous as they may seem in the short term, allow a season or two's worth of young to stay in the area. When the upwelling currents return they can feast on the sudden surge of productivity.

"The richness of the West Coast of the Americas is the benefit, the price of which is El Niño. It's as much a price as the fog is. You can't have one without the other," Barber said. "The idea that El Niño is just an act of God is just not so. To get this rich system you have to move water across the ocean, but you can't move it across constantly." The droughts and sudden drops in productivity that occur when this system breaks down, he argued, should all be expected.

"You plant wheat crops in a fertile, low-rainfall area and then you're surprised when every ten years there's a drought. There isn't any basis for being surprised with that sort of thing. Looked at as a whole we're delighted with El Niño because the cost-benefit ratio is eight to one: eight years of productivity for one or two years of problems. We build a whole fishing industry around it.

"It would be a big improvement for civilization if we took a more realistic view of nature instead of just treating droughts or hurricanes or El Niño events like something that just came in out of the blue." We could, he said, learn a great deal from native societies on that score. They expected those kinds of natural disasters and their folklore was often filled with stories and parables about them. Understanding the unpredictability of nature was essential to their way of life. In our own postindustrial society we have that information too, but we seem determined to ignore it. "We want to build a bigger breakwater instead of accepting the fact that there are storms," Barber continued. "It is a tremendous morality play about people and nature. A lot of things in nature are that way. The morality part of it is to come along and build a social system based on the good part and then act surprised when the bad part comes along."

• • •

Maps and charts of the seafloor off northern California fill Sam Clarke's office at the U.S. Geological Survey's Pacific Marine Geology branch in Palo Alto, California. The seafloor here is no more feature-less than the terrain on land. Clarke's maps reveal a continental shelf scored by faults and folds and submarine canyons. The San Andreas Fault is readily visible as well. North of Point Reyes its path is marked by the trace of a solid line that runs across the seafloor just offshore slicing through projecting points and headlands. As it nears Cape Men-docino, however, that solid line is replaced by a series of question marks.

The same wind that makes the waters off the cape productive also make them rough and dangerous. A week at sea can leave you feeling like you have spent the last seven days saddled to the back of a half-wild horse. With research ships in short supply and literally the whole ocean to choose from, most scientists choose to go elsewhere. That, coupled with the lack of any significant oil and gas deposits, has kept research on the area's marine geology to a minimum. Like the rugged forested hills that border it, the seafloor here is still very much a geo-logic frontier, and what little work has been done has not been enough to resolve its complex geology. Somewhere near Cape Mendocino the San Andreas Fault disappears, replaced to the north by the Cascadia Subduction Zone, which runs along the coast, and a series of spread-ing ridges farther offshore.

There is no clear consensus among geologists as to just where the San Andreas Fault goes near Cape Mendocino. Some have argued that the fault makes an abrupt right turn, heading straight out to sea to merge with the Mendocino Fracture Zone—a network of fractures and faults nearly a hundred miles wide that runs halfway across the Pacific in a path as straight and regular as a line of latitude on a globe. On the eastern side of the Pacific it seems to lead right to Cape Mendocino. Clarke, however, who has been studying and thinking about the area for years, believes that possibility is unlikely. "You can't simply snake the San Andreas Fault around in a right-angle bend some two thousand miles long," he said. "The geometry simply doesn't work." Instead what his research suggests is that the fabric of the fault "rolls over" as it passes Cape Mendocino, gradually transforming itself from fault zone to subduction zone.

South of Cape Mendocino the regional geology is dominated by faults that trend northwest and dip or slope to the southwest—like a ramp with its downhill side facing the Pacific. North of the cape that fabric of faults and folds still trends northwest but the slope is exactly the opposite—faults and folds dip or slope to the northeast—with their downhill side facing the continent and sloping down beneath it. Closer to the cape itself, faults and folds seem to dip in all directions, as if the regional fabric were rolling over, in Clarke's words, making the transition from fault zone to subduction zone like a ribbon with a twist in it.

As for the triple junction, Clarke believes that it may lie not offshore near the Mendocino Fracture Zone as geologists had traditionally thought, but possibly right under Cape Mendocino itself—right where the pattern of faults and folds begins to change directions. Near the cape, the faults and folds are almost horizontal—as if the layers of rocks offshore were sliding right over the top of one another like a deck of cards being shuffled. That triple junction, he added, may be not a simple, single point on the earth's crust but a broad zone of activity and change covering several square miles, a proposition that seems quite likely when one is contemplating the movements and features of objects the size of continents and ocean basins.

• • •

North of Cape Mendocino the start of the Cascadia Subduction Zone is no more clearly marked than the San Andreas Fault. While the appearance of volcanoes like Mount Shasta and Mount Lassen farther inland leave no doubt that subduction is taking place, there are few signs of its beginnings on the seafloor. In other parts of the world subduction zones are marked by deep gashes in the seafloor known as trenches where plates of the earth's crust are thrust down into the earth. Like the midocean ridges that drive the earth's plates around the globe, these trenches are some of the largest and most obvious geologic features on the seafloor. Off the coast of Alaska, for example, the Pacific Plate disappears into the Aleutian Trench, a thousand-mile-long gash on the seafloor that runs across the North Pacific in a broad, curving arc. While the floor of the deep sea is seldom more than twelve thousand feet deep, the floor of the Aleutian Trench is more than twenty thousand feet deep, a four-mile-deep gash in the earth's surface. Those depths are by no means unusual for subduction zones. In the South Pacific the floor of the Marianas Trench is more than thirty-five thousand feet deep.

Off the coast of the Pacific Northwest, however, the Cascadia Subduction Zone is all but filled by a wedge of sediments carried out to sea by the Columbia River. While the sands and silts of the Mississippi empty into the shallow waters of the Gulf of Mexico and feed a broad, meandering delta, those of the Columbia empty almost directly into the deep sea, forming a submarine fan that spreads outward from the edge of the continental shelf and into the subduction zone that lies just offshore. In places, it is more than two miles thick.

The Cascadia Subduction Zone is so well hidden, in fact, that some geologists have gone so far as to suggest that subduction is no longer taking place here at all. If it were, they argue, the wedge of sediments that now fills the trench would have long since been carried beneath the continent on the back of the descending Gorda Plate. Patterns of earthquakes (or, to be more specific, the lack of them) in the area seem to support this idea as well. Aside from a scattered series of quakes in the Cape Mendocino area over the past twenty years, the area has not been struck by a major quake—or even a small one—in more than one hundred years. Elsewhere in the world subduction zones have produced some of the most devastating earthquakes ever known—not mere 6.0s and 7.0s like those that have peppered the San Andreas Fault, but magnitude 8.0s and 9.0s that have released several thousand times as much energy and leveled whole cities.

In 1923 an earthquake centered along the subduction zone that lies off the coast of Japan killed more than one hundred thousand people and triggered fires in Tokyo that burned much of the city to the ground. In 1960 an 8.3 earthquake centered in the Peru-Chile trench that lies off the coast of South America killed more than five thousand people and caused $500 million in damage. Along the coast the quake was felt over a distance of more than seven hundred miles. Tsunamis or tidal waves triggered by that massive shift in the earth's crust rolled westward across the Pacific, killing more than fifty people in Hawaii and causing some $75 million in damage before continuing on to Japan, where they killed more than 130 and caused another $50 million in damage. After hitting Japan they bounced off the Asian coast the way a ball bounces off a wall and headed back across the Pacific to pound Hawaii again.

To some scientists the area's recent relative calm and its trench-filling sediments are a sign that massive quakes like these are unlikely in the Pacific Northwest today. Others, like Oregon State's Vern Kulm, who have studied that thick wedge of sediments in detail, however, have found that they suggest that the area is anything but quiet. Over the past few years Kulm has spent time studying the seafloor off the Oregon coast in deep-diving research submarines and on surface ships, mapping the area with sonar and studying its structure with seismic surveys. What he and his co-workers have found is that these trench-filling sediments are not flat and featureless but a kind of submarine badlands—a terrain of ridges and plateaus cut by a bewildering array of faults and folds that rise anywhere from several hundred to more than one thousand feet above the surrounding seafloor.

Most of those faults and folds are young and active, suggesting that the area is far from calm. In places, the signs of movement and strain are even more graphic—mud volcanoes several hundred feet high and cold-water seeps that support exotic species of clams and tube worms similar to those found by scientists in the research submarine *Alvin* in Monterey Canyon. They seem to be fueled by jets of mud and water that appear to be flowing out of the sediments themselves, as if they were being squeezed like a sponge. The force behind all this squeezing, of course, is subduction driven by the steady push of the spreading ridge that lies only a few dozen miles from shore. There is little doubt in Kulm's mind that the area is still active.

While there have been no major quakes in the area since the arrival of the first Europeans some 250 years ago, there are suggestions that the forces that shaped this wedge of sediments are not smooth and regular but more likely to occur in sudden bursts.

In the late 1960s, Gary Griggs, formerly a student of Kulm's and now a professor of geology at the University of California at Santa Cruz, found a troubling series of features in the fan of sediments that spreads out from the mouth of the Columbia River. That fan, Griggs found, had been regularly cut by a series of turbidity currents—underwater landslides similar again to those that have periodically coursed through Monterey Canyon. They were not small, steady slumps but massive flows that seemed to have been triggered by earthquakes. Dating revealed that they had occurred roughly every three hundred to four hundred years for the past twelve thousand years. Like the steady creep of the San Andreas Fault farther south that periodically rocks Los Angeles and San Francisco, there is no reason to suspect that the forces that have left these features on the seafloor have stopped. Like California, the Pacific Northwest is not immune to damaging earthquakes. The ground here is not quiet. It is merely the break between movements that is longer. The quakes themselves are even larger.

• • •

Unlike most people, Gary Carver knows precisely how many trees it took to build his home in the hills east of Trinidad: fifty-two. He built his house by hand, using trees cut from his six-acre homestead for the poles and beams. From start to finish the costs were covered by money that came, even on a professor's limited salary, out of pocket. He mined stumps for shingles and bought windows second-hand. Bolts were salvaged from an abandoned pier. No power line ties him to the local power plant. What little electricity his family uses today comes from a twelve-volt electrical power system and a small generator bought in deference to his two young daughters. He estimates that over the years he has graded several thousand papers by the light of a kerosene lantern.

His feel for the landscape is similarly intuitive. As we headed south he pointed out passing faults and folds and layers of rocks the way some people point out city parks and shopping centers to visitors from out of town: the scarp of a thrust fault running through a field; the trace of the Mad River Fault; a two-million-year-old layer of ash deposited from the eruption of an ancient volcano located near what is now Yellowstone National Park more than eight hundred miles away.

So much faulting and folding has squeezed the land here that the fifty-mile trip from Trinidad to Eureka is some twenty miles shorter today than it would have been just a few hundred thousand years ago. Thrust faults and folding have squeezed the earth's crust here like an accordion, stacking layers of rock on top of one another. Unlike the movements taking place on the San Andreas Fault farther south, the area is dominated not by two massive blocks of rocks sliding past one another but by thin sheets of rock that slide over the top of one another like the skin of an onion. In places, the features are all but identical to those found farther north in the wedge of sediments that fills the Cascadia Subduction Zone. Here on the northern edge of Cape Mendocino, according to Carver, that subduction zone seems to come right up on land.

For geologists used to working along the San Andreas Fault and thinking of things in terms of north and south, the area's features are subtle and hard to understand. Even after the advent of plate tectonics, geologists from farther south in the state tended to head north of Cape Mendocino looking for signs of northward drift and surface offset. When they failed to find them, they left with the mistaken impression that this northwestern corner of the state was so quiet that it had hardly moved at all. Little did they know that the features they were looking for were right under their feet.

"Faults here are so low angle they're almost unreal," Carver said. Movements are not northward but eastward, and the faults are inclined at shallow angles, barely five or ten degrees and often less. What looks like merely five or six feet of vertical offset at the surface is actually the product of five or six thousand feet of nearly horizontal movement below. Far from being inactive, the land here is even more restless than it is farther south. Trenching studies done by Carver and his students on the low-angle faults that slice through the area, like those done by Kerry Sieh and Tina Niemi on the San Andreas, have found that the slip rates have been as high as fifteen and twenty feet per quake.

• • •

North of the cape the Mad and Eel Rivers wind their way to the Pacific. For the space of about thirty miles the coast is wide and open, dotted with small farms and dairies. The ground, however, is seldom perfectly flat—it is cut by low rolling hills that are the products of faults and folds. In places, the fields and pastures that border the coast look almost like the terraced fields of Nepal or Tibet. The terraces, however, are not man-made but natural, products of the land's steady rise out of the sea and periodic fluctuations in sea level.

Over the past few hundred thousand years sea level has risen and fallen more than a dozen times, as massive sheets of ice have advanced and retreated across the face of North America in response to changing climate. Fifteen thousand years ago, for example, so much water was locked up in ice that sea level was more than three hundred feet lower than it is today. Farther north, Alaska and Siberia were linked by a vast subcontinent known as Beringia—a land bridge that linked Asia to North America. Today it lies beneath the Bering Sea.

On the West Coast these continual changes in sea level have scored the shoreline with a series of ledges and terraces. During high stands of sea level, ocean waves cut right into the base of coastal cliffs and rocks, carving out a notchlike bench or terrace. Later, when sea level fell, they were left high and dry. By the time sea level rose again, however, the same steady uplift that was building mountains along the coast had already lifted them well above the reach of the sea. Over the past several hundred thousand years repeated episodes of cutting and lifting have created a step-like pattern of terraces along the coast.

Geologists call these features marine terraces, and they are found up and down the West Coast from Mexico to Alaska. Here to the north of Cape Mendocino they rise up out of the sea like stair steps. Near Trinidad there are more than a dozen. The highest are more than two hundred thousand years old and more than one thousand feet above sea level. Near Arcata they run right through the center of town, a series of ramps and steps that lead up from the edge of Elkhorn Slough. The forces that have lifted them out of the sea are still taking place. Faults and folds are so active here that farther inland the terraces are no longer flat but tilted, like the panels of a sidewalk knocked off kilter by the roots of a tree.

••• •••

Halfway to Eureka we stopped at a roadside overlook above Clam Beach and waded through a thicket of scotch broom covered with bright yellow flowers to the edge of a steep, hill-like cliff overlooking the Pacific. The flat ground on top is actually a marine terrace, Carver says. In the face of the cliffs below you could see how it had been abruptly lifted out of the sea.

Carver led the way down to the beach, running down the steep slope and using the branches of willows and salal covering its flanks as pivots and brakes. In less than sixty seconds we were at the base of the cliffs wading across the mouth of the Mad River, ankle deep as it met the sea.

The water offshore was leaden colored, a mirror for the wet gray clouds overhead. South of the river the beach was narrow, littered with driftwood and blocks of dark brown rock the size of small cars that had broken off the base of the bordering cliff. Looking closely you could see the shells of clams and mussels, suggesting that they too had once been part of the sea.

Above that rocky base, however, the cliffs were capped not by layers of light gray sandstone like those near Trinidad but with successive layers of sand and peat. Like the broken and offset layers of rock found in the sides of trenches along the San Andreas farther south, these layers also have a story to tell about the timing and frequency of earthquakes. Here, however, the story is contained not in breaks and discontinuities but in the sequence of the layers themselves.

The lowermost layer of sand was a dirty gray—almost indistinguishable from the sand that makes up the beach today. Several hundred years ago, Carver says, this layer of sand in the cliff actually was a beach. Here and there you can see scattered pieces of driftwood and broken bits of shell inside. In the layer above it the sand is cleaner and more fine-grained. There is no more driftwood, but in places it seems almost laminated—cut by thin layers of sand that lean against one another like the cross-hatched pattern of a herringbone tweed. These finer sands, Carver observed, were left by a series of sand dunes that moved in and eventually covered the beach. Higher up on the cliff these dune deposits are overlain by a layer of peat and soil. In places, it is littered with fallen leaves and pine cones and even tree roots.

Taken together, Carver said, what these layers show is a classic marine transgression—revealing how this area was steadily lifted out of the sea. Uplift not only raised the ancient beach above the edge of the sea, it also exposed a new beach and broad stretch of sand farther offshore. Prevailing winds blew that sand inland to cover the ancient beach with sand dunes. In time, however, grasses and trees began to take root in those dunes, covering them with a thick coastal forest and giving rise to a thin layer of peat and soil. Here on the coast where the Mad River is cutting into the cliffs, that history has been laid out in plain view, as clear and straightforward as an illustration in a text-book.

What is so interesting to Carver, however, is not just that this sequence of events is recorded here but that it is repeated at least two more times in the layers above. Higher up on the cliff that lowermost layer of peat is covered by two more layers of peat-covered sand dune deposits. What they suggest is that there were at least two other episodes of uplift along this stretch of the coast. Sand exposed by these uplifts was then blown inland to form a new layer of dune and the cycle from dune to coastal forest was started all over again.

While the age of the middle event is uncertain, carbon-14 dating of pieces of driftwood fragments found in the lowermost layer of beach sand in the cliff reveal that it is roughly 1,200 years old, while the uppermost layer is just 300 years old. In all probability each episode of movement represented several meters of uplift—enough in any case to expose a broad strip of sand offshore that could be blown inland to form dunes—and those movements were undoubtedly accompanied by earthquakes.

For Carver, whose studies of the area's faults and folds have led him to believe that large quakes are possible here, these cliffs are another troubling piece of evidence. As the Gorda Plate slides under the edge of the continent, the ground here has been faulted and folded into a series of ridges and troughs. Clam Beach lies near the top of one of those ridges and the plate movements taking place here have steadily lifted it out of the ground. A few miles away, however, Elkhorn Slough on the edge of Humboldt Bay lies near the trough of one of those waves of rock—the Freshwater Syncline. While Clam Beach has been rising up out of the water, the slough seems to be sinking down into it.

• • •

By the time we reached Elkhorn Slough on the northern edge of Humboldt Bay, a light rain had begun to fall. It was low tide and down at the water's edge a series of mudflats stretched out from shore. Carver took a shovel out of the back of his Chevrolet Blazer and carefully picked his way out into the marsh, leaving a trail of footprints in the wet surface of the mud that quickly turned into pools.

Here and there the flats were covered with a fringe of pickleweed,

Salicornia grindelia. A rare plant in most of California because of its sensitivity to pollution, it thrives here on the edge of Elkhorn Slough. For geologists like Carver stands of pickleweed are also a useful reference point for measuring subsidence or sinking along the edges of bays and marshes because they grow only within twenty centimeters of the high-tide line. And, although they never grow far from water, they are also highly intolerant of it. If the surface of the marsh rises or falls below their limited range, the plants quickly die. The change is easy to see—submerged below water the bright green plants quickly turn to a rusty red. After the 1906 San Francisco earthquake, dead mats of these same plants had marked the floor of Bolinas Lagoon, showing in colorful detail where its bottom had subsided.

On the edge of a little channel Carver began digging out a small, square block of mud with his shovel. Each scoop sent a sour, sulfury smell wafting through the air—the smell of decaying plants and animals, the rich, sweet, stink of the marsh. Grabbing a handful of pickleweed he pulled the block free and cradled it in his hands. From the side you could see how the roots of the weed reached down into the mud. Below their base was another layer of mud several inches thick shot through with thin layers of gray sand. Below that was another layer of roots that were all but indistinguishable from those of the modern marsh above, except in one important respect: they were, according to Carver, more than three hundred years old, the remnants of a marsh that was probably flourishing here in northern California about the same time as the Pilgrims landed at Plymouth Rock. For at least the past two thousand years, faulting and folding have pulled the surface of Elkhorn Slough down into the sea in sudden drops of several feet or more.

• • •

Carver had been working in the area for several years before he thought of looking for signs of movement under the surface of Elkhorn Slough. In 1987 he had gone to a conference in Corvallis, Oregon, on the geology of the Cascadia Subduction Zone. One of the topics covered at the conference centered around the evidence of major quakes contained in the coastal marshes of both Washington and Oregon. In places like Netart's Bay on the central Oregon coast, scientists had found that the present-day marsh was underlain by as many as a dozen buried marsh deposits. They seemed to suggest that portions of the coast had periodically subsided or sank. From time to time the coastal marshes dropped out of sight. Afterward, uplift and new layers of mud and silt slowly built the marsh up again—only to be dropped down by a new round of subsidence. Those movements were not small. Most had involved drops of several feet or more that were undoubtedly accompanied by major earthquakes. Those drops were not random, scientists studying the area reasoned, but linked to plate movements on the subduction zone located only a few miles from shore. Like California to the south, the Pacific Northwest had apparently been rocked by major quakes of its own.

Even more disturbing than the size of the quakes suggested by these buried marsh deposits was that they seemed to have occurred at roughly the same time up and down the whole length of the coast—roughly every three hundred to four hundred years. What it suggests is that the area was prone not just to local quakes and tremors but to patterns of quakes and tremors that occurred almost simultaneously over distances of several hundred miles—a massive superquake running the length of the whole Pacific Northwest.

While Carver listened to his fellow scientists discuss the possibility of a major quake in the Pacific Northwest, it occurred to him that he had spent the past several years looking at evidence of the same thing in the faults, folds, and coastal terraces that bordered the coast north of Cape Mendocino. As soon as Carver's plane touched down in Eureka he hopped into his car and drove out to the edge of Elkhorn Slough. "It didn't take me more than thirty seconds to find what I was looking for," he recalled. Coring studies would later find evidence of as many as four buried marsh deposits. Farther up in the slough there were even traces of buried trees—forests that had sunk below the surface of the marsh when the ground was suddenly dropped downward. When they got the dates back from the laboratory they were similar to those from the cliffs at Clam Beach and on the same order as the dates geologists had been finding in Washington and Oregon.

For Carver the possibility of a major quake in the region suddenly took on a whole new dimension. The bits and pieces of evidence they had been looking at were not just signs of a large local quake but quite possibly the traces of a massive superquake or coseismic event reaching all the way from Cape Mendocino to Puget Sound. "How big would one of those quakes be? Think of one hundred San Francisco earthquakes all happening at once," he suggested.

Even more mind boggling than the reach of those quakes was the amount of damage they could cause. In Carver's mind the 1964 Alaskan earthquake offered a graphic demonstration of the kind of power contained in one of these superquakes. Triggered by plate movements along the subduction zone that lies off the coast of western Alaska, that quake registered an estimated 8.6 on the Richter scale. It released more than twice as much energy as the 1906 San Francisco earthquake and was felt over an area of more than half a million square miles. Centered under Prince William Sound it all but leveled

Anchorage, the state's largest city, more than one hundred miles away. While the Loma Prieta earthquake did not even break the surface, ground movements and faulting from the 1964 Alaskan quake were spread over a distance of more than six hundred miles. In places alongshore the land shot upward by more than fifty feet. Tidal waves spawned by this sudden massive shift in the earth's crust destroyed docks and canneries from Valdez to Kodiak Island—a distance of more than three hundred miles.

Coastal cliffs and buried marshes suggest that those kinds of quakes have also happened in the Pacific Northwest. The folklore and legends of coastal Indian tribes are replete with stories of giant earthquakes—an earthquake god who ran up and down the coast creating springs in times of drought by shaking the earth; a massive tidal wave that carried a blue whale seven miles up the Klamath River. While there are undoubtedly lessons to be learned from those ancient legends, few people seem to be listening, and what worries Carver is not the past but the future: "Now we aren't talking about shaking coastal marshes and forests but Seattle, Tacoma, Portland, Salem, and Eugene," he added. "Towns here aren't prepared for that. They aren't built to any kind of seismic code."

While the risks are very real, the preparations for a major quake in Washington and Oregon have lagged far behind those of California. Geologists have been able to unravel the details of the area's unstable past only in the past five to ten years, but the pattern of earthquakes in the area has also played a role: aside from a pair of magnitude 7.0 earthquakes in the Cape Mendocino area—one in 1970 that rocked Eureka and another in 1992 that leveled homes in Ferndale and Petrolia out near the cape—the area "hasn't budged for over hundreds of years," according to Carver. At present the entire seven-hundred-mile stretch of the Cascadia Subduction Zone is a seismic gap—just as the Santa Cruz Mountains segment of the San Andreas Fault was before the Loma Prieta earthquake. But in the Pacific Northwest, to date, there has been no regular pattern of quakes to galvanize the public or the area's political leaders. In the meantime local residents, of course, would like a precise prediction as to just when that major quake is likely to strike. But like Allan Lindh and Kerry Sieh, who work on the San Andreas Fault, Carver insists that the key to surviving that quake lies not with better prediction but with better preparation. Scientists, he argued, already know enough to say that that giant quake is likely to strike soon.

"What is the frequency of earthquakes here in the Pacific Northwest?" Carter suggested. "About every three hundred to four hundred years. When was the last major earthquake? About three hundred to four hundred years ago." The current sense of calm, he insisted, should not give anyone anything more than a mistaken sense of security. It is, after all, just what the buried record of ancient quakes suggests: long periods of quiet punctuated by massive earthquakes. Underneath the edge of the Pacific Northwest the rocks that slide into the Cascadia Subduction Zone are slowly gathering strength. It took the San Andreas Fault just eighty-three years to build up the energy released in the Loma Prieta earthquake. The Cascadia Subduction Zone has been quietly gathering strength for more than three hundred. "Oh sure it's quiet," Carver said. "But then periodically all hell breaks loose."

■   ■   ■

North of Trinidad the mountains return, running right to the edge of the sea. A few miles in from the coast you can walk through forests of giant trees—redwoods, hemlocks, and firs. Tall trees tower over the landscape, but the forest floor is open and clean, covered by small, low plants and clumps of lichen and moss that are capable of surviving for decades without ever seeing the sun directly. On a clear day solitary beams of sunlight reach all the way to the forest floor, illuminating the roots of a tree or the delicate bloom of a dogwood tree with a clear golden light.

Here and there the trunks of redwood trees seem to rise up like spires. Sounds are muted here, swallowed by the immensity of the trees and the layers of needles and moss that cover the forest floor. In the redwood groves the tallest trees can reach more than three hundred feet into air. There trunks can be more than fifteen feet in diameter. They are the largest living things on earth.

The first redwoods appeared roughly 120,000,000 years ago when dinosaurs still roamed the earth. At one time forests of these giant trees extended all the way from Mexico to Canada. Today they are found only along the coast in a narrow band stretching from the Big Sur area just south of Monterey Bay to the Chetko River in southern Oregon. Primitive trees, they lack the fine root hairs of more modern trees and depend on the heavy winter rains and persistent summer fogs that blanket the coast here to survive. While they have been found growing on mountainsides at elevations as high as three thousand feet, the densest groves are found along stream banks and river deltas. Their red, furrowed bark is extremely resistant to fire, a fact that helps them survive the fires that periodically cleanse the forest floor. They can live to be more than one thousand years old.

When the oldest groves of trees here first emerged from the ground, armies of knights and soldiers were marching across Europe toward Jerusalem to take part in the First Crusade. William the Conqueror had not yet invaded England. Two thousand miles to the south the pre-Columbian civilizations of Mexico and Central America were building great cities filled with palaces and pyramids. Columbus would not reach America for another six hundred years. The land here in the north, however, was not uninhabited—it was home to clans of coastal Indians who subsisted on the rich runs of salmon that swam through the rivers and streams—the descendants of the first early bands of men and women who migrated across the mass of dry land that once linked Asia to the Americas. In a lifetime spanning several thousand years these giant trees would have had to survive fires, diseases, droughts, and even earthquakes. Like the stones, perhaps they too have stories to tell.

The rocks underneath these ancient trees are old as well. Their history reaches back even further than that of the roots of these giant trees. North of Cape Mendocino the Klamath Mountains straddle the border between California and Oregon, running inland for more than a hundred miles. They are not part of the Coast Ranges that run through Los Angeles and San Francisco, but a province all their own. While the Coast Ranges of California are seldom more than thirty million years old, the rocks that make up the Klamath Mountains are more than three hundred million years old.

The Klamaths are exotic terrane, a displaced package of rocks whose history and origins are far removed from those that surround them. Geologists believe that they may be the remnants of a chain of volcanic islands that once lay off the coast of North America the way Japan now lies off the coast of Asia. Unlike the rocks to the north or south, they did not collide with the continent in response to the opening of the Atlantic, but they were part of an older episode of mountain building known as the Antler Orogeny that began more than 350,000,000 years ago and laid the foundation for the Rocky Mountains a thousand miles farther inland.

Today the Klamaths are an all but impenetrable wilderness of densely forested peaks, home to the mythical Sasquatch, or bigfoot, an apelike giant said to roam the forests of the Pacific Northwest. A tightly woven knot of low peaks, this rugged terrain is cut only by wide, wild rivers: the Klamath, Smith, and Rogue. Few roads find their way through the mountains. Their core is remote, sacred ground to the Yurok and Karok Indians. Jets bound for Seattle, Portland, San Francisco, and Los Angeles fly over the area daily, but the mountains here are a world apart. In an era of fax machines and cellular telephones, tribe members still strive to keep the traditional ways alive and maintain their ties to the land.

Underneath the roots of these ancient mountains subduction is still taking place. Oceanic crust is spreading outward from a series of volcanic rift zones on the floor of the sea and diving under the edge of the continent to fuel volcanoes farther inland. Here on the coast their presence is marked by occasional layers of ash in the coastal cliffs and soils—a record of past eruptions. Farther north in Oregon you can actually see flows of volcanic rock right along the shore. To the north the coast is not built up out of unstable shales and sandstones that have been scraped from the seafloor but from hard black lavas that erupted from the mouths of submarine fissures and volcanoes.

The appearance of these volcanic rocks along the coast is sudden and unexpected. Like so many other important geologic changes along the coast, their exact point of arrival is hidden from view. It occurs somewhere north of Coos Bay on the central Oregon coast. But from the edge of Coos Bay to Florence the coastline is covered by a blanket of sand that stretches for more than forty miles—the largest system of sand dunes in North America.

On a clear day you can stand at the tip of Cape Arago on the southern edge of Coos Bay and see the dunes—a vast wave of white sand rising up out of the sea. Narrow pockets of beaches and dunes occur up and down the West Coast, gathering in the shelter of rocky points and peninsulas, running alongshore in a narrow strip between the high cliffs and sea. In central Oregon, however, the sand comes into its own.

THE OREGON DUNES

A SEA OF SAND

THE BEACH WAS ALMOST WHITE IN THE MIDDAY SUN. IT WAS A STILL, clear, late fall day, midway between the steady winds of the summer and pounding storms of winter. The beach was flat and wide. Overhead the blue of the sky was marred only by the soft, curving shapes of mare's tail clouds. As we drove down the beach, flocks of seagulls ran with wings outstretched to gather speed before pushing themselves into the air with powerful wing beats that sent them curving over the waves offshore. A light wind blew through the open windows of the pickup truck. The late morning sun felt warm on the skin.

It was the second week of October and I was traveling along the beach that borders the Oregon Dunes with Gene Large, from the U.S. Forest Service. Large has been working here in the dunes for more than thirty years. At present he is the supervisor of dispersed recreation for the Oregon Dunes National Recreation Area, a loose term that includes keeping an eye on everything from hiking to mushroom collecting. After driving down to Florence that morning to take care of some paperwork, we had driven out across the high dunes on the edge of town to the beach. We were now on our way back to Reedsport, hugging the ribbon of hard, wet sand that lay just above the edge of the breaking waves.

The beach seems almost endless. It stretches for more than twenty miles in either direction, bordered to the west and east by the ocean and the wall-like expanse of the grass-covered foredune that rises more than twenty feet into the air. Behind it the dunes reach inland for more than two miles—a shifting sea of sand that covers the coast all the way from Coos Bay to Heceta Head above Florence—the largest system of sand dunes in North America. No sprawling developments of vacation homes mar the coastline here. The central Oregon coast is a world of sand, water, and sky.

Beaches have been public property in Oregon for more than one hundred years. In the pioneer days of the 1860s when Large's grandfather settled on the north fork of the Siuslaw River, the beaches were a public highway, a ready-made pathway for wagons and stagecoaches traveling between the small towns and settlements along the coast. Travel was far easier on the sand than through the rugged and densely forested hills farther inland. Although a paved highway runs along the edge of the dunes today, portions of the beach are still open to traffic. But while driving across the sand may be scenic, getting there is often easier than leaving.

"There are a lot of cars that have become permanent fixtures of the beach out here," Large said as he gingerly guided the pickup over the mouth of a small stream crossing the sand. Although the beach can be as hard and firm as pavement, in places it is as soft and unstable as quicksand. "We lose a couple of cars every year," he added in midstream. "They just disappear."

"One Sunday afternoon I was out patrolling the beach and some guy flagged me down and asked me to help pull his truck out of the sand," he recalled. "All you could see of his pickup was the shiny white glow of its roof through the wet sand. I tried to tell him there was nothing I could do to help, but he was pretty insistent. 'But I just got it Friday,' he said. 'You've got to help me get this thing out. I've got to go to work on Monday!' I don't know why, but it always seems to happen to new cars," he added.

The beach that day gave one the illusion of permanence. Built up by five months of gentle summer waves, it was more than one hundred yards wide. In a few weeks, however, the first winter storms would arrive and their high, sharp waves would begin eating into the beach. By December it would scarcely exist at all. At high tide the waves will cut right into the foredune, exposing the tangle of driftwood now buried near its base—not small sticks and branches but the sixty- and seventy-foot trunks of Douglas fir and hemlock washed out to sea from the forests that lie farther inland. "You just can't imagine what it's like out here in the winter," Large said. "Ten-foot waves and a few feet of storm tide with seventy-mile-per-hour winds can do a lot of damage. Those logs are thrown around like match sticks."

The only thing constant about the beach is change. As the tide rises and falls, incoming waves are constantly shaping and reshaping its face. Each wave moves grains of sand up and down the face of the beach. Those movements in turn are all part of an ongoing cycle of change that alternately builds and destroys the beach from day to day and from season to season.

Farther inland the dunes that rise above the beach are marked by their own cycles of change as well. There, however, the changes are driven not by waves but by wind. Wind keeps the dunes in constant motion, burying entangling plants and grasses under a choking layer of sand. Waves of wind-driven sand move through the dunes like waves at sea. In places they are advancing into the coastal forest at rates as high as fifteen feet per year.

Forty years of working here have given Large a healthy respect for the power of moving sand. A few years back when he was working as part of a crew planting beach grass in the dunes, they had just finished unloading two tons of fertilizer in eighty-pound bags when a winter storm sprang up. After covering the bags with a tarp they headed inland to wait out the storm. When they came back the next day their house-sized pile of fertilizer was nowhere to be seen. The wind had completely covered it with sand. More than twenty years later it has still not been found.

Originally the beach and its flanking dunes were part of a continuous system. Waves carried sand up the face of the beach where it was picked up by the wind and blown inland to feed the growing dunes. Today, however, that link has been broken by the foredune that now borders the coast—a high, grassy wall that has cut the dunes off from the beach and its essential supply of sand. The plants that anchor this natural barrier are thin-bladed bunches of European beach grass, or marram grass. They are not native to the coast here but were introduced to the area in the 1930s to check the flow of sand into neighboring harbors and rivers. That artificially introduced plant has done such a good job of controlling the sand that it now threatens to choke off the dunes entirely. With the influx of new sand held in check, plants and trees have been able to gain a foothold on the seaward edge of the dunes, establishing themselves in the lee of the foredune, turning once open dunes into a scrub-covered plain. Over the past twenty years the spread of plants has been so rapid that some scientists have suggested the dunes may all but disappear in the next one hundred years.

Others, however, believe that this troubling advance of plants is only temporary, part of a larger cycle of events linked to changes in climate and sea level that have seen the dunes advance and decline several times over the past few thousand years. These larger cycles span not just days or seasons but hundreds of years. In time a new wave of sand may break through the foredunes that now seal off the coast, bringing the high dunes back to life.

To the north and the south the coastline has been shaped by the earth's drifting plates. Underneath this vast sea of sand those movements are still taking place, but here on the central Oregon coast the surface of the land has been shaped by the wind and the sea.

• • •

At low tide the beach is broad and open. You can see approaching objects for miles. They rise above the flat slope of the sand and draw the eye: a tangled pile of seaweed, a solitary seagull, the shell of a crab, the trunk of a large tree. The steady roll of the surf fills the air, making it impossible to hear sounds more than a few feet away. The disparity between sound and sight gives one the feeling of disjointedness, of continually watching things from a distance. Sounds arrive suddenly and then disappear into thin air—the cry of a gull, the hiss of the water as it runs across the sand, scattered pieces of conversation.

Down at the water's edge you can see how each incoming wave shapes the beach. They break a few feet from shore and roll up the face of the beach, sending thin sheets of water flowing across the sand. Look closely and you can see individual grains of sand tumbling and rolling inside them. Higher up you can see the reach of the high tide in the debris that litters the beach—a tangled line of seaweed, shells, driftwood, and bits of plastic trash carried inland by the waves.

The debris that litters the beach, however, is deceptive. It gives the illusion of a rich and varied life just offshore. In fact, the beach is a barren world of unstable sand. While it may by a playground for swimming and sunbathing for us, the beach is a hostile world for plants and animals that offers little in the way of food or shelter. Waves break on the beach with a force of several thousand pounds. The sand itself is constantly shifting, making it virtually impossible for plants and animals to gain a foothold. The only plants capable of surviving here are the small bits of algae that grow on the sides of individual grains of sand. What little life there is here is carefully hidden below the surface—the scattered clams, crabs, and worms that lie burrowed in the sand. At the surface you can see signs of that buried life in the uprooted shells of clams and cockles that tumble in the surf and the tiny, pin-sized breathing holes of blood worms that dot the wet sand.

The term *beach* is usually synonymous with fine, white sand, but beaches can be made up of almost any loose material: fragments of pink coral in the Caribbean, black lava sands in Hawaii, even fist-size stones and cobbles on the storm-battered coasts of Alaska. All that is needed is a place for this loose material to collect in and waves to keep it moving. Here in the Oregon Dunes the beach is made up of coarse khaki-colored sand.

At first glance the grains look almost uniform. But pick up a handful and study it in the palm of your hand and you will find grains have been worn and chipped from more than a dozen different types of rock: clear, rounded grains of silica; small translucent chips of quartz; pale pink grains of feldspar; and tiny fragments of broken shell. Here and there are darker grains as well: micalike flakes of biotite and horneblende; small, red weather-beaten crystals of garnet; even flakes of valuable minerals like gold and chromite.

Like the more solid stones that lie farther inland, each beach has a story to tell. The composition of its sand is a reflection of eroded hills and mountains, or possibly a coral reef or an ancient volcano. The size and texture of its grains, in turn, are a product of the size and strength of the waves offshore. The high waves and storms that pound exposed points and shorelines build up a steep, coarse-grained beach, while the gentler waves of protected bays and coves build up a fine-grained one that can be as flat as a tabletop. The face of the beach is a reflection of the water and waves offshore.

■ ■ ■

From the top of the high grassy foredune you can see how the waves approach the coast. Here in central Oregon they do not come in head-on but roll in out of the northwest at an angle of thirty to forty-five degrees. On most days they move with a regular rhythm, their evenly spaced crests only a few seconds apart. In deep water the waves move without even touching the bottom and the sea is a clear, deep blue. That all changes as they approach the shore. When the water depth equals roughly one-half their wavelength (the distance between their crests) the waves begin to touch bottom. You can see that point of contact from shore, even from several hundred yards away—the water is suddenly a cloudy tan, filled with sand kicked up by the incoming waves.

Waves can travel for thousands of miles across the ocean with almost no change in size or shape, but as soon as they touch bottom they begin to change shape dramatically: slowing down and gradually growing in height. As those incoming waves slow down, they also begin to change direction. Like light passing through a lens, the shallow bottom bends and focuses the waves until they approach the shore almost head-on. This change in direction sets up currents in the water offshore—known as longshore currents—that move water parallel to the beach. While the incoming waves carry sand up and down the beach, these currents carry sand along it as well—parallel to shore often at speeds of several knots. At times the beach seems to be moving in all directions.

While all this is happening the waves continue rolling toward shore, slowing and growing until they reach some critical point of instability and break. High waves can break and reform several times before they reach the coast, but alongshore each breaking wave sends a surge of turbulent water flowing across the sand. And, like a fast moving stream, that turbulent surge of water carries sand up and down the face of the beach.

The beach includes not only the dry sand above water, but the sand underwater as well up to a depth of several meters. Walking knee-deep through the surf you can feel the pull of the waves against your legs and feel the sand shift under your feet. Waves and currents are continually carrying sand back and forth between these wet and dry worlds of the beach. While an individual grain of sand can move back and forth several times a day, the net direction of movement changes from season to season.

In the summer the net flow of sand is inland. Summer waves are low and rounded, only three to six feet high. They tend to break more gently than winter waves and roll softly up the beach, leaving sand behind the way a flooding river leaves mud and silt on a floodplain. By fall the beach can be more than one hundred yards wide.

In winter, however, the net flow of sand is offshore. Winter waves are sharper and more angular. They can explode on the beach with a percussive crack that shakes the sand under your feet like the tremors of an earthquake. More than twice as high as summer waves, they average some twelve feet in height, with storm waves that can top twenty feet. As they break on the beach they cut into the sand the way a rain-swollen creek cuts into its bank—capable of chewing up ten and fifteen feet of sand in a single afternoon.

That sand, however, is not really lost to the beach but merely carried offshore. Offshore, the sand from the summer beach gathers in shallow bars and banks that protect the coast behind them from the full force of the sea. High storm waves break on the shallow bars offshore, spending the brunt of their energy before they ever reach shore. In effect, they are filtered out and broken down into a series of smaller waves. When the gentler waves of summer return, that same sand moves back onshore again, building up a flat, wide beach.

The beach's mobility preserves and protects it. Problems arise when man steps in and expects the unstable sand to behave like solid ground: terra firma for vacation homes and beachfront hotels. A shoreline shored up by seawalls and jetties has no power to move and no ability to absorb the pounding of winter storms. Those who ignore the yearly cycles of change on the beach do so at their peril. Like the movements of rocks and continents that have built up this active edge of the continent, these movements of sand on the beach are part of the rhythm and pattern of things. Although the sand holds few clues about the history of continents and ocean basins, like them it too has been shaped by forces that are essentially global in nature. The waves that shape the beach spring from winds that are often half a world away.

• • •

Waves are a product of the interaction between wind and water. The critical factor in determining wave height is not wind speed but fetch, the distance over which the wind blows. It takes time to build up the large waves—and the longer the reach of the wind, the higher the waves. Starting from a flat, calm sea, there is an orderly progression of events and waves that leads from still water to a high, rolling swell.

Wind blowing across a flat, smooth sea hardly catches the water at all. In time, however, it begins to set up a faint pattern of ripples that grip the wind, gradually giving rise to short choppy waves. For the first ten hours those small waves grow rapidly, but after thirty hours they hardly grow at all.

In the open ocean waves come in all shapes and sizes, even different frequencies. Some are little more than ripples that have only just begun to form. Others are left from storms that may have occurred several days earlier. As the wind continues to drive the waves, however, these confused patterns slowly begin to sort themselves out. Some cancel out as the crest of one wave collides with the trough of another. Others, however, begin to slowly combine as crest meets crest, building up a chain of large waves with nearly identical shapes that move through the ocean with a regular rhythm: sea swell.

Once that regular pattern of swell is set up, waves can move through the ocean with little or no loss of energy. Out in the open ocean the movements of water molecules inside a wave are almost entirely vertical. Horizontally they hardly move at all. Their passage through the ocean is as effortless as ripples that curl through a flag blowing in the wind. That ability to travel long distances explains why you can go to the shore on a calm, windless day and find high waves pounding the beach. Waves in the ocean can travel for hundreds or even thousands of miles. Storms in the North Atlantic, for example, end up as surf in Morocco. Antarctic gales send high waves pounding into the coast of Alaska more than ten thousand miles away. Here on the Oregon coast, the summer waves that wash up on the beach come from storms in the Gulf of Alaska more than fifteen hundred miles away. Higher up on the beach the sand they carry toward shore is blown inland by the wind to cover the coast with a blanket of dunes.

• • •

The sand was almost featureless in the midday sun—a bright white sea broken only by scattered green clumps of American dune grass that rattled in the wind. The dunes themselves were all but invisible, their texture perceptible only underfoot: the corrugated feel of small ripples and ridges, the tilt of a steep slope leading up to a hard-packed ridge of sand and a sudden view of the Pacific.

Offshore, a purple layer of fog had begun to gather over the ocean. Here in the dunes, however, the air was sharp and clear, marred only by the heat waves that shimmered and danced across the sand.

The dunes are wide and open. You can pick out a distant point on the horizon and walk toward it almost straight across the sand, your path interrupted only by the steady rise and fall of the dunes. Sound carries a long way here. When the wind is light you can hear the wing beats of a raven flying a quarter-mile away. Walking down the side of a steep dune you can hear the metallic squeak of sand—the sound of grain rubbing against grain beneath your passing feet. Distances are deceptive. Objects that seem only minutes away take several hours to reach. After a half hour of walking you can stop on the top of a high ridge of sand and turn around and see your tracks stretching out behind you. Look closely and you will notice that they have already started to fill with sand.

Here and there are other tracks as well, small stories in the sand: the furrowed path of a beetle or an ant wandering through the open sand; the tracks of a deer heading out toward the deflation plain; the footprints of a bird that suddenly appear and pace across the sand and then vanish into thin air.

Like the beach, the dunes offer little in the way of food or shelter. Outside of the scrubby deflation plain that spreads out from behind the foredune, few plants grow here. Water is scarce. In midsummer the temperature of the sand underfoot can soar to more than 120 degrees Fahrenheit.

But while the dunes are capable of supporting few animals on their own, most living in the neighboring coastal forests visit the open sand from time to time. Mice and raccoons, for example, make brief forays out into the dunes. Birds fly overhead, stopping to examine a solitary bush or tree in the sand or a clump of grass growing near the crest of a dune. Deer forage in the scrub-covered deflation plain that borders the foredune. Even bear and mountain lion can be found here. With territories that sprawl over as much as one hundred square miles, the dunes are simply another facet of their far-ranging habitat.

At first glance the sand here looks no different from the coarse gritty sand on the beach, but pick some up in your hand and let it run through your fingers and you will find that it is as fine as flour. It works its way into shoes and camera lenses. Puffs of wind blow fine grains through the air like smoke.

As sand moves inland from the beach it is continually winnowed by the wind. Grains of sand do not float in the air as they do in the waves offshore; they bounce and skip across the dunes as they are driven inland. As one heads inland from the beach the sand becomes progressively finer-grained. Coarser and heavier grains are gradually left behind by the wind. Even a mile in from the beach on top of the high dunes that border the forest, however, the sorting is not complete. You can see it in the varying colors of ripples that run through the sand: crests are almost white, while the troughs are slightly darker, almost khaki-colored. The difference has to do with the composition of the grains themselves. The crests are made up almost entirely of light, clear grains of silica and quartz while their troughs are richer in grains of darker and denser minerals like magnetite, garnet, and horneblende passed over by the wind.

•••

By late afternoon the texture of the dunes was sharp and distinct. As the sun moved lower in the sky, the sand began to slowly change color, fading from white to tan, then bright gold. Shadows had appeared in the low light, setting the details of ridges and troughs off in bas-relief. Out toward the foredune the sand was almost flat, broken here and there by ripples whose shadows gave the open sand a texture that looked almost like corduroy. A few hundred yards farther inland the sand was built up into a series of low regular waves that rolled inland for more than a quarter-mile toward the towering, pyramid-like shape of the high dunes. Like waves at sea there is an orderly progression to the growth of these waves of sand in the dunes. Here, however, that growth takes place in not the space of a few thousand miles, but the space of a few thousand yards.

Behind the edge of the deflation plain that now marks the start of the dunes, the sand is almost flat. Like breezes blowing across a calm, smooth sea, the wind here seems to hardly move sand at all. Here and there tufts of grass and low scrubby plants have trapped piles of drifting sand, giving rise to hummocky mounds two and three feet high. In places ripples appear, offering a surface for the wind to grip.

A few hundred yards farther inland the wind has blown the sand into a series of low dunes as regular and orderly as the patterns of swell in the open ocean. Their crests are rarely more than six feet high and typically seventy-five to one hundred feet apart. The ridge of a single dune can run for more than a half mile. Seen from the side they look almost like teeth on the blade of a saw, a series of off-center triangles.

These transverse dunes face the summer wind with long shallow slopes canted at just ten or twelve degrees. Sand blows up that gentle windward slope and piles up until its height reaches some critical point of instability and tumbles down the backside or slip face of the dune. Slipfaces are steeper, invariably dropping off at a uniform thirty degrees—the loose sand's limit of stability. As sand falls down the slip-face the dune slowly moves inland, maintaining its shape as sand from its windward side slides down the slipface. They can travel as much as a hundred feet in a single season. Standing on top of the high dunes you can almost see how they roll through the sand.

As these dunes move inland they being to slowly rise up on top of one another—like waves at sea—as ridge collides with ridge. Crests run together to create long, winding ridges of sand. Troughs combine and intersect to form deep amphitheater-like bowls. The undulating shapes of the dunes in turn tunnel and shape the wind, creating vor-texes of turbulent air. In places, those fast-moving jets of air cut right through a high ridge of sand, forming blowouts. In less than a mile these waves of sand have grown into giant hills more than 150 feet high. The ridge of a single dune can stretch for more than a mile. Where they climb up the sides of the neighboring coastal hills, the crests of these dunes can be more than one thousand feet above sea level.

These high dunes are known as oblique dunes because their crests are oriented obliquely to the seasonal winds here. While winter winds blow the transverse dunes almost flat, the oblique dunes last from season to season.

Like the beach the dunes vary from summer to winter. Here in the dunes the same steady winds that drive upwelling in the ocean drive waves of sand. Summer winds are brisk and steady, blowing in out of the northwest at a steady ten to fifteen knots. In from the fore-dune and its fringing deflation plain they begin building up the famil-iar texture of the transverse dunes. The air is dry and the sand moves easily, filling holes and hollows and smoothing out sharp angles to give the dunes a rounded look.

Like the winter waves that pound the beach, winter winds are strong and unpredictable. They blow out of the southeast, driven by approaching storms. Out in the open sand wind speeds can top one hundred miles per hour. Other days are airless and still. Rainfall is heavier in the winter as well and it changes the texture of the sand immensely, cementing grains of sand together. The sand moves less easily and the shape and texture of the dunes become sharper and more angular. Down near the beach the transverse dunes are planed almost flat.

Up in the high oblique dunes this combination of wet sand and high winds can carve narrow, fifteen- and twenty-foot-deep notches in the sand known as yardangs. In places the wind blows probing fingers of sand back into the trees—parabola dunes whose northwest trend indicates the direction of the wind as readily as a weather vane. A one-hundred-foot-high dune can move several feet in a single storm. Standing on the inland edge of the dunes you can see how they move into the forest, choking off trees, plants, and flowers with a flooding wave of sand. Along the edge of the dunes the dead trunks of seventy- and eighty-foot fir trees rise up right out of the dunes like telephone poles, their bare trunks polished and worn to a bright silver by the abrasive action of wind-driven sand.

Here on the edge of the forest the sand seems almost unstoppable, but the battle between plants and trees is part of the ongoing cycle of things in the dunes. The wind is constantly shaping and reshaping the dunes. When the sand stops moving, the dunes quickly give way to a cover of plants and trees. Few things last for long in the open sand.

••• •

While dunes are typically thought of as dry, associated with hot spots like the Sahara desert of northern Africa and the barren, sunbaked terrain of the Middle East that borders the Persian Gulf, the critical factor for the formation of dunes is not dryness but a steady supply of sand. Here on the central Oregon coast more than eighty-five inches of rain fall each year, and in the dunes, water is seldom more than a few feet below the surface. Even during the hot, dry days of summer you can dig down into the dunes and find damp sand only a few inches below the surface.

Sand has been collecting on this stretch of the coast for at least several thousand years. The coastline here is gentler and more open, bordered by low rolling hills rather than high cliffs and mountains. Rugged points of land—Cape Arago to the south and Heceta Head to the north—define the reach of the sand. Not only does it stretch inland for the better part of two miles in places, it is also several hundred feet deep. Offshore this sea of sand runs out across the seafloor for miles.

In most parts of the West Coast the sand on local beaches is derived from local rivers and streams. The sand here in the dunes, however, is exotic. Its mineralogy—the composition of its constituent grains—does not match that of nearby streams and bays. Like the rocks that make up this mobile edge of the continent, the sand here in the dunes has come from somewhere else. Its grains are volcanoclastic: derived from the erosion of ancient volcanic mountains that once towered over the coast. In addition to grains of common minerals like silica and quartz, they also contain minerals like augite, epidote, hypersthene, garnet, zircon, and chromite. These minerals are not generally found in the rocks of the Coast Ranges that border the dunes today, but they can be found in those of the Klamath Mountains, which lie farther south.

The rocks that make up the Klamath Mountains today originally collided with the continent more than three hundred million years ago. Since that time they have been built up and eroded several times, giving rise to a flood of sand that has been washed out across the continental shelf. The most recent episode of uplift and erosion began just fifty million years ago. At one time they may have been as much as twenty thousand feet high. Today their highest peaks barely top eight thousand feet. Walking amid the dunes you can see the remains of ancient peaks that bordered the coast more than fifty million years ago.

In turn the processes that built up the Oregon Dunes have been going on for thousands of years. Centuries of wind have winnowed the sand, blowing fine grains of it in from the beach to build up the high dunes. Offshore the supply of sand is almost limitless. Carried inland by the waves it has kept the dunes alive, covering invading plants and trees with a choking layer of sand.

Over the past sixty years, however, that natural scene has been interrupted by plantings of European beach grass to check the flow of drifting sand. Today those plantings have grown into a grass-covered foredune that runs along the edge of the beach like a solid wall, dividing the beach and the dunes into separate worlds. Cut off from the beach's life-giving supply of sand, the dunes are in trouble. While the dunes are still moving inland, a carpet of plants has started to spread out from the base of the foredunes that border the beach. In the space of fifty years the dunes could all but disappear.

• • •

As we headed up into the high dunes outside of Florence, the pickup truck rose and fell over the rolling surface of the sand like a ship at sea in heavy swell. Wheels slid and slipped as we climbed up the steep slopes. The plunge down the other side was preceded by a brief feeling of weightlessness as the sand dropped out from beneath us. We roared through the long flat troughs between the successive waves of sand with the throttle wide open, gathering speed for another climb. Ruts and tracks rocked the truck like potholes on a city street. "Some people do this for fun," I thought to myself as the top of my head collided with the roof of the truck.

Gene Large would rather be walking than driving. Although traveling through the open dunes on foot can be slow and tedious, the details of the sand are easier to see and there is no need to shout over the whine of an engine. Near Cleavox Lake, a small freshwater lake on the edge of the high dunes, Large stopped the truck on top of a high ridge of sand to take in the view.

It had been a busy weekend, and after a week of still windless days, the sand was covered with tracks—as if a tank battalion had stopped by for a few days of maneuvers. Roughly half of the Oregon Dunes' twenty thousand acres of open sand have been set aside for foot travel only. The remainder, like this stretch of dunes near Florence, are open to off-road vehicles. Out in front of us the tracks of dirt bikes and all-terrain vehicles crisscrossed the dunes, some turning tight circles in the sand, others heading straight for the beach more than a mile and a half away. A few cut through stands of dune grass and up the flank of a scrub-covered slope topped with fir trees. At its base Large picked a large uprooted sign warning that the area behind it had been closed to vehicles.

After working in the dunes for almost thirty years, Large has seen a lot of changes. He first started working here in 1952 as a seasonal employee with the U.S. Soil and Conservation Service planting trees and beach grass in the dunes. They traveled around the dunes in a surplus army truck, a Dodge Power Wagon with oversized tires pirated from a DC-3. At the time, he recalled, he and the other soil conservation service employees were the only people who traveled regularly in the dunes. Out in the open sand they would often go for days without seeing anyone outside of their own work crew.

"At the time," he recalled, "the dunes weren't recognized as having any recreation or commercial value. The sand tended to drift into neighboring harbors and highways. Management of the area was largely synonymous with plantings of beach grass to hold the dunes in check. For a time the Forest Service even attempted to turn the dunes into a commercial forest, planting thousands of trees. Today, however, more than two million people visit the dunes each year and the primary concern of the Forest Service is not growing trees but preserving open sand. Plantings of beach grass that had once seemed so essential to manage the dunes now threaten to all but choke them off.

From our vantage point on top of the high dunes you could look out toward the Pacific and see how the plants had come to take over the dunes. A high grassy foredune hid the beach from viewing, running along the beach like a seawall. In places it was more than twenty feet high. Behind it a carpet of grass spread and scrub reached more than a quarter mile into the dunes. Thirty years ago, Large said, that foredune was nothing more than a low ridge of sand. "It wasn't all grassed in," he said. "We even did some work repairing the breaches in it." The open sand stretched all the way to the water's edge.

The culprit behind all these changes is European beach grass. A fragile plant along hot, dry coasts like Australia's, it thrives in the wet, mild climate of central Oregon. Its roots grow together like a woven basket, anchoring it in the loose sand. Above ground its spreading knee-high clump of thin blades checks the flow of sand off the beach the way a fence line stops drifting snow. Buried by sand it simply puts out new shoots. It seems to thrive under stress. Uprooted by winter storms, it drifts down the beach and puts down new roots wherever it lands.

European beach grass was first introduced here in the 1930s by the Forest Service and the U.S. Soil and Conservation Service to protect a chain of picnic areas and campgrounds that had been built on the ege of the dunes by the Civilian Conservation Corps. Other planting were designed to shield nearby harbors and highways from drifting sand. By the 1950s that program had expanded to include plantings of shore pine as well as plants like scotch broom, barley, and birds foot trefoil to provide cover for wildlife and help fix nitrogen and other nutrients in the loose sand. For the first thirty to forty years those plantings seemed to have hardly any effect at all, but by the time the dunes were protected as a national recreation area (NRA) in the early 1970s, the sand was disappearing at a rate of more than two hundred acres per year. Less than ten years after it had been preserved as a park, scientists were already predicting its demise.

The key to this sudden surge of the plants was the development of a high, solid foredune that sealed the dunes off from the beach and its essential supply of sand. At first it was nothing more than a low mound of sand, but once the beach grass established itself on the edge of the shore the foredune began growing rapidly in height as the grass on its flanks trapped sand blowing in from the beach. At times you could almost see it grow. "The coast guard had a communications line that ran along the top of the foredune," Large recalled. "It came across from the coast guard station just east of the Siuslaw and ran the full length of the beach to the Coos River, if I remember correctly. I saw two lifts of that line. From time to time they had to put in new poles. The sand just climbed right up to the wires."

Behind the foredune the changes were even more dramatic. While the high wall of grass cut off the supply of new sand from the beach, prevailing winds continued to blow the dunes behind it farther inland. Eventually so much sand was blown out of the area behind the foredune that the water table reached right to the surface—a deflation plain. What was once an area of flat, dry sand was now a wet, marshy plain where grasses and sedges and flowering plants like beach pea and monkey flower could take root and spread out across the dunes. Today the transition from sand to plants is so complete that in some spots, trees have even begun to take root in the deflation plain—thickets of willow and alder. Beaver have begun to move into the area from the nearby forest, cutting down trees and saplings to build small dams across the seasonal streams that run through the deflation plain. In some areas the plants are advancing out into the sand by as much as seventy-five feet per year.

After a lifetime of working here, Large is clearly pained by the possibility that the dunes may soon disappear. But his perspective on the problem is tempered by years of exposure to the power of moving sand. He is not inclined to judge the past by present standards.

"You know, I'm kind of defensive about it," he explained. "The Forest Service comes under a lot of heat from today's people for past management. They're blamed for covering all the sand. But I don't know how it could have been done any differently. If you're going to start out with this open block of sand today and try to live next to it and have harbors and towns and highways, I just don't see how it could be avoided. You manage in the best interests of the times. I think if they were going to start over again today they would do the same thing." Ever since the first strand of beach grass took root, "it's been kind of a race between the sand and the grass to see which could outgrow the other."

• • •

The first time I went to see John Gould, the supervisor of the Oregon Dunes National Recreation Area, I had to wait in the hall with a group of television and newspaper reporters from Coos Bay and Eugene. The dunes are not normally prime-time news, but the day of my visit happened to coincide with a California Mining Company's announcement that it planned to exercise its claim to a section of the high dunes near Coos Bay, tapping the fine, pure sand of the dunes for the making of precision glass. As the area supervisor, there wasn't much Gould could do except take the heat. The company had a valid mining claim on the area and, while it was within the national recreation area, it was without question perfectly legal.

Leaving the area open to existing mining claims was one of the many compromises that had been worked out to enable the area's preservation. Fortunately this claim was the only one still open on the lands administered by the national recreation area. Explaining the situation, however, did not lend itself to ten-second sound bites. While the company had a valid mining claim, Gould says, their land was entirely surrounded by parkland, and access was, legally speaking, impossible. What the company was really after, he later explained, was to swap its parcel of sand inside the NRA for a similar source of sand on publicly owned land outside the park. The possibility of mining in the dunes, however, had captured everyone's attention.

"People were outraged and they had a right to be," Gould said.

"But we have a law and they followed the law and I don't think the government has a right to tell people that they can't do what the law says they can do if they follow the law. If society doesn't like the law, maybe they should change the law."

Gould is a wildlife biologist by training, and a few days after my initial visit we spent the better part of an afternoon talking about the plants and animals of the dunes as well as the steady advance of the scrub-covered deflation plain. While past supervisors spent their time finding ways of stabilizing the dunes, Gould spends time thinking up ways of keeping the dunes moving—a goal that is largely synonymous with removing the beach grass his predecessors worked so hard to plant.

In contrast to the doomsday predictions of the 1970s that predicted an imminent demise of the dunes, Gould is cautiously optimistic—at least in the short term. "While I don't have any scientific evidence to back it up," he said, "I've been here for eight years. I'm not sure why, but after looking at the aerial photographs I don't think we're seeing a rapid expansion of the deflation plain anymore." In part, he explained, that may be because the plants have largely covered over the flat sand portion of the dunes. Now they are faced with the much more difficult task of establishing a foothold in the high dunes that lie farther inland. Away from the coast the dunes are pushing into the neighboring coastal forest by as much as fifteen feet per year, burying the trees beneath a choking layer of sand. "One thing's for sure," Gould said, "the dunes aren't sitting around waiting for the plants to arrive."

Down at the beach, however, the problems are still very real. Gould would feel much more comfortable about the long-term future of the dunes if he could find a way of bringing the beach grass that borders the shore under control. It took almost forty years for that grass to check the flow of sand off the beach. Solving the problems it has created may take even more time.

In theory it is possible to simply remove the beach grass plant by plant, but in practice the prospect of weeding the entire forty mile reach of the foredune by hand is so prohibitively expensive that it has never been seriously considered. Shortcuts have been tried, but so far there has been a noted lack of success.

For a time the Forest Service experimented with herbicides like Round-up and Dowpon on a small section of the foredune, but the chemicals—widely used to control weeds on farms—seemed hardly to faze the plants at all. Before they could get any conclusive results, the program was quickly brought to a halt by a court injunction obtained by environmental groups that objected to the spraying. The Forest Service was not the only agency disappointed by the setback. The spraying had been approved, even encouraged, by the state's wildlife agency.

Seeking to provide more nesting ground for the endangered snowy plover that had historically used the flat sand portions of the dunes now taken over by the fordune and deflation plain, the agency had reasons of its own for trying to control the spread of European beach grass.

To date, the one technique that has proved successful is simply removing the foredune—sand, grass, and all. When CETA (Comprehensive Employment and Training Act)—a federally funded job training program—came calling in 1984 to look for a place to train heavy-equipment operators, Gould put them to work on the foredune. In the space of a few days they had opened up a 150-yard-wide breach in the foredune, using bulldozers to push the wall of sand and grass bordering the beach into the sea.

What had started out as a simple plan to train would-be bulldozer operators turned into an ongoing scientific project as the Forest Service called up researchers at Oregon State University to see if they might be interested in tracking the movement of sand through the breach. Previously, some experts had predicted that the sudden invasion of plants was in part due to a shortage of sand; that theory was soon put to rest. In the space of two years, however, more than thirty thousand cubic yards of sand shot through the narrow breach—enough to bury a football field twelve feet deep in sand. Expand that gap to fifty miles, Gould says, and you didn't have to think too hard to understand why dunes cover this stretch of the coast—or that knocking down the foredune could quickly bring them back to life.

Seven years later that breach is still not completely healed. But although the breach proved that the problems created by the foredune could be quickly cut down to size, there are not plans to start removing more parts of that grassy wall. Once word of the flood of sand that had shot through the gap began to get around, it didn't take long for questions to be raised about the effects this new wave of sand might have on local harbors and highways. Finding a technically and politically feasible way of freeing the sand remains as elusive today as it was twenty years ago. In the meantime, the battle between sand and beach grass is one Gould expects to lose. "I hope I'm wrong, but in my opinion the death knell for the Oregon Dunes was sounded when the first beach grass plant was introduced to the Oregon coast," he said. "Whether it takes one hundred years or five hundred, the Oregon Dunes National Recreation Area will eventually become the Oregon Dunes National Forest. Natural succession will take place."

••• •

While Gould is pessimistic about the long-term outlook for the dunes, others, like Chuck Rosenfeld, one of the scientists from Oregon State

University who came down to study the Forest Service's breach in the foredune, sees things in a different light.

Years later Rosenfeld can still remember the first day he arrived at the beach. The bulldozers were already at work, driving mountains of grass and sand into the sea. It didn't faze the water at all, he recalled. Almost as soon as it was pushed into the sea, the sand floated off, carried away by the tides and waves. The only thing left was the 150-yard-wide breach in the foredune. For Rosenfeld, an associate professor of geography who had been studying beaches and dunes up and down the coast for several years, it was a chance to see the past, to see how sand had moved through the dunes before the arrival of European beach grass. Over the next few years the mountain of sand he saw shoot through the gap gave him a very different perspective on the future of the dunes than either Large or Gould.

Over the next few decades, Rosenfeld believes, the beach grasss that now anchors the foredune will slowly give way to sand again. "People who have seen the short-term change in the past fifteen to twenty years are panicked by it," he said. "But things balance out. I don't really expect to see a lot of freezing up of the open dune sequence. The grass has peaked out in total area as far as I can tell. What people don't realize is that it only propagates when there's fresh sand being dumped on it. When its roots are exposed by deflation it dries out and dies. It's only a transient species. Once it stabilizes and there's no longer that high nutrient loading for the beach grass it just tapers off. It becomes less virile." In short, the beach grass's success in anchoring the sand may prove to be its undoing. Elsewhere along the coast, Rosenfeld added, these kinds of changes are already taking place. "I've done a lot of work up on the Klatsop Plains on the coast near Portland. They first planted beach grass there in the early 1900s. Now American beach grass, the broad-leafed native plant of the area, is coming back and coming back fast."

Rather than an inevitable march of plants across the dunes, what Rosenfeld believes will happen is that the foredune that now blocks the flow of sand into the dunes will break up into hummocks, or "whale backs," as he calls them, as the grass gives way. Sand will begin shooting through these gaps just as it did through the breach the Forest Service had been able to open in the early 1980s, and bring the dunes back to life.

All this, he cautioned, is not likely to happen in one human lifetime, but it may occur over the course of several decades, perhaps even a century or more. Rosenfeld believes, however, that the changes are inevitable. The surge of plants across the dunes that has accompanied the arrival of European beach grass is merely part of an ongoing cycle of change in the dunes—one that has seen the dunes advance and retreat over the coast here several times during the past few thousand years. In time the dunes will begin building again as a new wave of sand washes up on the beach and begins moving inland.

• • •

From the top of the high dunes the view is panoramic, a study in contrasts. You can see both forest and sand. To the west you can look out over the wide reach of the dunes toward the soft, blue swell of the Pacific. To the east you can look out over a green coastal forest of Douglas fir, spruce, and hemlock that stretches all the way to the crest of the Coast Ranges. In places, you can stand on top of the high parabola dunes that punch back into the forest and look right down into the tops of the trees. Here and there the dunes are bordered by small lakes. They lie almost right on the edge of the dunes, black pools of water amidst the deep green of the trees. As the sand has moved inland it has overrun not only trees but streams and rivers, blocking their path to the sea to create small lakes.

Three days of wind had blown the dunes clean and the sand was as trackless as a field of freshly fallen snow. The wind was still blowing. It sent thin sheets of sand dancing across the dunes like the mists that play across the surface of a lake in early morning. Behind the edge of the sand, the curving reach of the Pacific was streaked with whitecaps. Offshore a freighter headed north, hugging the coast.

There above the Umpqua River, the dunes seemed almost endless. Here and there, however, you could see scattered stands of trees. They seemed to rise up out of the sand like ships at sea. They were not part of the deflation plain, but tree islands—the remnants of an ancient forest that has been all but buried by sand.

Although it seemed only a few hundred yards away, the walk out to the nearest island took more than two hours. Travel was slow, a journey across several miles of open sand that moved and drifted with the wind. From the tops of exposed ridges you could actually hear the sand sliding across the dunes—the soft scrape of grain against grain as it moved with the wind.

In the deep hollows between dunes the air was so still that you could hear the rattle of dry leaves blowing across the sand and the buzz of a solitary dragonfly. Halfway out to the island three ravens flew overhead, their shadows on the white sand an almost perfect reflection of their black shapes. The sand seemed barren, almost lifeless.

Near the trees, the air was suddenly full of sounds: the cries of squirrels running from treetop to treetop and the chatter of goldfinches. Out in the open sand the trees were an oasis of life. From the top of a nearby dune you could look right down into the tops of the sixty- and eighty-foot fir trees of the island and see chestnut-sided chickadees flitting from branch to branch; small black-capped birds, their gray backs and white bellies were accented by rust-colored streaks running down their sides. Down at ground level, however, the island was an almost impenetrable thicket of salal, huckleberry, and willow. A narrow path leading into the heart of the trees disappeared in less than ten feet.

From a distance the island had seemed like a small point in the sand, but the walk around its base took more than half an hour. The sand surrounding the island was crisscrossed with the tracks of deer. On its seaward side I surprised a gray fox that sprinted off toward the edge of the deflation plain a few hundred yards away. Here and there around the edges of the island the troughs of the surrounding dunes cut right down to the water table and the damp sand was dusted with tufts of green grass and small yellow and orange flowers that looked almost like orchids.

In places, the base of the trees was level with the surface of the dunes. Elsewhere, low scrub-covered cliffs of rust-colored sandstone rose right up out of the sand. In places you could make out layers of bedding beneath the covering plants—thin layers of rock only a few fractions of an inch thick that leaned against one another like a house of cards or the familiar diagonal lines of a herringbone tweed. Scraping off a few grains of sand from the cliff and studying them in the palm of your hand, you could see that their color was the product of fine layers of rust, the result of dissolved iron in the groundwater here. The shape and form of the grains themselves, however, were almost indistinguishable from those of the surrounding dunes.

In fact, the layers of sandstone that lay beneath the tree island were actually made up of dune sand—the remains of an ancient dune sheet that had been transformed into soft rock. What its appearance here under this island of trees suggests is that the dunes have moved across this landscape not just once but at least twice in the recent past—once to build up the ancient dunes that now lie beneath these trees, and a second time to overrun the forest that had covered them to create this island of trees. Like the faster movements of sand seen in the dunes out on the beach, the changing boundary between forest and sand is part of an ongoing cycle of change that has seen the dunes advance and retreat through the coastal forest several times in the past few thousand years.

■   ■   ■

Geologists have no clear idea as to just how old the dunes are. While they talk with certainty about the trajectories of continents and the beginnings of ancient seas, the dunes are so recent and active that there is no clear consensus of opinion regarding their age. Dunes have existed here and there around the world for several hundred million years. On the Oregon coast they may have appeared as early as the Pleistocene some 1.8 million years ago at the start of the Ice Age.

As the earth's climate began to cool, ice spread out of the Arctic into North America, at times reaching as far south as Kansas. At its peak, so much water was locked up in ice that sea level was more than three hundred feet lower than it is today. Off the coast of Oregon that large drop exposed wide stretches of the continental shelf—a sandy plain covered with the eroded rocks and sands of the ancestral Klamath Mountains. Evidence from such diverse sources as the orientation of bedding layers in ancient dunes in the deserts of eastern Oregon, and the wind-blown ash deposits of ancient volcanoes like Mount Mazama, which blew apart several thousand years ago to create what is now Crater Lake in southern Oregon, suggests that the region was swept by the same seasonal wind patterns that shape the dunes today. With a broad continental shelf exposed offshore, those winds would have blown sand toward shore, building up a wide sheet of dunes.

When the glaciers melted and sea level began to rise, again encroaching waves would have pushed sand farther inland, driving the dunes in front of them like the blade of a plow. Later when sea level dropped as the climate cooled and the glaciers advanced, those dunes would have been left high and dry as both the waves and the beach headed back out to sea. With their supply of new sand cut off, the dunes would have slowly given way to plants—a steady succession of grasses, shrubs, and trees that led gradually back to a rich coastal forest. A few thousand years later when the glaciers began to melt again the cycle started anew. Over the past two million years sea level has risen and fallen more than a dozen times.

Geologists believe that the most recent episode of dune building began some seventeen thousand years ago, about the time that the first humans were making their way across the continent-sized land bridge that once linked Asia to North America. That building reached a peak here some six thousand years ago—although there have been several smaller advances and retreats since then. At one time the dunes reached much farther inland than they do today. The rolling forested hills that lie between the dunes and the Coast Ranges today are, like the tree islands, the remains of those ancient dunes. Long before the first blade of beach grass was planted here on the Oregon coast, the dunes had advanced and retreated across the coast several times, alternately covering the trees and being covered by them.

••• •

Rocks and stones are not the only evidence of change here on the coast. In 1977 when the Forest Service was enlarging a parking lot at Tahkenitch Lake, a small freshwater lake on the inland edge of the dunes, workers uncovered a series of archaeological artifacts that offered a human perspective on the evolution of the dunes.

Few archaeological sites on the Oregon coast are more than three thousand years old; those that are, presumably, were submerged by the rise in sea level that followed the close of the last Ice Age. Artifacts buried near Tahkenitch Lake, however, range from three thousand to eight thousand years old—the remains of what archaeologists believe were once an ancient hunting camp and later a village. From the start this site was recognized as unique and valuable from an archaeological standpoint. Later it was realized that it had important geologic significance as well. Collections of shells and bones from that village trash heap, or midden, suggested that the landscape surrounding the ancient site was far different from the dunes that border it today.

The oldest layers of midden were filled with the remains of fish and animals that had undoubtedly been hunted for food. They were not, however, fish and animals one would expect to find on a freshwater lake in the dunes. They were marine: the bones of cod, hake, and flatfish like flounder, fish one might expect to find in a deep coastal bay but not in a small lake in the dunes. Scattered here and there among them were other bones as well: those of seals, cormorants, and ducks. Like the fish, they too are common on large marine bays. Outside of the thick deposits of bone, other artifacts were scarce in this lowermost layer. They consisted mostly of crude stones used for chopping and scraping, suggesting that this site was a hunting or fishing camp. It is not likely that these native peoples would have caught these birds and fish at a bay along the coast and then carried them inland to a small camp on the edge of a lake in the dunes. They had no need. Eight thousand years ago, scientists believe, this dune lake was a deep ocean bay.

The record of artifacts did not end with these leavings of an ancient hunting camp, but continued in the layers above. By five thousand years ago the site seems to have been transformed into a village. Artifacts are more prevalent and varied: stone knives, hammers, tube beads, a bone whistle. Deposits in the village midden suggested that the surrounding landscape had changed as well. Instead of being filled with the bones of fish and birds caught in a deep marine bay, layers from this area are filled with the shells of clams, mussels, cockles, and limpets. The change, archaeologists believe, suggests that the deep bay that once lay here had begun to fill in, transforming itself into a shallow lagoon or estuary where shellfish could be easily harvested from the broad tidal flats reaching out from shore. Fish bones, of course, are still found here, but they are much less prevalent, suggesting that they were no longer as readily available—although it is possible they were no longer as highly favored. Interestingly enough, the midden of this five-thousand-year-old village also contains the bones of a blue or humpback whale (scientists are unsure as to its exact identification). The find is unusual. In all probability the whale washed ashore, a gift from the sea and a convenient source of food for the village. Unlike the ancestors of the Inuit and Aleut who live on the edge of the Bering Sea farther north, there is no evidence to suggest that the prehistoric peoples who lived here on the central Oregon coast had developed the skills or technology necessary for hunting whales. Surrounded by rich bays and forests, they had little need to stalk the whales that traveled along the coast.

By three thousand years ago the village seems to have almost disappeared. Artifacts at the site begin to taper off in those more recent layers. Clam shells and fish bones are rare, suggesting that food was less plentiful here or that the village's residents were spending much of their time elsewhere. Eventually the site was abandoned altogether. The timing of this decline coincides almost perfectly with the age of Tahkenitch Lake. Like uppermost layers of the midden, the oldest fossils of freshwater diatoms found on the floor of the lake are some three thousand years old. Their appearance is thought to mark the creation of the lake, the time when it was finally cut off from the ocean by the encroaching dunes. The changes that have taken place here over the past eight thousand years are all related to the movement of sand.

Rising sea level would have slowly driven sand toward the coast, transforming the deep bay that was once located here into a shallow estuary before sealing off its exit to the sea and transforming it into a freshwater lake. For early people the dunes would have offered no more in the way of food or shelter than they offer to plants and animals today. No doubt the change from open bay to open sand occurred so gradually that successive generations gave it little thought. When the dunes began to cover this stretch of coastline village residents slowly moved elsewhere, seeking sites where the food and game were more plentiful. Like the sand, they too were a part of the cycles of change that have shaped this unstable edge of the continent.

• • •

I stayed out at the tree island until late in the afternoon, heading for the beach as the sunset approached. Beyond the island sandy hummocks of grass gradually gave way to the scrubby reach of the deflation plain. Meadows of grass reached back from the edge of the foredune, dotted with scrubby stands of willow and shore pine. Like the tree island, the deflation plain was full of life—sparrows jumped from bush to bush and sang from the tops of trees. Tracks of deer and raccoon covered the sandy trails that led through the high grass and scrub.

The pools and streams of water that run through the deflation plain in the winter and spring were dry, but climbing the grassy slope of the foredune I came across a stand of ferns surrounded by clumps of marram grass and sand—a reminder that this seemingly dry stretch of coast receives more than eighty-five inches of rain per year. Here and there among the grass you could see the vinelike shoots of wild strawberries running through the sand and the thick, green leaves of sea rockets, tipped with bright red seed pods. From the top of the foredune the high dunes that rose above the forest's edge seemed to reach up toward the sky like mountains.

Down at the beach the tide was falling and pools of water filled swales and curves of the beach. Offshore the surf rolled toward the beach with a sharp curl, each collapsing wave trapping a pocket of air that exploded like a pistol shot as it broke on the beach. Above the steady beat of the surf you could hear the cries of gulls and the soft hiss of water running across the sand. At the edge of the falling tide tight knots of sandpipers—small, plump, gray and white birds—skirted the edge of the breaking waves, probing the sand with their short stout beaks. Behind them, the higher waves offshore seemed to rise up like a solid wall. Each surge of the surf sent the sandpipers sprinting back up the beach in a stiff-legged gait just inches ahead of the incoming waves.

Higher up on the slope of the beach the wet sand was an almost perfect mirror for the sky, reflecting the clouds offshore in soft blues and whites that slowly turned to gold as the sun moved lower in the sky. As the sun began to set the beach seemed to catch fire—glowing red, yellow, and orange, as if the sand had suddenly been transformed into molten rock. Several hundred million years ago these grains of sand were part of a chain of volcanic islands. Wind and water have reduced them to fine grains and built them up into a wide sheet of dunes bordered by a broad, flat beach.

The last few seconds were merely an illusion. The sun was already below the horizon, but its image still hung in the sky—a bright orange disk that slowly disappeared into the sea. As the top of the sun sank below the edge of the ocean, the clouds anchored offshore turned bright red—as if they were reflecting the rift zone that lies just fifty miles from shore and more than a mile deep in the sea.

Twenty miles to the north the dunes give way to hard rock. Cliffs of black lava rise up right out of the ocean, the products of rifting and a succession of volcanic events that have shaped the coast of the Pacific Northwest all the way to Puget Sound. On a clear, bright day you can look north from the dunes and see them rising out of the sea.

As the last of the twilight faded away, the darkness on the beach was almost total, broken only by the lights of fishing boats working offshore. In another hour's time the moon began to rise over the dunes. I walked back through the dunes in the moonlight. The sand seemed almost endless in the soft, blue light—waves of silver and black that stretched along the edge of the coast for miles.

THE PACIFIC
NORTHWEST

A FIRE ON THE
FLOOR OF THE SEA

NORTH OF FLORENCE THE DUNES BEGIN TO NARROW, PINCHED OUT BY cliffs that rise up out of the sea in a shallow ramp leading north-ward. From their tops you can look back across the long curving arch of the beach, a strip of sand that reaches all the way to Coos Bay more than forty miles away.

As the road climbs up the cliffs it begins to bank and turn. Passing around the edge of a sharp curve the dunes suddenly disappear. Up ahead the view is taken over by sheer cliffs of black rock more than five hundred feet high. They are not made up of the same brown and gray layers of soft shales and sandstones that anchor the coast farther south but hard, black volcanic lavas that erupted from a collection of submarine rift zones and volcanoes that once bordered the coast here.

Today these remnants of ancient volcanoes and rift zones run all the way from central Oregon to the Olympic Mountains on the edge of Puget Sound. After the bright, open world of the dunes the sudden change in the landscape is striking; it gives one the feeling of having entered a different world, of having traveled several hundred miles in the space of a few hundred feet. The land here is dominated by dark, solid colors—the black of the rocks, the green of the trees, and the solid blue of the sea.

As you round the top of the cliffs Heceta Head comes into view— a dome shaped mass of rock that rises up out of the sea like a small volcano. A lighthouse lies on its seaward flank. On a bright, clear day its white walls and red roof seem to glow in the sun, set off by the emerald green backdrop of the surrounding forest that runs right to the edge of the cliffs. Geologists believe that fourteen million years ago the headland was an active volcano, one of perhaps several dozen that once sat here along the coast of the Pacific Northwest. These volcanoes were not part of an isolated event but merely the final phase of a series of volcanic eruptions that had been building up this edge of the continent for more that fifty million years.

Down at the narrow beach that fringes the headland the black flows of lava look so fresh that they could have cooled yesterday. In the sides of the cliffs you can make out the roseate patterns of dikes and tubes where the hot rock rose through the cooler rocks around them. Where the headland is fringed by a narrow marine terrace of reddish-brown sandstone, the lavas seem to have cut through its softer rocks the way a hot knife passes through butter. Here and there in the face of the cliff you can see volcanic bombs—chunks of lava rock the size of watermelons that blew out of the mouth of the volcano. Some are so shot through with holes that they look almost like sponges—the tiny holes are vesicles, the products of hot gasses once trapped inside. Other pieces of volcanic rock poke through the narrow sand beach along shore, small projections of blue-black rock that look almost fused, like the slag left from a blast furnace.

Like the displaced rocks farther south along the San Andreas Fault, these piles of volcanic rocks are a product of plate tectonics. Unlike them, however, they did not drift in from the south or west, but originated right here. They came up right out of the ground, flowing from the mouths of a series of submarine volcanoes and rifts that once bordered the coast here. Instead of being a product of motion, they are a product of plate tectonics' fiery engine, the rift zones and volcanoes that drive plates of the earth's crust around the globe. Today a similar rift zone lies just one hundred miles from shore—the San Juan de Fuca Ridge, the center of a string of oceanic ridges that begins off of Cape Mendocino and runs northward all the way to the northern tip of Vancouver Island. Spreading along its crest is slowly driving the floor of the sea under the edge of the coast.

Waves pound the coast here, but the rocks alongshore are not lifeless. Look closely and you will see that they are carpeted with barnacles and mussels and clusters of purple and orange starfish. Above the reach of the tide the sheer cliffs and offshore rocks are alive with nesting seabirds. Over the steady beat of the waves you can hear the barks of seals and sea lions. The small islands of rock that dot the volcanic coast of the Pacific Northwest are rich and full of life, a safe haven where seabirds and sea mammals like seals and sea lions can breed and raise their young.

From the high grassy slope above the lighthouse you can look north and see point after point stretched out along the coast—small headlands that reach out into the sea like tiny fingers, their tops marked with the dark silhouettes of Douglas firs and spruce trees. Shelves and jetties of rock push out into the ocean one after the other like the images reflected in a hall of mirrors—a chain of ancient volcanoes running along the coast.

• • •

Beams of sunlight reached through holes in the soft, gauzelike layer of fog that hung over the sea, illuminating the water below with shafts of warm yellow light. Steep hillsides carpeted with huckleberry, salal, and thimbleberry led down to the sheer basalt cliffs. The ocean was more than five hundred feet below. From the top of the cliffs you could hear the waves pound the rocks, each one sending a white surge of foam over the rubble of stones that lined the base of the cliff.

It was the second week of August and I was standing at Cape Foulweather on the Oregon coast with David Bukry and Parke Snavely, Jr., two scientists from the U.S. Geological Survey in Menlo Park. Bukry is a micropaleontologist who specializes in using microscopic fossils known as coccoliths to date marine rocks. Snavely is a field geologist who has been working on this part of the Oregon coast for the better part of forty years.

Cape Foulweather rises more than five hundred feet above the sea, reaching out into the Pacific like the prow of a ship. Named by Captain James Cook in 1778, it was his first site of land since discovering the Hawaiian Islands, then known as the Sandwich Islands, six months earlier. Bad weather kept Cook and his men aboard the *Resolution* and *Discovery* from landing. Instead they pushed northward, complaining of the "grate stinking sea fogs" that hung over the coast. In the heavy fog, they sailed right past the mouth of Puget Sound without even seeing it. They finally made landfall at Vancouver Island.

Winter winds of one hundred miles per hour are not uncommon here at the cape, but in late summer the air was clear and relatively still. A thin layer of fog hung over the sea, but above it you could see up and down the coast for miles.

Snavely was not so much interested in the scenery as the rocks below. In the face of the cliffs he pointed out fluted columns of basalt. Down on a ledge at the base of the cliff those rocks seemed to be laid out in a series of concentric circles—"the nicest ring dyke you're ever likely to see," Snavely says—the center of a channel of lava that once shot through the ground. Like Heceta Head, Cape Foulweather was once an active volcano. There are no fossils here for Bukry to date. Fourteen million years ago these rocks came out of the ground at more that two thousand-five hundred degrees Fahrenheit. Underneath them are several miles of other, older volcanic rocks. From the coast you can drive inland, back through sixty million years of history to the base of the volcanic rocks that began building this rocky reach of the coast. From central Oregon to the mouth of Puget Sound, the coast of the Pacific Northwest has been built by a fire on the floor of the sea. For the better part of fifty million years the coast of the Pacific Northwest has been fringed by a rift zone and active volcanoes.

• • •

Parke Snavely, Jr., first started working on the Oregon coast in 1946. Since then his work as a geologist has taken him to Africa, Asia, South America, and the South Pacific. Although Snavely's success has saddled him with an increasing load of bureaucratic and administrative responsibilities, he still manages to get back to Oregon almost every year. "Delegate in the summer and get out into the field. You die professionally if you don't get out into the field," he had told me one afternoon at his office in Palo Alto. Spending a few days working in the field with Snavely leaves one with the impression that he intends to live forever.

Snavely's government-issue Jeep Cherokee is a laboratory on wheels. As we drove back into the hills, I took stock of the materials piled up inside. In back was a collection of metal field boxes filled with aerial photographs of the Oregon coast and copies of every topographic and geologic map of the region. Others held a small library of scientific papers and journal reports on local and regional geology. A rolled up cross section of the Oregon coast lay curled up on the dashboard. Fastened to the visor above it with a fat rubber band was a collection of marking pens in twenty-four different colors for marking maps and photographs. Between the seats was a Tupperware container stuffed to the gills with pens, pencils, and a collection of Koh-i-Noor Rapidograph pens. Behind it was the box from a bottle of Glenfiddich scotch filled with a collection of tobacco and pipe cleaners for his ever-present pipe.

Snavely himself was dressed in a blue tennis shirt, light-blue denim pants, and a light-green canvas vest the pockets of which were stuffed with still more pens and pencils. Around his neck was a string cord with a hand lens and a tiny army-issue folding can opener—a relic from his earlier days in the field when he fed on C and K rations. On his head was a faded-orange cap bearing the logo of the U.S. Geological Survey.

Parke's view of the world is shaped by rocks. As we headed inland he pointed out features in road cuts alongside the highway: a layer of thirty-eight-million-year-old basalt, an ancient marine terrace, the trace of a fault scarp. As we followed the Siletz River he pointed out the house used in the filming of Ken Kesey's award-winning *Sometimes a Great Notion*. "It wasn't a bad movie," he said. "My daughter rented it one time, but she finally asked me to leave the room because I kept pointing out the geology in the background."

• • •

Away from the coast and its thin layer of morning fog the air was already hot and dry. Off to either side of the road the trees were thick and green. It was impossible to see any further than a few feet into the forest. Roads, rocks, and homes were all hidden by trees. Outcrops of rock are miles apart here, found in road cuts and stream banks and the sides of logging roads that wind back into the hills. "Without logging roads our maps wouldn't be half as good," Snavely says. "I always felt there ought to be a roving geologist in the Pacific Northwest to describe things in detail when they make a new cut to realign Highway 101 or I-5, because a new cut goes to pot in two or three years." Road crews have no respect for rocks. "They just spray grass seed on them!" he says with disbelief. "God they just ruin the thing."

A half hour in from the coast we turned up a steep narrow road leading back to a rock quarry hidden in the trees near the highway. A collection of rusting equipment for working the quarry filled the clearing. Getting it to work again seemed like something that would require divine intervention, or at least a major act of faith, but Snavely assures me that it all works quite well when the need arises. Over the years he has gotten to know the quarry owner rather well, but that comes as no surprise. Here on the central Oregon coast Snavely knows almost everyone whose property contains interesting rocks.

Snavely looks up at the cliffs of black rock with an affectionate smile. "There's more money in quarry rock than oil and gas," he says. "Some people love gold. Some people like oil. But a good rock quarry is worth more than an oil well in places." He reckons that he could have retired several times over just telling quarry operators where good hard rock like this could be found. As it was, companies could get all the information they needed off of his published reports and scientific papers for free. As a government scientist, his work is all public property. "If I had taken a royalty of twenty-five cents per ton I would have made millions. It's just Geology 101, but they don't know that."

In the walls of the quarry you could see the shapes of the cooling tubes of lava. They rose up out of the ground and then branched outward like a cluster of lilies. Elsewhere the cliffs were marked with car-sized blocks of rock.

The rocks in the quarry walls were part of a formation known as the Siletz River Volcanics, the base of a thick sheet of lava rocks found up and down the Oregon coast. At first glance they look no different from volcanic rocks found on the coast at Cape Foulweather and Heceta Head. But these rocks are not fourteen million years old; they are between forty and sixty million years old.

Handing me a chunk of basalt from the quarry wall and his hand lens, Snavely told me to take a good look at the small white specks filling the holes that pocked its surface. The white specks, Snavely says, are zeolites, traces of carbonates from a coral reef that once fringed these rocks. Forty million years ago these rocks of the quarry wall were not in the midst of a forest, but part of a chain of volcanic islands that once sat near shore.

• • •

If you could have stood on top of the Cascade Ranges seventy million years ago and looked out toward the Pacific, you would have seen not the Willamette Valley where Portland, Salem, and Eugene now stand nor the Coast Ranges and the rocky shore beyond them, but water—a deep ocean basin that extended offshore for several miles. Over time you could have watched it slowly fill in with flows of lava that erupted on the floor of the sea and a periodic stream of sand and silt eroded from hills and mountains that rose and fell alongshore.

The oldest volcanic rocks appeared here fifty-eight to sixty-two million years ago, a group of rocks known as the Siletz River Volcanics that erupted from an ancient rift zone on the floor of the sea. Today those rocks reach inland for more than one hundred miles in places and are found several miles offshore as well. Altogether they amount to some fifty thousand cubic miles of basalt, enough to cover the United States under a blanket of lava some seventy feet thick.

By forty-three million years ago these ancient lavas had been built up high enough to form islands offshore. Looking closely, perhaps, you may have been able to see the rocks that now make up this forest quarry in the cliffs of an island several miles from shore. When these volcanic rocks first appeared, their chemistry had been almost identical to that of the basalts found on midocean ridges today. By the time these islands were forming, however, the chemistry of the lava was almost identical to that of lavas erupting in the Hawaiian Islands today.

While these islands appeared offshore, farther inland the Klamath Mountains were being uplifted, built into mountains nearly twenty thousand feet high. It took less than ten million years to plane them flat. By forty million years ago their eroded sands had filled in the basin that had once separated these islands from the mainland, linking them to the continent. Today those filling sands and rocks are preserved in cliffs and outcroppings of sandstone and shale scattered among the harder darker lavas of the coast.

Thirty-eight million years ago a new episode of rifting began, opening another deep basin on the seafloor. In places, these new rocks were all but indistinguishable from the lavas that preceded them. Elsewhere, however, their composition was unusual, laden with high-temperature, high-pressure rocks like camptonite and nepheline syenite, suggesting that they had come almost straight from the earth's mantle.

Later that basin was slowly filled in by deposits of plants and animals that fell out of the ocean above and sands and rocks washed offshore from the neighboring continent—a collection of deep water shales that slowly graded into near shore deposits laden with the shells and fossils of clams and fish as the basin shallowed. The rifting which opened this new basin and the sediments that later filled it were an almost exact reproduction of the rifting and filling that had begun nearly thirty million years before. Accurately dating and identifying the rocks here often hinges on the skills of specialists like Bukry who can identify the tiny microscopic fossils of the sandstones and shales that interfinger with the lavas here. "Rocks of different ages can look almost more alike than rocks of the same ages here," Snavely explains. "The coast here has gone up and down like a yo-yo." By twenty million years ago the Willamette Valley was finally above water, but the Coast Ranges farther west were still hidden from view, waiting for another episode of volcanic eruptions and faulting that would lift them out of the sea.

The final phase of volcanism here happened just twelve to fourteen million years ago. Alongshore it built up prominent points and headlands like Cape Foulweather and Heceta Head. Thrusting associated with the subduction zone and plate movements offshore would complete this picture, finally lifting the coast ranges out of the sea. For Snavely, the repetitive cycle of geology here is one of its most intriguing properties. "There are," he says, "so many places where you can stand on the ancient shoreline."

• • •

It was a clear, cloudless day in early June at Yaquina Head outside of Newport, but the wind blowing in from offshore at thirty knots made it feel more like midwinter. Fifteen million years ago these black rocks of the headland were an active volcano. That day its surface was alive with thousands of nesting seabirds: black swanlike cormorants and parrotlike puffins. A few hundred yards from shore more than thirty thousand common murres carpeted the top of Colony Rock, a pluglike island of black lava, like stubble. Rafts of terns, surf scooters, sea ducks, and auklets floated in the water near shore.

Seabirds are a miracle of adaptation. Aside from their short three- to four-month breeding season, many of these birds spend their lives entirely at sea, dozens or even hundreds of miles from shore, feeding and roosting on the open water. Like desert animals they can derive what little fresh water they need from the food they catch at sea. Even here on the coast they do not drink from freshwater lakes or streams.

They seem as at home in the water as in the air—capable of skimming the surface of the waves or even, in some cases, diving beneath them for fish and krill. Some, like the murres that carpet the top of Colony Rock, can even fly underwater, using their wings to propel them like a skate or manta ray. Northern relatives of the penguin, they have small nictitating membranes that cover their eyes while underwater, improving their vision while diving. They have been found diving to depths as great as five hundred feet, well beyond the safe limits for scuba divers and even some submarines. Their relationship with the land is so tenuous and temporary that here on the coast they do not even bother fashioning a nest; instead they lay their eggs right on the rock, cradling them between their webbed feet and soft white belly as they wait for them to hatch. While they frequently travel in small groups out in the open ocean, here they gather by the thousands in small compact colonies, often with other seabirds right nearby or even in their midst.

Here at Yaquina Head there are typically more than a dozen different types of birds nesting on the cliffs and offshore rocks. It seems logical that the competition for privacy and space would be fierce, but from the rubble of fallen rocks that lies at the base of the cliffs to the thin layers of soil that lie at their top, each species has carved out its own particular niche in the landscape. Black oystercatchers picked their way through the rubble that lay at the water's edge, their jet-black feathers blending almost perfectly with the rocks. Only their bright orange bill and sudden movements betrayed their presence. Higher up on the rocks pigeon guillemots nested in the dry crevices between boulders and stones. Halfway up the rocks you could see pelagic cormorants sitting on nests fashioned out of seaweed and excrement, clinging to the sides of the sheer black cliffs. Up above them the rocky shelves and the bare tops of the rocks were covered with common murres and Brandt's cormorants. Where the wind and rain had not washed the grass and soil away from the clifftops and offshore rocks, storm petrels, auklets, and puffins were hidden away in burrows beneath the grass. Up above the rocks the air was filled with the comings and goings of hundreds of birds.

• • •

The scene being enacted at Yaquina Head was also taking place at sea cliffs and offshore rocks up and down the coast. Each spring the Oregon coast is home to an estimated 1.2 million seabirds, more than in California and Washington combined. The most prevalent are common murres, whose numbers total an estimated six hundred thousand. The thirty to forty thousand that gather on Colony Rock amount to roughly 5 percent of the state's estimated total population.

The rocks at Yaquina Head are part of the Oregon Islands National Wildlife Refuge, a chain of more than fourteen hundred small offshore rocks and islands. By definition the refuge includes all rocks and reefs and islands off the Oregon coast above the mean high-tide line separated from the mainland. The term *islands*, however, is somewhat misleading, according to Roy Lowe of the U.S. Fish and Wildlife Service. Most are too small to appear on even the most detailed maps and charts. In effect, the refuge includes almost every offshore landmass in the state.

From his small office at the Hatfield Marine Science Center in Newport, Lowe is the refuge's sole employee, responsible for overseeing not only its far-flung rocks and islands but three other coastal refuges as well—Bandon Salt Marsh and Three Arch Rocks farther north and an old growth forest on Cape Meares. With responsibilities spanning more than two hundred miles of shoreline, Lowe's office is often the least likely place you will run him to ground.

The critical factor behind the state's thriving seabird population, according to Lowe, is habitat. While seabirds can survive a full gale in the open ocean with ease, they are almost defenseless on land. The walls and islands of black lava rock that border the coast north of the dunes are a near-perfect nesting site for the vulnerable birds, protecting them from would-be predators like raccoons, foxes, and coyotes. People are the birds' only true threat.

• • •

Seabirds have been coming to the Oregon coast for thousands of years. Like the salmon that spawn in the coastal streams here, most have returned to the same sites year after year and generation after generation. Almost every prominent point on the coast that is used by seabirds is also an archaeological site, littered with the artifacts left by the area's earliest inhabitants—knives, scrapers, and piles of shells as well as the bones of seabirds. Aboriginal peoples here not only collected their eggs, they also hunted them for food. Undoubtedly those raids on the nesting sites had an impact on the birds, but when the first Europeans arrived here in the early 1800s the birds were still thriving. In less than one hundred years they would almost disappear.

From 1850 until the turn of the century, San Francisco was the economic engine that powered the West Coast. Like Point Reyes farther south, the land here was put into harness to feed and build the city. Forests were cut for building timbers and paper. Coastal dairies provided milk, butter, and cheese. Seabird eggs were gathered as well. They brought as much as a dollar per dozen, wholesale, in city markets. At Port Orford alone in southern Oregon more than a million seabird eggs were gathered from the offshore rocks near town. In California things were often worse. Eggers took more than fourteen million eggs off of the Farallon Islands near San Francisco. Each season the eggers spent their first few days smashing every egg they found, Lowe says. That way they could be sure that every egg harvested thereafter was fresh. By the turn of the century the birds had all but vanished from the Farallons and the numbers in Oregon were beginning to sag as well.

Not only did commercial exploitation play a roll in crippling the birds' numbers, but the era's perception of sport and entertainment played havoc with them as well. In the late 1800s a popular Sunday afternoon pastime at Tillamook Bay in northern Oregon was to charter a tugboat down at the harbor and fill it with people and ammunition and then run out to Three Arch Rocks and spend the day "blasting away at the birds," according to Lowe. Dead birds washed up on the beach by the thousands. Not everyone was amused, particularly weekenders who had come from Portland and other inland cities for a few days at the beach. One of those who happened to witness the slaughter was William O. Finley, a prominent photographer and conservationist and friend of President Theodore Roosevelt. In Washington he began lobbying Roosevelt to protect the area as a wildlife refuge. In 1903 Three Arch Rocks became the first federal wildlife refuge west of the Mississippi River.

Today the numbers of seabirds along the coast have been steadily increasing. While seabird rookeries are no longer raided for their eggs, they are threatened by new problems, the effects of which can be equally devastating.

Gill netting and oil spills in both California and Washington have killed tens of thousands of seabirds over the past decade. So far, Oregon has been able to avoid those kinds of problems, partly by luck (there have been no large oil spills off the state) and partly by design— gill netting is prohibited by a series of treaties with Canada to protect the valuable stocks of salmon and steelhead feeding offshore. What threats are present here are often unintentional, coming in the shape of boats, jets skis, and low flying aircraft that pass too close to the rocks and frighten the birds. When a low-flying aircraft passes over the bird colonies, for example, the birds "just come raining down off the rocks," Lowe says, leaving their eggs and chicks unprotected or even knocking them into the water. Most people simply don't realize how fragile and vulnerable the nesting birds are or, in the case of low-flying aircraft, how easily panicked seabirds can break through the windshield of a plane or foul a prop or engine.

While pilots and boaters have been fined and cited for passing too close to the refuges that line the coast, the real key to the problem, Lowe believes, is educating people about the beauty and fragility of these birds. "If you want people's help in protecting this resource, you have to find a way of allowing them to use it," he says. "In the case of seabirds that means watching them from a distance."

For the moment seabird population numbers along the Oregon coast are increasing. At Yaquina Head, for example, the number of murres on the rocks have increased from twelve thousand to more than thirty thousand over the past decade. Other birds seem to be making a comeback as well. Once-rare rhinoceros auklets—small, plump birds resembling the parrotlike puffin—now number more than a thousand along the coast. For the birds, the critical issue is still habitat. With care, there is still enough room on the coast for both birds and people.

• • •

By late June the guarding and incubation of eggs has been replaced by the care and feeding of young chicks. Watching the birds on the offshore rocks through a pair of binoculars or a spotting scope you can see the heads and beaks of young murres nestled under the protective bellies of their parents. The cliffside nests of pelagic cormorants are filled with as many as a half-dozen young birds that look like little more than round balls of feathers. A few feet away, on the opposite side of the fence that runs around the perimeter of the headland, gulls sit attentively on nests in the midst of the green grass of the clifftop, surrounded by their newly hatched chicks. They look almost like young chickens, balls of bright yellow feathers speckled with brown dots. For the gulls, the chicks of the neighboring seabirds are a readily available source of food. When the gulls are perched quietly on their nests however, they seem to observe a kind of uneasy truce.

For most seabirds it takes both parents to care for the vulnerable young chicks: one to watch over them and another to gather food. Predators—gulls, crows, and hawks—are continually working over the colony in search of an easy meal, and a crow flying overhead quickly elicits quavering cries of protest from the seabirds gathered below.

The birds may be firmly anchored to these rocks and small islands near shore during the nesting season, but their lives are still focused on the sea. Most feed not in the surrounding coastal waters but several miles offshore. From sunrise to sunset the air is filled with the steady comings and goings of birds. Puffins circle the cliffs with the tails of small fish dangling from the sides of brightly colored, parrotlike beaks. Murres plunge off the windward side of the rocks and glide down toward the water, furiously beating their slender wings to build up speed before heading offshore. Their search for food can take them away from the rocks for hours or even days.

Like their nesting sites on the cliffs and offshore rocks, the feeding areas of these seabirds are carefully partitioned. Different species rarely compete for food. Seagulls, for example, skim the water's surface near shore in search of food while terns dive after prey several inches or even feet below the surface. Nearby, among the rocks that dot the base of the coastal cliffs, pelagic cormorants dive into the surf searching for food.

Some birds nesting on the near shore rocks and cliffs feed only several miles from shore. Murres, for example, hunt for small fish fifteen and twenty miles from shore. They feed near the surface when possible but are capable of diving several hundred feet below and swimming with their short, slender wings when the need arises. Other birds are found out here as well—but each one has its own particular source of food. Unlike the murres, tiny cassin's auklets feed on swarms of euphausid shrimp, the same food sought by whales migrating along the coast. Some, like the storm petrels, venture even farther out to sea, flying out to the open ocean anywhere from fifty to two hundred miles from shore to feed on zooplankton, the tiny insect-size predators that graze on the tiny floating plants that are the base of the marine food chain. Small, black birds the size of swallows, they feed right on the surface, fluttering and hopping over the waves with swift, short wing beats, their small webbed feet pattering across the surface of the sea beneath them. They seem to almost dance across the water.

The timing of the hatch is no accident. The arrival of the chicks coincides with the arrival of upwelling currents offshore. There is little margin for error in the open ocean, and the arrival of these rich currents of water are literally a matter of life and death. When an El Niño event strikes, the results can be disastrous, killing literally thousands of birds. Sometimes, during an especially severe El Niño event, they fail to even nest at all. Others leave thousands of birds starving to death. In a bad year you can walk down the beach and see the birds staggering ashore to die. When that happens, Lowe's office is often flooded with calls about paralyzed and poisoned birds washing up on the beach. "They aren't paralyzed, they're starving to death," Lowe says, and unfortunately there isn't anything anyone can do except wait for the wind and the upwelling currents to return.

Like the early summer months in northern California, June and July can often be the coldest months of the year here on the Oregon coast—or at least feel like it, with a cold, damp, thirty-knot wind. While the Willamette Valley that lies between the Coast Ranges and the high volcanic peaks of the Cascade Range can spend the summer baking under a hot dry sun, the coast can be blanketed with fog for days on end. As that hot, dry air rises in the valley, it amplifies the prevailing winds even further, pulling cold air in off the Pacific. "If it's one hundred degrees in Corvallis, it's going to be foggy and windy here, I can guarantee it," Lowe says. "A lot of people on the coast say, 'Oh jeez. It's going to be windy again.' But without that system we don't have anything. The fisheries and everything collapses." Like fish in the ocean offshore, the lives of seabirds are keyed to rich currents of water that arrive with the wind.

• • •

By August the rocks are bare. The only birds that remain in the area are the gulls that flock to nearby beaches and the cormorants that gather from time to time on the seastacks offshore, drying their outstretched wings in the sun. The only sign of the thousands of birds that had gathered alongshore only a few weeks earlier is the white stain of guano that coats the black rocks like spilled paint. Feathers and small pieces of down stirred up by the wind float through the air like old, dry leaves.

Murres are the first birds to leave the rocks, departing as early as the first week of July. For the mother, the work of raising the chick is now over. For the next six to eight weeks, fledging the young bird—teaching it how to survive on the open water—will be the sole responsibility of the father.

When they leave the nest the young chicks are still incapable of flying. They leap off the rock and float down toward the water more than one hundred feet below, an act of instinct, or faith. Once in the ocean they can dive almost instantly, but the art of flying is still several weeks away. The young birds are little more than plump balls of down at this point, and it is critical that they link up with their parents as soon as they reach the water.

The murres do not stay in the area long. Banding studies suggest that they head north almost immediately, young chicks and all, to Washington and British Columbia. Only three to four thousand murres nest on the Washington coast, but by late August there are several hundred thousand in the waters of Puget Sound and the nearby Straits of Georgia off British Columbia. The protected waters of the straits and the sound are a training ground where the young birds can learn the skills they will need to survive on the open ocean.

While common songbirds like robins or sparrows depend on sheer numbers to survive, laying four or even half a dozen eggs at a time two or even three times in a single breeding season, most seabirds depend on training and skill to survive. Age differences reflect a similar bias: while most songbirds rarely live to be more than four or five years old, seabirds like murres can live to be more than thirty. Murres lay only a single egg during their short breeding season. Every egg and every young bird is important and represents a major investment in time and effort by the parents. Once it hatches, the young bird's survival will hinge on its ability to learn and its father's ability to teach.

A casual observation of birds leaves one with the illusion that their lives are effortless and perfectly in tune with the natural world around them. However, survival in the open ocean requires aggressiveness and skill, whether the task is diving for fish or surviving a three-day storm. Errors in flight or navigation—failing to find food quickly enough, misjudging the height of a wave or the strength of the wind—can be fatal. Storms, disease, and a lack of food all take their toll on the young birds. In September and October when the young are left to fend for themselves, the beaches are often littered with the bodies of young seabirds. What happens to the birds once they leave the nest is a big black box, Lowe says. There are many ways to die in the open ocean.

From their nesting grounds on the Oregon coast these seabirds travel around the world. While the murres frequent the waters off Washington and British Columbia, the puffins and pigeon guillemots that nest on the surface of these ancient volcanoes travel to the Gulf of Alaska. In midwinter the storm petrels that nest in small burrows amid the grass on top of the coastal cliffs can be found near the equator, dancing across the waves several thousand miles from shore.

The following spring the birds' wanderings will come full circle, bringing them back to these ancient volcanic islands and lava flows that line the coast north of the dunes. The first few tentative visits come as early as December, by gulls from California and British Columbia that come by to stake out territory on the top of coastal cliffs. By January the murres are already coming in from offshore to make short visits to survey the nesting grounds they have been using for thousands of years. Storm petrels and puffins are often the last to arrive, reaching the coast by late March. By April the rocks and cliffs of the coast will be covered with a swirling cloud of birds.

•   •   •

The stump was gigantic, more than seven feet in diameter. Notches cut roughly into its sides with an axe led to its top several feet above the surrounding thicket of huckleberries and salal. Ray Wells of the USGS stood on its top, talking about the geology of the landscape below.

We were standing in a clearing near Cape Meares on the northern Oregon coast—the view was wide and panoramic, stretching for miles to the north and east both up the coast and inland over the forested hills of the Coast Ranges. It was a bright August day. Rafts of rounded cumulus clouds floated in the sky like giant ships, their hard, flat, gray bottoms capped by soft, white tops that rose up into the sky like sails. As they drifted through the sky they cast shadows on the ground below, patterns of black and green that moved across the forest.

In the foreground you could see Netarts Bay and the sheen of tidal flats that fringed the green marsh bordering its shore. A spit of sand ran across its mouth shielding the shallow water behind it from the Pacific. Offshore in the ocean the waves broke in a clean line as they rolled toward the spit. Marsh deposits buried beneath the surface of the bay have played a prominent role in the developing theory of coseismic quakes—the idea that the Pacific Northwest has been periodically struck by massive earthquakes. The bay below was one of the first sites in the region where those deposits were discovered.

The marsh, however, is not the only sign of movement and change here. The surrounding ridges and hills are the products of ancient volcanoes and lava flows that span more than fifty million years. The two high ridges bordering the bay, Wells says, are part of the same Siletz River Volcanics that Snavely, Bukry, and I had looked at the day before. But while they are both made up of the same type of rock, their origins are quite different. One high ridge was formed by the eruption of an ancient volcano; the other was formed by lavas that filled an ancient valley. When those former hills eroded they left this ridge of rock behind, like a casting formed by a mold.

Out on the coast the rocks are even younger, ten to fifteen million years old, part of the same flows, or at least the same type of flows that made up Cape Foulweather an hour's drive to the south. Those younger volcanics at Cape Meares and elsewhere on the coast are part of a "stupendous story," Wells says. The lavas, or at least ones chemically and structurally similar to them, have been traced all the way back to the Columbia River basalts in eastern Oregon that lie along the banks of the Columbia River and reach inland all the way to south-central Idaho, following the path of the Snake River—a distance of more than five hundred miles. Some geologists believe these rocks are all part of the same flow and originated several hundred miles inland.

Others, however, believe that they came out of a series of fissures and cracks spread across the Pacific Northwest that were fueled by a massive pool of magma below.

Snavely and Bukry listened to Wells, taking it all in. We had driven up for the day from Otter Rock to visit Wells and his rocks here a few miles to the north. Wells, now just in his thirties, is an expert on paleomagnetism. A few years back he was hired on at the survey by Snavely. This summer they are neighbors after a fashion. Wells has been mapping the coast a few miles north of the area where Snavely has been working.

Unlike the rocks to the south in California, these coastal lavas do not seem to have drifted northward at all. In fact, for the past twenty million years or so they have been almost motionless with respect to north and south. Instead they seem to have turned in place, spinning and rotating like giant blocks.

The clues to these movements are hidden in patterns of magnetism locked in the rocks as they cooled. Like the patterns of magnetic stripes locked in the lavas that make up the floor of the sea, these patterns reveal how these rocks have moved and shifted with time.

· · ·

From the clearing we headed inland, up into the Tillamook State Forest, a block of state-owned land extending over several hundred thousand acres on the coast west of Portland. Snavely turns on the CB-radio tuning it to the channel numbers scrawled on the sides of large stones and pieces of plywood scattered here and there alongside the road. Logging trucks here communicate by CB-radio, charting their progress through the forest as they careen down the narrow dirt logging roads at thirty, forty, and fifty miles per hour with a full load of logs. Listening to the CB gives one a chance to get out of the way.

The forest here is all second and third growth, filled with trees forty and fifty feet tall. In 1933 a forest fire charred several hundred thousand acres, burning out of control for days. Fires struck again in 1939 and 1945. Today the charred, black stumps of old trees rise up out of the forest floor like stones.

Wells guides us through the forest largely by memory, looking for specific exposures of rock, steering us across high ridges so narrow there is barely enough room for Snavely's government-issued Jeep Cherokee to pass through. There are no street signs here, only junctions with other roads that lead into the trees and sudden dead ends on the edge of a landslide or washout. Rocks here are hidden away under a carpet of trees and plants. They crop up here and there alongside the road surrounded by thickets of salmon berry, alder, and sow berry— clumps of black-gray weathered rock. Most, Wells says, are volcanic breccias, tufts of lava from the eruptions of ancient volcanoes, so reshaped by exposure to rain and open air that they are almost indistinguishable.

In terms of decay the roads themselves are an even match for the rocks, so bad that they are often impassable even in our four-wheel-drive. The state apparently wants neither the expense nor the responsibility of maintaining the roads here. In today's twisted world of legalese, maintaining them would imply liability for the accidents and mishaps of those who choose to travel through the forest.

The branches of salmon berries and low scrub all but cover the road in places, scraping along the sides of the truck with a high-pitched squeal. During the 1981 eruption of Mount St. Helens, Wells was working in an area farther north up in Washington. Ash, he says, coated trees and scrub lining the forest roads like snow. As they drove through, it peeled the paint off of their truck like paint thinner.

By late afternoon we had found the rocks Wells had wanted us to see. They were not volcanic but light-gray marine shales found in a small cliff alongside the road rising up through the trees. The thin layers were broken into small chips. They looked almost like cracked pieces of porcelain. Their smooth surfaces scratched as easily as unfired clay. They were oil shales, according to Wells. Holding a piece under your nose, you could just barely detect the pungent odor of crude oil. Wells lit a match and held it to the surface of the rock. In a few seconds the flame started to sputter and dance as the oil inside the rock caught fire. There is not enough oil in these rocks to make anyone think of putting an oil well here, but it is unusual that oil occurs here at all in the midst of the region's predominantly volcanic rocks. It offers an interesting perspective on the area's geologic history.

The same rifting that gave rise to the lavas that now make up this edge of the coast also created a series of deep basins on the seafloor. Like those farther south in southern California, some of these basins slowly filled up with sediments that were eventually turned into oil—or at least oil-rich rocks. Like in southern California, plate movements here broke the rocks above into blocks that were slowly rotated in place. Looking at the paleomagnetic patterns of the volcanic rocks and lavas here, Wells says, you can see how these blocks were slowly rotated.

The rotation that occurred here, however, was not related to a plate that was moving alongside the edge of the continent but to one that was moving underneath it and into the Cascadia Subduction Zone. As the plate moved, these overlying rocks spun in place like a record on a turntable. Their movements seem to have been fairly steady, according to Wells, and the further back in time you go, the more the rocks have been rotated. Those from the lower Eocene some fifty-five million years ago have been rotated by as much as seventy degrees, while those from the late Miocene, a mere five to ten million years ago, have been rotated by only sixteen degrees. For the last fifty million years these rocks seem to have been spinning in place.

• • •

The same volcanic rocks that run along the Oregon coast extend several miles offshore as well. "I'd always kind of wondered what was going on out there," Snavely said, but everything ended right at the water's edge. Fishermen would bring him rocks dredged up from the seafloor, but there were only small clues that teased the mind without really revealing much at all. Much of the area's geologic features were rooted in things that had taken place offshore—subduction, the spreading on the nearby ridge, and faulting and folding. Keeping his feet on the ground, Snavely began to believe, was too limiting. "In effect it was like playing with half a deck," he said.

When Snavely was asked to start the USGS's newly created branch of pacific marine geology in 1966, one of the first areas he wanted to explore was the seafloor off of Washington and Oregon. The lava rocks that anchored the coastline, he had begun to suspect, did not stop at the water's edge but extended offshore for several miles. Like the rocks on land they had their own history too, but that history varied greatly on whether they were inside or outside of the Fulmar Fault, a strike-slip fault paralleling the Cascadia Subduction Zone and located only a few miles from shore. To the west of the fault the rocks had been carried northward as freely as rocks farther south along California's San Andreas Fault—some of them as far north as Alaska. East of the fault, however, the rocks had moved northward only slightly—some one hundred miles or so.

Their drift north, Snavely says, seems to have been stopped by the hard wall of Vancouver Island, "a buttress for the folding of the Olympics." Straighten all the kinks and folds in the rocks that have been stuffed into the Olympic Mountains near the mouth of Puget Sound and you end up with about one hundred miles of rock.

Nothing on the West Coast is ever wholly what it seems. The subducting plate offshore approaches the land not head-on but from a slight angle. While it pushes rocks under the edge of the continent, it is driving them northward as well. How that subducting plate is coupled to the rocks above, however, is not well understood. "We have a pretty good handle on the surface geology, but we don't know what the deep structure is like," Snavely says. "We need deep geophysics in Washington and Oregon to understand the boundary between the North American Plate and the San Juan de Fuca Plate."

Patterns of faulting and folding on the seafloor suggest subduction has been turned on and off several times in the past fifty million years on the Oregon coast and has come from different directions as well—sometimes head-on, sometimes at an angle. The forces that drive it, however, take place on the spreading ridge just one hundred miles from shore, a fiery rift on the floor of the sea. Lavas similar to those that once built up the coast of the Pacific Northwest still rise to the surface, flowing out of the mouths of submarine volcanoes and rifts. As these rocks cool, they spread outward—new fresh pieces of the seafloor, driving the rocks in front of them toward the coast.

• • •

It was the third week of January, and outside Bob Embley's home on the Oregon coast the town of Newport was weathering the brunt of a good, hard blow. The local chamber of commerce likens winter on the central Oregon coast to standing on the prow of a ship at sea. At the time, that description seemed particularly apt. Sitting around the fireplace in Embley's living room you could hear the waves pounding the rocks offshore. Sheets of rain danced across the windows and over the roof. Outside the temperature was hovering at fifty degrees, but it felt much colder with the gusts of thirty- and forty-knot winds blowing in off the water.

Embley is a senior scientist for the National Oceanic and Atmospheric Administration (NOAA). Since 1984 he has been working with the agency's Vents Program, studying a series of vents and springs on the spreading ridge that lies offshore just two hundred miles from his office at the Hatfield Marine Science Center in Newport.

We were talking about what it feels like to drop down several thousand feet below the surface of the sea. "I always wanted to be an astronaut," Embley said, moving back on the couch to take a sip from his tea. "But I couldn't because of my eyes. My vision wasn't good enough. So I guess going to the bottom of the sea was the next best thing. I guess there are a few places in the world where you can go and see a place that no one has ever seen before. Maybe in Antarctica. . . . But every time you go down to the bottom of the sea you're seeing something that no one's ever seen before. You're seeing something for the first time. It's exciting. Or at least it's exciting to me. It's a privilege to go down there. I try never to forget that."

• • •

The ocean is the earth's last frontier and it is a large one, covering nearly three-fourths of its surface. As recently as the 1950s the floor of the sea was thought to be a flat and featureless plain. The development of sonar to combat submarines during the Second World War had slowly changed that perception, but it took more than a decade to piece into a coherent picture the information that sonar made possible. That all changed in 1957 when Bruce Heezen and Marie Tharp of Columbia University's Lamont Doherty Geological Observatory published a revolutionary new map of the floor of the sea.

Unlike the maps that had preceded it, Heezen and Tharp's map was based not on imagination and guesswork but on reams of data gathered by sonar and seismic surveys of the seafloor. The floor of the sea, their map revealed, was not flat at all, but scored by a belt of mountains and submarine volcanoes that circled the globe like the seams of a baseball—a submarine mountain range more than forty thousand miles long. The idea was so revolutionary that it was featured on the front page of the *New York Times*.

While pioneers like Alfred Wegener had suggested in the 1920s that the earth's continents had moved, these submarine ridges and rises were the piece of geologic evidence that would finally open the door to plate tectonics. Geologists soon discovered that these submarine ridges were volcanic. Patterns of magnetism locked in the rocks as they cooled suggested that lavas had spread outward from these submarine rift zones, driving the continents across the face of the earth.

In the space of a few years the discovery of these ridges on the seafloor would revolutionize the field of biology as well. In 1979 a group of scientists from the Woods Hole Oceanographic Institute were exploring a series of vents on the Galápagos Rift Zone off the coast of South America. They found that these deep-sea ridges were not barren but carpeted in places with giant tube worms and clams that thrived in the midst of vents of superheated water streaming out of the rift zone. The animals were eerie, almost surreal, bone-white with blood-red interiors. The color, it was revealed, was due to hemoglobin, blood cells similar to our own. But that was as far as the comparison went.

Not only had these animals never been seen before, they had never even been imagined before and were unlike any other form of life on earth. They actually thrived in the high-pressure, high-temperature world of the rift zone, deriving the energy they needed to survive from the breakdown of hydrogen sulfide in the superheated vent waters.

Up until the discovery of these strange animals biologists had believed that all life on earth was based on the breakdown of molecules composed of carbon, oxygen, and, ultimately, energy that came from the sun and had been harnessed by plants through photosynthesis. The energy that fueled these clams and worms, however, was not solar but chemical—it came not out of the sky but right up out of the earth.

It was a revolution in biology's perception of the world. Not only were these animals a new form of life, but biologists came to believe that they may have been the original form of life on earth, a family of bacteria and perhaps even small animals that were capable of subsisting on the gasses and fluids that poured out of the earth's molten interior long before the earth had developed its oxygen atmosphere. Since their initial discovery in 1979, scientists have discovered and identified more than sixteen new families and more than fifty genera of these unusual animals. If the ridges represented the beginnings of oceans and ocean basins, it is also possible that the primitive animals living there represented the beginnings of life on earth.

• • •

Today one of the most promising sights for the study of ocean ridges lies on the seafloor of the Pacific Northwest on the San Juan de Fuca Ridge. Off the coast of Washington and Oregon scientists have found what has been described as a submarine Yellowstone, a system of superheated springs and vents on the seafloor covering an area of more than one hundred thousand square miles.

North of Cape Mendocino the San Andreas Fault is replaced by a series of spreading ridges paralleling the coast—the Gorda Ridge off northern California, the San Juan de Fuca Ridge off Washington and Oregon, and the Explorer Ridge north of Puget Sound off the coast of British Columbia. These ridges are offset by several miles by a series of east–west-running faults. Seen on a map, they seem to be staggered along the coast, leading northward like stair steps. The heart of this new terrain lies near the junction of the San Juan de Fuca Ridge and the Blanco Fracture Zone, the fault zone that separates the Gorda and San Juan de Fuca Ridges.

Midocean ridges are some of the most difficult features on earth to study. As their name implies, most are located in the middle of ocean basins. Not only are they often several thousand miles from shore, they are typically more than ten thousand feet below the surface. The San Juan de Fuca Ridge, however, is less than two hundred miles from shore and less than seven thousand feet beneath the surface. Transit time is measured in hours, not days and weeks, and scientists in Oregon, Washington, and British Columbia can study the ridge on a fairly regular basis with research submarines as well as surface ships towing camera gear and sonar equipment.

Geophysical data suggest that the San Juan de Fuca Ridge is spreading at the rate of some six centimeters per year. Its top is scored by a central valley about one kilometer wide and one hundred meters deep. Cutting through that is a narrow cleft, according to Embley, just wide enough to drive a submarine through—the actual center of the rift zone. Through the windows of the submarine you can see the edges of two plates—the Pacific Plate to the west and the San Juan de Fuca Plate to the east. The cleft is the actual eruptive center of the ridge. Flows of lava along it are fresh and glossy.

• • •

Here and there the surface of the ridge is marked by spires and chimneys spewing smoke and superheated streams of water. The basalt that lies just below the surface comes out of the ground at more than twelve hundred degrees Fahrenheit. Rocks here are often porous, and seawater flows through their cracks and pores, emerging from vents as superheated fluids laden with sulfur, copper, zinc, and other minerals picked up from the rocks. These vents can quickly weather basalt to the point of unrecognizability, Embley says, leaching out manganese and leaving behind a rock that is strangely rich in silica and aluminum. Studying these vents is precarious work for submarines. The water coming out of the ground can be three to four hundred degrees Fahrenheit—hot enough to melt the submarine's Plexiglas windows in a matter of seconds.

The vents come in all shapes and sizes. Some seem to spew right out of cracks in the rocks; others are surrounded by chimneys five, ten, and even thirty feet high. At seven thousand feet below the surface, water temperature in the ocean hovers at only a few degrees above freezing. When the superheated waters of the vents hit the cold water, minerals like sulfate, anhydrite, barite, and sphalerite precipitate out of the water, building up narrow chimneys, or stacks. Some plumes look almost milky. Others are almost black, rich in dissolved minerals like iron, copper, zinc, and sulfur. Some are rich in gas, bubbling away on the bottom and causing the water above them to shimmer almost like the heat waves that play across the surface of the desert at mid-day.

In spite of their striking appearance, these vent fields can be hard to find. The floor of the sea is perpetually dark. Visibility extends no further than reach of submarine lights. One of the largest vent fields discovered by NOAA scientists had been completely overlooked by a group of French scientists who had studied an area in detail less than a mile away.

Near the vents, the ridge is surprisingly alive, carpeted with clams and tube worms feeding in the jets of rich, hot water pouring out of the rift. They look almost like freaks from a sideshow—worms two and three feet long that sway like reeds in the currents kicked up by the vents like reeds. In some places there are giant clams, often more than a foot across, surrounded by billowing mats of bacteria. That flourish of life has in turn attracted other animals, predators that live on the fringe of the vents—rattail fish, long-legged spider crabs, and a collection of barnacles and sponges that coat the nearby rocks. They don't just survive here, they thrive. The temperatures and chemical content of the water would kill most other animals. The levels of sulfide inside their bodies is two times higher than what is commonly considered toxic.

Other strange features have been found here as well. One of the more startling finds was made in 1988 on the San Juan de Fuca Ridge at the urging of Cindy Lee Van Dover, a graduate student at the Woods Hole Oceanographic Institute. After observing that shrimp living near vents on the mid-Atlantic Ridge appeared to have eyelike organs on their backs, she theorized that there must be some type of light actually coming from the vents that enabled these deep sea shrimp to see. To test her idea, scientists studying a system of high-temperature vents on the San Juan de Fuca Ridge outfitted the *Alvin* with an array of ultrasensitive cameras and descended to the floor of the rift. As they hovered a few inches from the mouth of a superheated vent, they turned off the sub's lights. When they truned on the cameras they found a faint, flame-like glow coming out of the vent, a kind of black body radiation—light from the earth's fiery interior.

• • •

The discovery of midocean ridges opened up a new window on the world, but even after thirty years of work, scientists have only recently begun to peer through it. One thing that has become clear, however, is that it is impossible to understand the earth without understanding the spreading ridges that circle the seafloor.

Sixty percent of the earth's surface has been created at ocean ridges from cooling magma. But while scientists have been studying volcanoes on land for centuries, we know almost nothing about these volcanoes on the floor of the sea. Although their lava exceeds the total amount of lava on land by several orders of magnitude, no one has ever actually seen one of these submarine rift zones erupt. Rift zones are the key to understanding plate tectonics, the final and largest piece of the puzzle.

Scientists believe that here in the Pacific Northwest they have a very good chance of finally seeing one of these eruptions firsthand. The vents that litter the ridge are thought to be a sign of hot magmas just below the surface and possible markers of recent eruptions.

In 1986 and again in 1988 scientists were startled to find giant plumes of mineral-laden hot water floating through the ocean. The largest was more than fifteen miles wide and more than two thousand feet high—submarine ash cloud floating through the sea with concentrations of minerals that were five times higher than those of the vents and plumes they had been observing on the ridge. The amount of heat and the concentration of chemicals involved was so great that scientists came to believe that these plumes could even be affecting such things as chemistry of seawater and the global heat budget of the oceans.

In 1989 scientists finally discovered the site where the mega-plume may have come from, a seven-mile-long crack on the seafloor not far from a vent site they had been studying for more than a year. Near the crack geologists detected fresh flows of lava, the youngest yet encountered on the ridge. Like an ash cloud on land, the plume may have been the signature of a volcanic eruption of a rift zone on the floor of the sea.

For Embley it was one of the most interesting things he had seen in several years of work on the seafloor. "What I'm beginning to think is that the geology and the big plumes up in the water column may be telling us that this is an active phase of seafloor spreading. It may be that we're getting rifting and extension with large bursts of hydrother-mal activity."

The key, of course, is seeing an eruption as it happens. Chances are slim that a research submarine would be prowling the ridge when one of those eruptions occurred—or that anyone would necessarily want to be there several thousand feet below the surface when it did. Instead what scientists are hoping is that a network of monitors and cameras on the seafloor will enable them to actually see that spreading take place.

"It's been repeated many times in the history of geology," Embley said. "If you don't see the process, you'll never understand it. It's very difficult to understand a process unless you see it going on. Nobody has ever observed an active phase of seafloor spreading yet. They see the immediate aftermath of it, perhaps, in very fresh lava flows, and people have occasionally seen earthquakes on the ridge. We think something is going on, but we've never been able to document it or get out there and look at it. When you think about it, we know more about the surface of the moon than we do about the floor of the sea."

• • •

Surf scooters and pigeon guillemots rode the waves below, rising and falling with each passing ridge of water, cresting the top of each wave just before it broke and speeded toward the rocks alongshore. At the base of the cliffs below Sea Lion Caves a few hundred yards south of Heceta Head, more than one hundred stellar sea lions lay in the sun, scattered like logs over the rocks. Their dry fur was as tawny and gold as the hide of cougar or lion. They seemed almost motionless. Through binoculars you could see them lazily open an eye to look over the scene around them or wave a black flipper in the air.

Offshore, other sea lions streamed through the water like dolphins, turning and diving with quick graceful movements. From time to time a pair or two would speed toward the rock ledge where the others were hauled out, timing their arrival to coincide with the surge of a wave that would help lift them up onto the rocks. They were wet and cold as they emerged from the sea, and you could hear the angry barks and bellows of complaint rise up from the herd as they trundled over the tops of sleeping animals, nudging them awake. If you were to look closely at the sides of the sleeping animals, you could see cuts and scrapes from the rocks and the teeth marks of sharks. A few wore bits of plastic fishing nets and trash around their necks like necklaces.

The sea lions that were on the rocks below are part of a population of some three thousand that gather here along the offshore rocks and reefs that fringe the Oregon coast to breed and raise their young. Some come from as far away as the Gulf of Alaska. By late summer they will be joined by an estimated two thousand California sea lions—all males that head north for the summer from their breeding grounds in Baja and southern California. There were harbor seals offshore as well, floating amid strands of kelp or sleeping on low-lying rocks only a few inches above the waves. While the sea lions typically range several miles from shore, the harbor seals, as their name implies, typically stay closer to the coast and are often found in bays and estuaries. Their total population is thought to number somewhere near five thousand.

While seals and sea lions are often found resting in close proximity to one another, they typically feed and haul out in different areas. Seals are rarely found more than five miles from shore. Sea lions rarely feed closer than five miles to it, preferring the more open reaches of the ocean farther away from the coast.

On the coastal rocks seals are seldom more than a few inches above the water, while sea lions may climb to the top of high rocks ten and twenty feet above the waves. The difference has to do with biology. The sea lions' hind flippers are hinged, and the animals are able to bend them underneath their bodies to actually walk across the rocks in a rough fashion. Seals, however, are more rigid. Unable to bend like the sea lion, they are reduced to inching their way across the ground like snails.

• • •

While seals and sea lions spend much of their life at sea they still depend on land to breed and raise their young. While seals can be found hauled out in bays and estuaries and on low rocks up and down the coast, sea lions are more restricted. There are only half a dozen known haul-out sites on the coast and only two breeding grounds where pups are raised—Orford Reef near Port Orford and Rogue Reef off the mouth of the Rogue River, both several miles from shore.

The males arrive as early as April to stake out terrain. The females arrive in early May and give birth within a week or two of their arrival. After the pups are born the females spend the first few days nursing and setting up a feeding schedule, feeding offshore for a few days and then returning to nurse and raise their pups.

While land animals can have their young anytime from early spring until late fall, marine mammals like sea lions are more restricted. With only a limited number of suitable breeding and pupping sites they come together not only to raise their young, but also to find each other and breed. Birthing and breeding are compacted into the space of two months instead of nine or ten. How that is all possible is due to an ingeneous biological trick.

Females give birth in the spring almost as soon as they reach the rocks. Within two weeks they are in heat again and begin breeding with the dominant bulls. Once fertilized, the sea lion ovum does not implant itself in the uterine wall immediately but enters a dormant blastocyst stage for approximately two months. The delay enables them to give birth the following spring when they return to the rocks to breed rather than in the colder, rougher waters of Alaska.

A collection of dominant bulls rule the breeding ground, but instead of claiming and defending specific females, they defend specific territories, breeding with any and all females passing through. Like all real estate, location is critical, and the choicest areas are those closest to the water along the paths used by the females as they come and go from the rocks. While the females periodically leave to feed and fish in the waters offshore, males remain on the rock for two to three months without leaving or eating. They can mate with as many as sixty females in a single season, but the strain takes a quick toll. An eight-hundred-pound bull can lose as much as two hundred pounds in a single breeding season. Most die young. While females typically live to be twenty or twenty-five years old, most males seldom make it past sixteen.

As with seabirds, care and feeding of the young pups is critical for their survival. Most sea lions have only a single pup, and there is a strong bond between each mother and her pup. While seals are often weaned in a week and can swim almost immediately and often travel on their mothers' backs, baby sea lions are remarkably vulnerable. The newborn pups weigh just twenty to forty pounds and typically do not have enough blubber for buoyancy and insulation. Drowning is actually the leading cause of death among pups. They can also be crushed to death by the comings and goings of other animals in the colony.

Congenital birth defects claim others. The nursing pups grow quickly. In three months they generally weigh as much as year-old adults. Training, teaching them to hunt, swim, and survive, can last for as long as six months.

Roughly seven hundred stellar sea lions are born on the Oregon coast each year, five hundred at Rogue Reef and two hundred at Orford Reef. While the numbers are small, they have been increasing for the past decade. At present it is the only breeding population of stellar sea lions in North America outside of Alaska—and Alaska's population is roughly ten times that of Oregon's. However, the population here is stable, while Alaska's has been declining rapidly, by as much as 50 percent from 1981 to 1985, due mainly to the arrival of large factory ships fishing for pollock, the sea lions' principle food.

••••

Seals and sea lions come back to the same sites year after year and have been using these offshore rocks and reefs as breeding grounds for several thousand years, according to Rob Brown of the Oregon Department of Fish and Game. Coastal Indians hunted seals and sea lions long before the arrival of the first Europeans, and in some places their hunting appears to have been so successful that the animals were locally eliminated. While these animals can move with agility and grace in the open ocean, they are nearly helpless on land and would have been easy targets for spears and clubs. Midden deposits on Seal Rocks south of Newport, for example, are littered with the bones of stellar sea lion pups and northern fur seals—neither of which are found there today.

Settlement and development of the area in the 1850s proved as disastrous for seals, sea lions, and other marine mammals as it did for seabirds. Seals and sea lions were hunted for their pelts or shot for sport. Within a few decades the northern fur seals were all but eliminated in the waters off the Oregon coast. By the early 1900s sea otters had disappeared from Oregon for good, having been hunted for their valuable pelts, which brought as much as $100 apiece. Taking advantage of the strong bond between mothers and their young, hunters made a point of killing the pups first because the mothers would then stay nearby to see what happened to their young.

For others, extermination was no accident. At Rogue Reef, fishermen were known to scare the sea lions off the rocks and then set off dynamite charges in the water to kill them. Others in Alsea Bay set up land mines with trip wires to kill the seals as they came ashore. By the 1960s surveys predicted that the total number of seals and sea lions along the coast amounted to just five hundred animals. As recently as

1972 bounty hunters could kill seals and sea lions in Oregon and receive as much as $25 a head for helping control what were widely perceived as vermin.

• • •

The roots of this ongoing war with seals and sea lions are basic, Rob Brown says. "They eat fish, and fishermen catch fish. Traditionally fishermen don't like them as competitors no matter what fish they're fishing for." The passage of the Marine Mammal Protection Act in the late 1970s changed all that. While fishermen with special permits can still shoot problem animals, the law makes it a federal crime to shoot, harass, or annoy any marine mammal. Protection has dramatically increased their numbers, but it has also created a highly controversial issue, particularly for federal and state employees like Brown and Roy Lowe. "If I'm in uniform,"Lowe says, "that's the first thing I get nailed for: 'God, when are we going to start controlling those marine mammals?' You try to talk to them about it and it doesn't do any good." Brown has had similar experiences of his own.

While seals and sea lions can and do steal fish and foul nets and gear, in the end they may be more of a help than a hindrance to populations of locally valuable fish like salmon. Although conventional wisdom has it that sea lions and seals are decimating local salmon runs, studies have shown that that view is far from accurate. When Brown was a graduate student at Oregon State University he worked with a team of scientists studying the feeding habits of seals along the coast. They spent more than one hundred hours studying the animals in the field, watching them feed, examining their feces, and even shooting a few to examine their stomach contents. What they found was that the number one food for local seals was lamprey, which are parasites on salmon. Salmon made up only 1 or 2 percent of their diet.

"What I try to tell people when I get involved in these kinds of discussions," Brown said, "is that the seals we are familiar with have been around for five million years and salmon and steelhead have been around for more than five million years and that there has always been some kind of balance. To say that they're decimating runs of fish just really doesn't wash."

The problem with seals and sea lions, so far as it exists, Brown says, does not lie with the total number of seals but with problem individuals. Up in Seattle, for example, a sea lion that the local media quickly named Hershel learned to spend time in front of the fish ladders by the locks leading inland from Puget Sound and was able to wipe out an estimated 60 percent of the salmon run there.

Like bears that learn to congregate around a campground or a city dump, there are a few crafty animals that learn to cause problems.

Some seals and sea lions, Brown admits, have learned to follow fishing boats and steal salmon and steelhead. The solution is to deal with those problem animals on an individual basis just like bears—by either killing them or relocating them.

·   ·   ·

It was a low gray day at Otter Rock, a few miles north of Newport on the central Oregon coast. Walking out among the shelves of rocks that reach out from shore you could see rich tide pools filled with starfish, anemones, and a host of scurrying crab. Waves hardly broke at all on the rocks, their passing nothing more than a slow and steady rise and fall of the water.

A few feet from shore a harbor seal slept on a low shelf of rock only inches above the water. As I drew near he slowly opened an eye and rolled on his side, to watch me. His fur was bright silver, mottled here and there with large brown spots. As I drew closer he bounced awkwardly across the rock and then slid into the water, gliding off to a tangle of kelp to regard me with large, wet brown eyes.

On land, seals and sea lions seem almost awkward and ungainly, with bodies shaped like plump, furry sausages, with only flat, soft flippers to propel them over the ground. What seem like disadvantages on land, however, are really adaptations to living in and moving through water.

Life on land is ruled by gravity. Its pull through the open air demands structures that are anchored by rigid bones and shells. Weight exerts a constant strain. In the water, however, the operative word is *turbulence*. The buoyant effects of water can make even the heaviest objects almost weightless. The soft, massive body of a whale, with its relatively delicate bone structure, would never survive on land. In fact, beached whales die not from a lack of water but from suffocation as their heavy bodies flatten their lungs and internal organs.

In the water, the critical factor is not weight but resistance. Streamlined objects, whether they are a seal or a submarine, move more easily through the water. The rounded shape of a seal that seems so awkward on land is actually highly efficient in the water. Limbs are less useful here. Instead, the power of movement comes from waving a whole body or tail. Shoulders are reduced, almost nonexistent—hips too. Arms and legs in the seal have become small flippers, planes, or fins used to turn and maintain balance.

The layer of blubber that helps streamline the seal also helps insulate it from the cooling effects of water, which is a remarkable absorber of heat. Fifty-degree water can kill an unprotected swimmer in a matter of minutes, and seals and sea lions stay in the water for hours. Seals and sea lions have other adaptations to the cooling effects of water as well. Flippers and tails have what is known as a counter-current mechanism. Arteries leading out to the flippers are encased in a network of veins so that the heat of the blood coming out from the heart is not lost in the bare, uninsulated flipper but is passed on to the returning blood in the veins.

Seals and sea lions also have amazing endurance. Trolling in a fishing boat or research ship several miles from shore you can suddenly find a seal or sea lion swimming alongside, carefully watching you pass by. Although they can dart through the water with quick bursts of speed, they are foragers and typically move more slowly in the open ocean. Researchers have tracked seals swimming at three and a half knots nonstop for as long as nine hours.

But while we often think of these animals at the surface, their ability and agility underwater is phenomenal. Some seals, such as the elephant seal, routinely dive as deep as twelve hundred feet in the ocean. Others, like weddell seals in the Antarctic, can dive for as long as an hour and a half. Underwater seals and sea lions have what is known as the mammalian dive reflex, which shunts blood away from the extremities toward the brain. Other chemical adaptations within their bodies help as well. Their muscles are richer in myoglobin, which binds with oxygen. Fats in their bodies are also more capable of absorbing nitrogen then ours, so they do not get the bends—a painful and sometimes fatal build-up of nitrogen gas in the blood—from a deep dive. Below the surface, the water pressure can be phenomenal— as much as six hundred pounds per square inch down at the depths reached by an elephant seal. Rather than fighting the pressure, body cavities simply give into it, shrinking and collapsing as depth increases.

Light is also limited at depth, and it is not certain how seals and sea lions navigate below the surface. Unlike whales and dolphins, they do not echolocate—that is, they do not use the reflected noises of grunts and squeals to judge distance and location. Sight, however, is thought to play a role because they have unusually well-developed eyes for low-light conditions. Like cats, their eyes have a tapetum, a reflective surface just behind their retina that bounces light passing through the retina back through it again. In effect, the rods and cones in the back of their eyes are exposed to light twice, once on the way in and once on the way out. (A cat's tapetum is what makes its eyes so visible at night.)

About the only adaptation these animals have not made to life in the sea is breeding and raising their young, which forces them back to the coast each year. Each spring they come back, part of the cycle and rhythm of life on this volcanic edge of the continent.

. . .

When Parke Snavely first started working on the Oregon coast in the late 1940s the timber industry was just beginning to take off and trees were not carried to the mills by truck but by train. Narrow-gauge railroads ran almost everywhere in the foothills, and heading out into the field for a day's work Snavely would often hitch a ride uphill and then walk the tracks back down. In Snavely's opinion, even better than the actual ride uphill was that, because the trains were built with shallow grades that cut through the hills and ridges, he could see layer after layer of the rocks below in the sides of deep road cuts.

After joining the U.S. Geological Survey, Snavely spent his first field season in the area in the summer of 1946, paired with Harold Vokes, a professor of geology from Johns Hopkins University. They mapped sections of the Coast Ranges west of Eugene. Volks wore a white suit while they worked out in the hills and was, in Snavely's estimation, "a little too citified for a real field man." Whatever his taste in clothing, however, Vokes worked quickly and he and Snavely mapped a 7.5-minute topographic quadrangle—an area of more than thirty square miles each month. As the summer wore on, Snavely let it be known regularly and often that he preferred working alone. By September of that year he got his wish. The powers that be in Washington put him to work in the coastal hills near Neskowin on the northern Oregon coast.

It was a wet fall. For the next two months it rained every day. Snavely worked alone, following the streams back up into the hills, looking for good outcrops of rock and carefully marking them on his maps. At times he waded through spots as deep as his waist. The salmon "were as big as logs," he recalled. "I used to tell people I could walk across streams on them." It was, he said, kind of lonesome up on those drainages. For emergencies he carried a service .38 revolver loaded with pyrotechnic shells—although it was never quite clear who would see his signal in the rain. After a full day in the field he would come back to his rented house near the beach to eat dinner with his wife while his wet field clothes dried on an inside line.

When Francis Wells, a senior geologist at the USGS who had befriended Snavely, heard that he was still out in the field in Oregon, he made a point of stopping in to see Snavely's boss in Washington, D.C. "What are you trying to do? Kill that young man?" he asked. "Have you ever spent a winter out there? He might drown." It was a reprieve. Around Thanksgiving a telegram arrived ordering Snavely to Johns Hopkins University. He and his wife loaded up their car and drove inland across the Coast Ranges, swinging south through California on their way back east. Crossing the border into California they finally left the rain behind. "When we got to Mount Shasta the sun was out and we just got out of the car and looked at it," his wife told me one morning. "We hadn't seen the sun for thirty days."

Today most of Snavely's fieldwork on the Oregon coast takes place in the summer. The field season, however, is still very much a family affair. His wife takes care of their rented apartment in Otter Rock and prepares their meals, while Snavely searches for new exposures of rock. For a time, his son—Parke Snavely, III, now a geologist with Exxon in Houston—spent the summers working with him as an assistant. Now his grandchildren have taken over that role, coming out to spend time with him in the field looking at rocks. Visitors are fairly common as well—geologists from the United States, Japan, and Europe come by to hear about Snavely's perspective on the terrain and see the rocks through his eyes. In the late 1960s when astronauts were first preparing to go to the moon, they came to Oregon for a crash course on volcanic rocks, with Snavely up on the fields of basalts that line the Columbia River. If the moon was volcanic, NASA wanted to make sure they knew what they were looking at.

• • •

After a full day in the field with Ray Wells we were out the next morning by 8:00 collecting rock samples that Bukry would be using for dating up in the hills behind Lincoln City. After chipping off a few chunks of rock from an outcrop alongside the road, Snavely sat in the Cherokee making notes and marking his maps. "This is the bear outcrop. See, it's even marked on the map," he said, pointing out the small silhouette of a bear drawn near our location on the map. "My granddaughter put that on the map. Only a girl would do that," he added with a sigh of feigned disgust. Doodles on the Holy Grail. "We saw a bear here two weeks ago coming out of the woods and boy did he take off like a rocket."

A few miles up the road we stopped at another outcrop for more samples and more careful notes and markings. After just a few hours in the field you get the feeling Snavely knows the area here so well he could drive through it blindfolded and never miss a stone. "Gee, Parke," Bukry said as he walked around the outcrop, "there are hammer marks all over this outcrop and they look really fresh."

"Yeah, that's right," Snavely said rather sheepishly. "I'd say they're about three to four weeks old." (The marks are undoubtedly Snavely's.)

"You wouldn't happen to have any idea what the elevation is here would you?" Bukry asked.

"Well. . . " said Snavely, rolling his eyes up toward the sky, "I don't have my altimeter with me, but offhand I'd guess it's about 542 feet above sea level."

We had lunch on the edge of an old clear-cut farther back in the forest. The hillside was terraced by the winding path of a logging road and the trunks of dead trees were scattered helter-skelter down the hillside like toothpicks. Stumps and snags burned charcoal-black from the slash fire rose up above the scrub and fireweed that covered the hillside like the slats of a picket fence.

Down at the bottom of the slope the burn faded into an orderly planting of Douglas fir. Scarcely higher than the scrub that surrounded them, their green, feathery branches were only a few years old. Down below a small creek flowed through the hills in a series of S-shaped curves. Small farms and dairies with bright tin-roofed barns lay along its banks, bordered by fenced-in pastures and fields laid out like a checkerboard. Looking west you could follow the winding path of the creek stream all the way to the Pacific less than five miles away. Rocks the size of small islands lie just offshore, shaped like giant eggs.

As we delved into our sack lunches Snavely thought back over more than forty years of work here on the Oregon and Washington coasts. Almost all of it has centered around field geology, getting out and seeing the rocks and trying to understand how they came to be and how they have been faulted and folded. Over the years he has mapped some twenty quads of Oregon and Washington coast and has gotten to know their details firsthand. The fieldwork has been constant. Only the goals have changed, ranging from finding oil and gas to predicting earthquakes. For Snavely, getting out and seeing the rocks is what it's all about.

"It's all field geology: hydrology, oil and gas work, earthquake hazards, minerals. Field geology is the backbone of the whole science," he said. "Most geologists today are interested in writing five- to ten-page potboilers. There's still a lot to be done with fieldwork, but it's becoming a lost art in an age of specialists. It's been lost at the university level, the corporate level, and at the USGS. There's always new techniques and new technology, but there's no substitute for a guy putting on a pair of boots and wading up a creek and making direct observations, writing notes, and collecting samples. There's no panacea for that."

With several decades of fieldwork under his belt, Snavely could easily retire and sit back and spend the rest of his days writing papers and proposing new theories. I had to ask: what keeps him coming back?

"Well there's always some refinement you can make on your observations," he said after a few moments' thought. "Another piece of the puzzle to discover. I suppose you could say that after forty years of work I'm beginning to run up against the law of diminishing returns.

But then they put in one hundred miles of new logging roads this summer and that's exposed a lot of new rocks. After two or three years the plants will cover them over again and then nobody will be able to go in and check those areas for another one hundred years until they decide to cut the trees again. I don't have any hobbies like gardening or painting. Geology is my avocation and my profession. It is not work for me."

•••

North of the Columbia River the volcanic rocks continue to run right along the coast. In southern Washington the shoreline is often dominated by broad bays and inlets—Greys Harbor and Willapa Bay—but the rocks are still present. Farther north they rise up out of the ground again in the form of small cliffs and headlands. Offshore, spires and seastacks of hard black basalt rise up out of the water like ancient monuments, some shaped like giant heads, others like ships, topped with thickets of spruce and fir for sails. They fringe the coast like a stone fence. Small pocket beaches between rocky headlands are littered with the trunks of giant trees, driftwood washed downstream from the neighboring forest. Inland, for a time, the coast is bordered not by a range of coastal peaks but by low rolling hills, as if the land were saving itself for one final push.

North of the Quinault River the Olympic Mountains appear. They sit on the northern end of the volcanic cliffs and headlands of the Pacific Northwest like the dot at the top of an upside down exclamation point. Unlike the Coast Ranges farther south in Oregon, they are not low, rolling mountains topped with trees but high, rocky peaks capped with snow and ice. Like the high cliffs farther south, the rocks here are volcanic as well. The mountains, however, are not the sight of a volcano that once bordered the coast—they constitute a package of basalts and sandstones from the seafloor that have been driven northward by faulting and lifted up onto the edge of the continent to build mountains more than a mile and a half high. The site of a national park today, the mountains are the center of a wedge-shaped peninsula of land that shields Puget Sound from the open Pacific.

From the high peaks and ridges on the western side of the park you can see the Pacific, but the mountains themselves seem not coastal but alpine. So much snow falls here that the tops of the high peaks are perpetually covered with it, and in places that snow has been slowly transformed into ice. Small glaciers cascade down the rocky flanks of the highest peaks. The terrain is sharp and angular—cut by steep ravines and high, narrow ridges. Higher up the deep-green forest of spruce and fir is broken by meadows that bloom with white avalanche lilies and purple lupines when the winter snows recede. Herds of deer graze in the grass. On the faces of scree slopes near the crest of the peaks you can see the white shapes of mountain goats clambering over the rocks.

Streams run out of the mountains in all directions like the spokes on a wheel. The land here is wet and green. River valleys that face the Pacific receive more than two hundred inches of rain per year and are filled with dense rain forests of moss-covered trees. Herds of Roosevelt elk roam through the park. Although it is less than one hundred miles from Seattle, the park is a different world. Here and there in exposures of rock, you can see the twisted and folded layers of lava that have been driven up out of the sea to make these high peaks. In places, they are folded almost double. The lavas may have cooled several thousand years ago, but the ground here has not lost its ties to the earth's interior. In places within the mountains there are hot springs, streams of hot water rising up from inside the earth.

On the high ridges behind Hurricane Ridge on the northeast corner of the park you can look north across the Straits of Juan de Fuca more than a mile below. Vancouver Island seems to lie off the coast of British Columbia like a giant ship waiting to dock at the edge of the continent. Flotillas of small islands surround it. Ridges of high, forested peaks run the length of the island, stacked up row upon row. You can see the higher peaks of the mainland rising up behind them, solid walls of granite rising up right from the water's edge. Farther inland you can see the cones of volcanoes, masses of red rock covered with snow. Ten thousand years ago it was all covered with ice. Today fjords and *U*-shaped valleys lead back into the heart of the peaks.

Between the island and the mainland lies the Straits of Georgia, the start of an inland sea that leads north all the way to Alaska. The coast of Alaska and British Columbia is a landscape unlike any other in North America—a world of high mountains and small islands shaped by moving rocks and ice. Farther north the ice runs right to the sea.

SOUTHEASTERN ALASKA AND BRITISH COLUMBIA

EXOTIC TERRANE

O UTSIDE THE WINDOWS OF THE PLANE YOU COULD SEE NOTHING BUT ICE
and rock and water. They stretched for more than a hundred
miles in either direction—glaciers that streamed down the sides of the
high peaks and ran all the way to the water's edge. As we turned in a
tight circle over the surface of Icy Bay, Mount St. Elias appeared on
the left side of the plane. Ice fields and snow fields sat on its flanks. A
thin white cloud seemed to cut its almost perfectly triangular shape in
half. Higher up you could see streamers of snow blown into the air by
the high winds that played across its summit. Down below the deep
blue waters of Icy Bay were filled with the white, angular shapes of
icebergs.

Somewhere below the boundaries of Alaska, British Columbia,
and the Yukon Territory all come together, but political distinctions are
meaningless here among the high peaks and ice. Massive alpine
glaciers stream out of the high peaks of the St. Elias Range on the
U.S.–Canadian border—and head downhill toward Alaska and the
coast, coalescing into a single sheet that leaves only narrow ridges of
rock and the sharp points of the highest peaks exposed. The result is a
moving plain of blue and white ice several thousand feet deep covering
hundreds of square miles. Close to the coast it flows out of the mountains like milk streaming down the sides of an overfull bowl.

Flying just one thousand feet above ground level you can see the texture and structure of the ice. Up on the instrument panel of the plane the air-speed indicator says we are traveling at 140 knots, but the scale of the landscape is so large that we seem to be hardly moving at all. Mount St. Elias looks just ten miles away—in reality it is more than thirty. It rises from sea level to more than eighteen thousand feet in less than fifty miles. Its summit is nearly a mile higher than the highest peaks in the Rockies. From base to summit it is the tallest mountain in the world, five thousand feet higher than Mount Everest in the Himalayas, which does not spring up out of the sea but sits on top of a three-mile-high plateau.

Beneath the ice you can see the black and gray rocks of the peaks, stacked up in layers like the pages of a book. Near the coast those rocks are young and marine, a collection of shales and sandstones lifted out of the sea just 5 million years ago. Farther inland, however, the rocks that lie in the core of the mountains are more than 350 million years old, the remnants of ancient volcanic islands and oceanic plateaus. Their origins are exotic. They did not come from the nearby sea but may have drifted halfway across the Pacific, possibly from as far away as Peru. While the Himalayas were formed through the collision of India with Asia some 42 million years ago, the high peaks that border the coasts of British Columbia and Alaska were formed when these drifting blocks of rock collided with the continent some 50 million years ago. Today they rise up right from the water's edge, blocking wet winds of the North Pacific and squeezing so much moisture out of the clouds that the mountains here are perpetually covered with snow. From Puget Sound to the tip of the Alaskan Peninsula, the coastline has been shaped by moving ice.

•   •   •

I was flying over the glaciers and ice fields that cover the coast of southeastern Alaska with Bob Krimmel from the U.S. Geological Survey's Ice Climate Project in Tacoma, Washington, and Jan Lundquist of Sweden's Institute of Quaternary Studies. We had been in the air for almost two hours, flying between Cordova and Yakutat, a distance of some two hundred miles. The flight was part of a field trip to the glaciers of southeastern Alaska organized by the American Geophysical Union. Somewhere in the air behind us were three other small planes filled with a collection of scientists from Europe and North America.

We had started out early that morning in Cordova following dirt and gravel roads through the forest to explore the rolling moraines that lie at the base of Sheridan Glacier. We then went to the banks of the Copper River to watch blocks of ice calve off the face of the Child's

River Glacier and float into the fast moving current. Ice seemed to cut through the landscape like a knife, carving out *U*-shaped valleys and leaving behind rolling moraines. The size of things was so large that it was hard to get a feel for what held them together.

Up in the air, however, the landscape began to take shape. The fronts of ice that had sliced through the forest rose up through winding valleys to a vast white sea of ice and snow that ran along the coast for miles. The snow in those high bowls and plains was white and trackless, broken only by the narrow ridges of rock and the sharp-sided shapes of high peaks that had been carved by the ice. Smaller glaciers led out from this high icy sea, fingers of ice reaching down into the land below.

Beyond the cover of snow, the ice was not featureless but full of texture and color. Bands of rock scraped from the sides of the peaks ran across the surface of the ice, bent into curving lines and folds by the steady creep of the ice. Where it plunged down the side of a steep valley its surface was cut by crevasse fields, giant fissures and cracks more than a hundred feet deep. One hundred fifty feet below the surface ice moves plastically, flowing smoothly downhill. On the surface, however, its movements are brittle. Studying the patterns below you could almost see the ice move. At the base of these ice falls, the surface of the glacier was colored with winding bands of light and dark ice known as ogives that looked almost like taffy.

In places, these rivers of ice reached all the way to the sea, littering the milky water of a bay or inlet with icebergs. Elsewhere they ended in broad moraines—rolling fields of brown and gray rock—feeding silt-laden streams that ran through the landscape below like thin, gray threads.

From time to time Krimmel would call out the names of the glaciers passing below us: Sheridan, Sherman, Goodwin, Child's, Miles, Bering, Mount Miller, and Yahtse. Our pilot, Gail Rainey from Yakutat, weaved through the peaks and fields of ice. Occasionally we pointed out features on the fields of ice and snow below: a pattern of ogives on the ice, an amphitheater-like cirque carved in the side of a high peak. Beyond that we hardly talked at all, lost in a silent contemplation of the ice.

There are more than one hundred thousand glaciers in Alaska, and some 5 percent of the state's 586,000 square miles are covered by ice. Most of that terrain is all but unexplored. While the landscape passing beneath us has been photographed hundreds of times, most of its peaks and glaciers are still unnamed. The size of the landscape is so large it is almost impossible to comprehend: Bering Glacier sprawls over an area as big as the state of Delaware. A few miles farther down the coast Malaspina Glacier covers an area the size of Rhode Island. A little more than a century ago these glaciers were all but unknown.

It had been cool when we took off from Cordova, but as we circled over the ice in the bright sun it was almost sweltering inside the tiny plane. Within minutes of taking off, we had stripped out of jackets and sweaters down to shirtsleeves and T-shirts. It was warm down on the ice as well. Flying across the broad, icy sweep of Bering Glacier we had crossed hundreds of small pools on the surface of the glacier, round ponds of water whose white bottoms reflected the almost perfect blue of the sky overhead like small chips of turquoise strewn across the ice. As we threaded our way through the Guyot Hills we could see a stream pouring out of the glacier's side. It shot out of the ice like a huge fountain, spraying several hundred feet into the air. On dry land its roiling flow would have been enough to feed a river. Here it plunged back into the ice through the mouth of a cavern that was several hundred yards wide.

In the summer, the lower ends of these glaciers have water running and flowing not only over their surface and collecting in pools but running through them as well—a network of underground rivers and streams that drains water off the ice: glacial plumbing. Near the glacier's terminus a path of roiling, boiling water marked the spot where that river within the ice reached the sea. The calving glaciers that ringed the bay littered its surface with angular blocks of ice, transforming it into an icy, inland sea. One hundred years ago this bay on the Alaskan coast was not even here. The ice extended more than five miles offshore out into the open Pacific. Over the past ninety years the glacier has retreated by more than twenty miles.

For the past three million years the earth's climate has alternately cooled and warmed causing glaciers to expand and contract. Some of these cold spells have sent glacial ice as far south as Kansas. Others, however, have been far smaller: small changes of a few degrees in the earth's climate that have nonetheless set the ice in motion. The most recent of these small glacial periods ended as recently as 1850. Since that time the earth's climate has been gradually warming. In response, the glaciers here in southeast Alaska have been slowly retreating, pulling back up into the mountains and leaving moraines and rock-covered glacial outwash plains behind. Sites where glaciers once reached the sea are now marked by bays and inlets and mountainous fjords.

As we flew into Yakutat Bay we could see rolling fields of rocks and stones left by the retreating glaciers. Alongshore the ridgelike pattern of moraines marked the reach of the ice. It is not always easy to tell where the ice ends and the land begins. As the ice slides downhill it is often covered with layers of rock and soil that have been scraped and dropped from the sides of mountains and hills. What looks like solid ground is often ice—ice-cored terrain—a massive block of ice hidden by a thin patina of boulders and stones. In places, the cover is so complete that forests of spruce and hemlock have sprung up. Later when the ice below begins to melt, the forest floor collapses, leaving the trees above tilted and leaning at precarious angles—a drunken forest. From a thousand feet in the sky they look almost like craters left by meteors that have been haphazardly covered by trees. Elsewhere in the forest are small round ponds, pools of water left by the melting ice below.

Rainey set the plane gently down on the gravel strip that parallels the main runway. The stall horn sounded just as the plane's fat tires touched the ground. We rolled to a stop and then taxied down to a parking lot near her small office on the edge of the airport. One by one the remaining planes came in, sweeping in from the bay and the ice beyond it.

The next day most of the group would be heading on to Glacier Bay National Park. An hour after we landed, however, I was on my way back to Cordova with Rainey's son and Austin Post, a glaciologist with the U.S. Geological Survey, to meet up with another group of scientists traveling by boat up to Columbia Glacier on Prince William Sound. With the darkness rapidly approaching we flew along the coast, skirting the edge of the mountains. For Post the details of the landscape were all familiar. He has worked here in southeast Alaska for the better part of two decades, studying the rapid advance and retreat of its coastal glaciers. As we headed back toward Cordova, he pointed out the features of the glaciers and ice fields that lay farther inland.

Along the coast the land was flat and green, covered with a dense forest of spruce and hemlock. Since the retreat of the glaciers the land here has been slowly rising through a process known as glacial rebound. Freed from the weight of the overlying ice it is rising up out of the water like a ship being unloaded at a pier. Today a spit covers the mouth of Yakutat Bay. Two hundred years ago when Spanish sailors made their way along this icy reach of the coast Yakutat Bay was open to the sea. The forest that now covers the coast is the product of more than two hundred years of plant succession, a process that begins here with bare rock and progresses through small bits of lichen and moss to tall trees.

Off in the distance you could see the glaciers that have shaped this green landscape. As we glided past the mouth of Icy Bay, Mount St. Elias was clear and cloudless, its top mantled in snow and ice. Around it, row after row of high peaks stretched off into the distance, colored with a thousand shades of gray, blue, pink, and red like the soft layerings of color in serigraph painting. In midsummer twilight lasts for hours.

As we passed over Berg Lake, the ice beyond it seemed to glow with an inner fire. Up ahead the Copper River Delta came into view. Fed by glacial streams and littered with blocks of ice from the glaciers that run right to its edge, it carries more than three times as much sediment as the Yukon River farther north, which drains an area five times as large. Glaciers are some of the most efficient erosion mechanisms in the world, grinding the rocks that lie along their base into sand that is finer than flour. Here in the delta that riverborne silt and sand stretches out into the bay in curving, islandlike lobes of sediment that divide the river mouth into a maze of channels and marshes. Looking down at the delta below I could see the long-legged shapes of moose grazing out in the shallow marshes near shore. A flock of geese flew over the water in a perfectly shaped wedge. It seemed strange to be up in the air looking down on them.

Up ahead the spinning green and white beacon of the Cordova airport came into view. Aside from the scattered lumber camps along the coast, its light was the first sign of human life we had seen in nearly two hundred miles. By the time we parked the plane it was nearly midnight. As we drove back into town you could make out the dark shapes of moose feeding in the marshes alongside the road. In the distance between the black shapes of the trees you could see the glaciers rising up toward the sky, blue and white in the starlight.

· · ·

By 4:00 the next morning we were sailing up Prince William Sound toward Columbia Glacier. The sound is three times the size of San Francisco Bay and bordered by mountains well over a mile high. Glaciers flow down the sides of the peaks, to border the sides of bays and inlets with ice. The Alaskan oil spill had happened just five months before, but the only hint of it along our route was the promenade of oil tankers that wallowed up and down the shipping channel escorted by tugboats and clean-up barges—a belated effort at protection. The oil had all drifted elsewhere, fouling fjords and inlets to the southwest of us.

As we headed up into the sound the water was full of life. Rafts of sea otters floated on their backs a few yards away from the boat, using their fur-covered bellies as tables for shellfish gathered from the bottom below. Small icebergs drifted by, floating terra firma for harbor seals and seagulls. Islands lay scattered across the surface of the sound like small green jewels, covered with thickets of sitka spruce and hemlock. Bald eagles and ospreys perched in the tops of tall, dead trees. As we turned the corner of a channel we could suddenly see the face of Columbia Glacier, a wall of blue and white ice more than six miles wide and three hundred feet high rising up out of the water and reaching back into mountains that ringed the sound.

• • •

We dropped anchor off of Heather Island. In 1975 the front of Columbia Glacier reached all the way out to this island, but today its front is more than two miles away. Since the mid-1970s Columbia Glacier has been retreating by as much as a quarter of a mile per year. Ice from its rapidly retreating front breaks free and drifts out into Prince William Sound. The *Exxon Valdez* was allegedly trying to avoid drifting ice from the Columbia Glacier when it ran aground on Bligh Reef.

Today all that remains of the ice's former reach is its terminal moraine, now a shallow underwater bar. All but the smallest bergs from the glacier run aground on the submerged moraine, turning the water behind it into what is in effect an ice-choked lake.

It was late June but it had been cold enough last night for a thin layer of ice to form near shore. As we headed toward the glacier in a small zodiac we plowed through a thin skin of ice, landing on a narrow bar of sand on the glacier's eastern side. Behind the bar was a protected inlet of water filled with hundreds of icebergs so covered with dirt and gravel that they looked almost like floating stones. In the salt water of the open sound they would have melted long ago, but here in the protected pools of fresh water fringing the glacier, these trapped bergs tend to last for years, refreezing into a solid sheet each winter, the equivalent of pack ice in the Arctic. Sun and wind had carved them into fantastic shapes—they looked almost surreal, like the rounded, curving sculptures of Henry Moore or the figures of polar bears and whales carved by Inuit artists in the Arctic. Walking along the edge of the lake you could hear the groaning shift of berg against berg and the steady drip of water from melting ice.

Up close the faces of the icebergs were as varied as finely bedded pieces of sandstone. Within the ice were layers of varying texture and color. Some were as clear and smooth as glass; others were almost granular, cut by the faces of tiny crystals. The differences have to do with the movement of ice within the glacier and rarely correspond to the original snowfall. Here and there were small stringers of pebbles and layers of sand, the products and debris that fell on top of the ice as it moved downhill or was scraped up from below. In places those layers had even been folded and bent. Up close you could see variations in the color of the ice itself: it could be clear, white, gray, and even blue. While the gray ice was often due to ash and grit, the deep blue color seen occasionally in the largest icebergs was actually due to the color of the ice itself. Like water in the sea, the water molecules inside a chunk of ice reflect blue light. And where the ice is thick enough it can look as blue as water in the ocean seen while snorkeling or scuba diving fifteen or twenty feet below the surface.

The process of transforming snow into glacial ice can take serval years. Higher up on the snow fields and glaciers that lie amid the high peaks, so much snow falls out of the sky that it does not melt during the short, warm summers but lasts from year to year, collecting in drifts and piles that are slowly transformed into dense glacial ice.

New snow is lightweight. So much air is trapped in its intricate form that its density is often only one-tenth that of water. A bucket filled with ten inches of snow often weighs no more than one filled with an inch of water. By the time that newly fallen snow makes it through the summer melt season it becomes what glaciologists call firn. Over the course of that first year, the weight of newer snow and ice above has squeezed so much air out of this old layer of snow that it is often four to eight times heavier than when it first fell. More time is still needed to transform the firn into glacial ice. As the weight and pressure of the overlying snow continues to work on the firn, what were once individual snowflakes begin to flatten and merge together forming single crystals of ice that can be more than a foot long—the start of dense glacial ice. Bubbles of air squeezed out of the firn become trapped in ice, sometimes with pressures as great as 750 pounds per square inch. Several years or even thousands of years later when that ice begins to melt you can hear those bubbles of air break open—a chorus of crackles and pops known as bergy sizzle. On an icy inlet strewn with chips of glacial ice that bergy sizzle can fill the air with sound. Aside from these small bubbles of air, however, glacial ice is almost airless. Its density is nearly nine-tenths that of water. When icebergs break off or calve from the front of a tidewater glacier and float off in the sea, barely ten percent of their surface lies above water.

• • •

Near the center of Columbia Glacier's front or terminus, walls of ice more than two hundred feet high rise right from the water's edge. On the side where we had landed, however, that glacier had already retreated up onto dry land, leaving glacial moraine behind—an open landscape covered with angular stones the color of coffee with too much milk in it. I spent most of the morning walking along the edge of the ice with Bruce Molnia a glaciologist with the U.S. Geological Survey. As we walked, we talked about glaciers and moraines, stopping from time to time to study the features left by the retreating glacier.

The ice rose up from the ground in a gentle ramp and you could walk from the water's edge across the moraine right up onto the surface of the glacier itself.

Up from the shore it was remarkably quiet. A cool, stiff breeze was blowing off the glacier, but out amid the bare rocks of the moraine the wind moved without a sound. You could feel it brush against your face, but there were no rustling grasses or swaying trees to mark its path. The sky was clear and the sun was bright. Heat waves danced across the surface of the stones.

Out in the moraine you could see for miles. The terrain was open but not flat and featureless. In places its carpet of stones was cut by piles and mounds of rock. Staggered across its surface were a series of recessional moraines, piles of stone that had been pushed downhill in front of the ice and then left behind when the glacier retreated. Each successive ridge marked a periodic stand of the glacier as it withdrew toward the higher ground farther inland. Between them were other smaller features as well. Here and there eskers wound across the ground like giant mole tracks, low ridges of stone and gravel left by streams that once flowed beneath the ice. In places there were small mounds of rock known as kames that had collected in crevasses and holes in the glacier's surface, only to be dumped in a heap when the ice melted. Elsewhere were craterlike depressions known as kettles created by melting blocks of ice left by the glacier's retreat. Giant boulders and stones littered the moraine—glacial erratics—their path marked by flutes and grooves in the carpet of stones that covered the ground.

Out on the moraine it was not always easy to tell where the land ended and the ice began. Walking across what seemed like solid ground, the air would suddenly be filled with the sound of water running beneath your feet. Chipping away at the seemingly solid stones beneath your feet with an ice axe you would find that they were underlain with solid ice. The streams were all underground, running a labyrinth of pipes and mills within the glacier.

• • •

While the faulting and folding associated with plate tectonics is pushing the land up, glaciers and streams are slowly pulling it down. As glaciers move they carve the ground beneath them, carrying mountains of rock and sand downhill. Beneath the ice, rocks and stones are plucked out of the ground and slowly dragged downhill. That steady movement coupled with the weight of the overlying ice can grind the stones below as fine as flour. Along its leading edge a glacier pushed a mound of boulders and stones in front of it like the blade of a bulldozer. What drives these movements of ice is nothing more complex than gravity. The glacier's weight pulls it slowly downhill toward the sea. When glaciers retreat they leave a rock-covered plain behind—a glacial moraine.

The moraine is a barren world of stone and sand, as stark and lifeless as a salt flat in the desert. Here on the fringe of the Columbia Glacier, for example, the landscape is less than five years old. It has only recently been freed from the grip of the ice. In the space of an hour's walk the only plants or sign of life you will see are small bits of lichen and moss clinging to the sides of the rocks.

In midsummer the moraine's bare stones bake in the sun. In midwinter they are buried again with snow. And yet for all the harshness of the land here, the plants are tenacious and resourceful, gaining a foothold wherever it is offered. Less than a mile from the edge of the ice you can find flowers and small trees rising up from what was ice-covered ground barely a decade before. Walking back toward the beach we passed through a rolling terrain of kettles and kames covered with a carpet of fireweed and low green plants. The fireweed was in full bloom: its stalks of bright purple flowers were nearly waist high. Nested among them was a colony of glaucous gulls. As we drew closer they rose up into the air with angry cries, diving down toward our heads to ward us away from their nests. Looking closely among the stones and the fireweed you could see small trees: tiny sitka spruce barely an inch high, low clumps of alder and willow blown so flat by the wind that they seemed to crawl across the ground like vines. I moved farther away from the gulls and stood on top of a bare ridge of cobbles looking across the purple blooms of the fireweed toward the blue and white ice of the glacier. Out on the barren rocks a hummingbird flew by, bound for the fireweed up ahead.

• • •

After stopping for lunch at Growler Island a few miles away we headed back to the glacier, landing on its western edge where the ice runs right up against the solid rock walls of the neighboring peaks.

Although Columbia Glacier has been retreating by several hundred yards per year for the past several years, what is so interesting here on the western side of the glacier is how much the ice has thinned. As Molnia and I walked along the western edge of the glacier, you could see the trimline that marked the former height of the ice more than one hundred feet above our heads. Above it were thin stands of sitka spruce. Below it the rocks had been stripped clean and looked almost polished.

Since 1975 Columbia Glacier has retreated more than three miles. Glaciologists, Molnia says, believe that this retreat, spectacular as it is, could increase fourfold, releasing some fifty cubic miles of ice into Prince William Sound—enough to seal off the Port of Valdez. This rapid retreat has to do with more than just the slowly warming climate of the past seventy-five years. While it is true that a warming leads to a general retreat of the ice, superimposed on this melting are sudden surges and rapid retreats of the ice that have more to do with the stability and configuration of the glaciers than climate. In College Fjord, for example, two glaciers that sit side-by-side are moving in opposite directions: Harvard is slowly advancing while Yale is slowly retreating. And while the icebergs that calve off the front of a rapidly retreating glacier like Columbia Glacier can create serious problems for shipping, advancing ice can be no less problematic.

Two hundred and fifty miles from Prince William Sound, in Yakutat Bay, Hubbard Glacier has been slowly pushing out into the bay. In early 1986 it actually sealed off Russell Fjord, turning it into an ice dammed lake. By fall the lake was more than seventy-five feet deep. When the dam finally broke on October 8, the built-up water emptied into the neighboring bay in less than thirty hours. Hubbard Glacier is still advancing and glaciologists expect it to seal off Russell Fjord again in the not too distant future. For nearby rivers and streams the results could be disastrous, flooding them with a surge of water that could seriously damage the area's valuable salmon and steelhead fisheries.

Whether a glacier is retreating or advancing, ice is always moving downhill. For tidewater glaciers like the Columbia Glacier, however, there is a delicate balance to the speed at which the ice advances and the stability of the ground beneath it. Ice breaks up easily once it reaches open water, but if a glacier is firmly anchored to its terminal moraine it can actually be quite stable. If the glacier advances slowly, it can push its moraine out in front of it, keeping its base resting on the rocks below. If it advances out over the edge of its terminal moraine, however, the results can be catastrophic—"like pulling a stopper out of a bottle," Molnia says. The unstable ice begins to break up and in most cases it rapidly retreats all the way back up to bedrock. Once back on hard rock it can begin building out into the water again, but much more slowly than it retreated. While the glacier can lose several miles of ice in a decade, replacing it can take more than a thousand years. Although the movements of ice taking place at Columbia Glacier seem catastrophic on a human time scale, they have been happening here for thousands of years. "Glaciers," Molnia said, "have gone up and down these valleys hundreds of times in the past few hundred thousand years."

• ▪ ▪

When the first European explorers reached this ice-bound coast of Alaska, they assumed they had found the edge of the polar ice cap.

They did not realize that they were looking at alpine glaciers (they called them snow cliffs) that streamed out of the high coastal peaks, or that the Arctic was still several hundred miles to the north. Most, like the English navigator James Cook, were not looking for a new land to settle but for a northwest passage, a pathway through the Arctic seas to the Far East. Early explorers left little more than their names behind: Heceta Strait, Malaspina Glacier, Cook Inlet.

The first European to realize that the "snow cliffs" bordering the coast of southeast Alaska were actually alpine glaciers was Jean François Baloup de la Perouse, a Frenchman who sailed into Lituya Bay in 1786 and noted the glaciers that lay on the flanks of the surrounding high peaks. By the mid-1800s the Russians had established towns and trading posts in Juneau, Sitka, Ketchikan, and St. Petersburg. Alaska was the center of the Russian Empire in the New World. Maps and charts by the Russian-American Company and the Russian navy by this time now accurately marked the location of the region's so-called tidewater glaciers, those that reached all the way to the sea.

That era of Russian exploration and study came to an end when Alaska was sold to the United States in 1867. For the next several decades the region was all but forgotten. By the turn of the century, however, a host of new scientific expeditions were organized to study the region's coastal glaciers. One of the largest and most elaborate was organized by the railroad magnate and financier, Edward Henry Harriman, in 1899. Intending the trip as a vacation for his family, Harriman chartered the steamer *George W. Elder* in Seattle and then invited some of the most prominent scientists in America along to study the Alaskan coast. Included were the U.S. Geological Survey's Grove Karl Gilbert and the Smithsonian's William H. Dall. Seven years later Gilbert would lead the research team that studied the aftermath of the 1906 San Francisco earthquake. Dall would eventually lay the groundwork for the study of biology and geology in both Alaska and the Pacific Northwest. Edward S. Curtis, who would later win fame for his photographs of American Indians, was the expedition's official photographer. John Muir, the founder of the Sierra Club, was the ship's naturalist. In Prince William Sound they explored College Fjord, naming its glaciers for prominent East Coast colleges: Columbia, Harvard, Yale, and Smith. They also discovered a previously unknown inlet, which they named Harriman Fjord. In two months they logged more than nine thousand miles. Twelve separate volumes of studies were published as a result of the expedition.

Other groups and research institutes mounted their own expeditions to the area, including the Smithsonian Institute, the National Geographic Society, and the U.S. Geological Survey. Even G. Allan

Hancock got into the act in the 1920s (however, he ran his steamer the *Oaxaca* aground in the Wrangell Narrows).

· · ·

Strange as it may seem today, the notion that glaciers and ice actually move was once as controversial as Darwin's theory of evolution or Alfred Wegener's conception of continental drift. Sheets of glacial debris in Scotland and Switzerland were taken to be the residue left over from the Great Flood of the Old Testament. Glacial erratics, the giant boulders and stones carried by glaciers, were thought to be features that had been left behind by comets or even "devil's stones"—rocks scattered across the earth by Satan to trick God's people into questioning the Bible's absolute truth and its static view of the earth. When Harvard geologist Louis Agassiz published his theories on glaciation based on the studies of actual glaciers in Switzerland, his work was met with skepticism and sarcasm.

Today we know without question that ice has moved quite freely across both North America and Europe. Scattered across the United States you can see the same features that are found today on the edge of Columbia Glacier. At one time the sheets of ice that spread outward from the Arctic planed much of Ohio, Indiana, and Illinois almost flat, leaving behind a hummocky terrain of kettles and kames and terminal moraines. In Wisconsin and Minnesota, the grooves they carved in the ground became the basins for lakes and ponds. On the East Coast the reach of the ice was often marked by giant moraines—high, long mounds of stone and gravel like Cape Cod and Long Island. Today they are separated from the ice by several thousand years and several thousand miles. In Alaska, however, those processes are still taking place and the ice that created them is only a few feet away.

The ice ages that sent glaciers southward from the Arctic were the culmination of a general cooling trend in the earth's climatic history that began some thirty million years ago. The last three million years of that trend have been characterized by repeated cool and warm periods lasting anywhere from several hundred to several thousand years.

In the Pleistocene 1.8 million years ago, ice covered some 30 percent of the earth's surface compared to only 10 percent today. Glaciers once reached as far south as Kansas and the climate was much cooler. Driven southward by the advance of the ice, boreal forests of pine and fir like those found in northern Canada today were found as far south as Florida. At times, as has been mentioned earlier, so much water was locked up in ice that sea level was as much as three hundred feet lower than it is today. Broad reaches of the continental shelf were exposed. In the Arctic, North America and Asia were linked by dry land—a subcontinent known as Beringia that now lies beneath the Bering Sea.

No one knows for sure why ice ages happen. It has been suggested that changes in the earth's axis of rotation or its path around the sun may be responsible. That change in solar radiation, in turn, causes changes in the earth's weather systems. As snowfall increases, so does the earth's reflectivity. Instead of being warmed by the sun, the white mirror of the snow reflects even more of the sun's heat back into space.

These changes have been particularly rapid in the last few thousand years. During the Holocene, for example, the most recent twelve thousand years of the earth's history, there have been at least four separate periods of cooling and glacial advance. The most recent began in the early 1500s and lasted until the 1920s, a period of time that saw many of Alaska's glaciers advance. Six thousand years ago the climate was three to five degrees warmer than it is today.

The ice sheets in Alaska, however, were not linked to those that spread out from the Arctic. They flowed out of the high coastal mountains and reached only as far south as Puget Sound. Between these alpine glaciers on the West Coast and the continental ice sheets of the Arctic, the interiors of northern Alaska and northwest Canada were ice-free, characterized by broad grassy plains inhabited by woolly mammoths, camels, and horses that had migrated across the land bridge from Asia.

During the ice ages more than 50 percent of Alaska and parts of its continental shelf were under ice. Deposits of rock and glacial debris found offshore suggest that the coast may have been bordered by an ice shelf like the one that currently extends off the coast of Antarctica. By twleve to fifteen thousand years ago that ice had started to melt. Equilibrium with present sea level was reached just four thousand years ago, but the land here has not remained stable. Freed from the weight of the overlying ice, the coast of Alaska has been rising by as much as an inch a year in places—a process known as glacial rebound.

That movement of land in response to the retreat of the ice, however, is merely part of the story. The same forces that built up these coastal mountains are still raising them up. Glaciers are such efficient eroders that if the land here had not been continually uplifted it would have long since been planed flat. While the glaciers here slide down to the sea, the mountains are slowly rising up out of it.

• • •

"You can't really understand the glaciers of southeast Alaska without understanding the area's geology," said George Plafker. "The reason there are so many glaciers in southeast Alaska is the tectonic activity," he explained one fall morning at his office at the USGS western headquarters in Menlo Park. "Tectonic activity has built those mountains up so high that they form an almost impossible barrier to the weather systems blowing into Alaska. It's a real anomaly in Alaska that the farther north you go, the fewer glaciers you have and the farther south you go, the larger they are. The reason for this is that all of the precipitation is wrung out of the weather systems that blow through the area. In the interior of Alaska north of the Alaska Range glaciers are very rare. In fact, the north of Alaska is an Arctic desert by definition. The precipitation is less than eight inches per year."

The coast of southeastern Alaska and British Columbia is one of the most geologically active and complex areas in North America. As the Pacific Plate heads north toward the Aleutian Trench, it sideswipes the coast, piling up stray pieces of rock and bits and pieces of islands and oceanic plateaus on the edge of the continent—a maze of high, wet peaks and mountainous islands cut by channels and fjords. In the south this rugged terrain is covered by a thick coastal forest of spruce and hemlock. Farther north it is covered almost exclusively with snow and ice.

The amount of rain that falls along this mountainous northern coast is phenomenal. Places like Yakutat, halfway between Anchorage and Juneau, for example, receive more than 120 inches of rain per year, while the high peaks of the St. Elias Range behind it receive more than 300 inches of precipitation, most of it in the form of snow. Meanwhile, less than one hundred miles north of the peaks, the high plains and plateaus of the Yukon Territory in northwest Canada are drier than Los Angeles, receiving less than ten inches of rain per year.

• • •

Plafker joined the Alaskan branch of the U.S. Geological Survey in 1952. Except for a six-year stint in South America, he has spent his career working almost exclusively in Alaska. In the 1950s and 1960s working here was unlike working anywhere else in North America, with the possible exception of Arctic Canada. Much of the land had never been mapped, much less studied in detail. Intellectually and physically, Plafker and the handful of other geologists who came here in those early days had the land almost entirely to themselves. Work was remote. "It was kind of an ultimate Boy Scout–type thing," Plafker recalled. A plane or helicopter would drop them off in the bush with food and supplies and the assurance that someone would come back in a few weeks to pick them up. They worked frantically, covering as much ground as possible, up and over entire mountain ranges. "You had a real chance to know the country," he said. "Every foot of it. In great detail."

High up among the peaks in places like the Wrangell Mountains they would find the shells of clams and the remains of coral reefs. While the fossils left no doubt that the rocks here had been lifted up out of the sea, how those warm-water corals had gotten to the rim of the North Pacific was a bit of a mystery. Offshore on the floor of the sea other scientists were beginning to piece together the evidence that would lead to the theory of plate tectonics. But while that theory saw continents and ocean basins as mobile and active, no one had yet thought of applying that notion of motion to individual mountain ranges. Mountains simply rose and fell. There was no unifying theory to explain why it happened.

At the time, the lack of a theory didn't bother Plafker at all. "You know we geologists are very imaginative," he said. "As one geophysicist told me, 'we can find a trend in a keg of nails.'" The 1964 Alaskan earthquake, however, would suddenly make him see things in a different light.

•••

"No one had ever seen an earthquake on that scale," Plafker explained. Registering an estimated 8.4 on the Richter scale, it released ten times more energy than the 1906 San Francisco earthquake and was felt over an area of more than half a million square miles. The quake had occurred along what was later proved to be a subduction zone running along the coast of southern Alaska—although no one knew that at the time. Fault movements extended for hundreds of miles. In places, coastal terraces shot up out of the ground by more than twenty feet. Plafker, who had recently been put to work on studying the offshore geology was quickly reassigned to the quake. "I had no real background in earthquakes certainly, and I don't think anyone else in the branch did either, but we just got involved, and, as happens very frequently, emerged as experts," he said. After the 1989 Loma Prieta earthquake, Plafker would play a key role in writing the USGS's summary report of that damaging quake as well.

From the standpoint of plate tectonics, the timing of the 1964 Alaskan quake was critical. While marine geologists talked about spreading ridges and magnetic anomalies on the seafloor, geologists on land at the time were looking for some concrete proof of large-scale plate movements—some dynamic sign that such large movements were possible. The massive quake in Alaska provided it in graphic detail. "It really was a very constructive laboratory experiment," Plafker said.

For Plafker the quake and the credence it lent to plate tectonics planted a seed as well. It gave him the idea that the rocks he had been studying were not necessarily in place, that they could have moved not just vertically, but horizontally as well. Definitive proof of that idea would eventually come from samples of volcanic rocks in the Wrangell Mountains first collected by Plafker and his co-workers. Patterns of magnetism locked in the volcanic rocks of the Wrangell Mountains suggested that they had originally been deposited several thousand miles farther south, possibly as far away as Peru. Their movements, however, were not linked to those of the continent as a whole. Instead they seemed to have moved across the ocean independently—a small, mobile island of rock. In fact, it seemed that the Wrangell Mountains had actually been islands at one point in their history. Although its rocks were more than two hundred million years old, they had only collided with the continent some fifty to sixty million years ago. Plate movements had carried not only continents and ocean basins around the world, but also volcanic islands and oceanic plateaus: pieces of rock and debris scraped from the floor of the sea. It was plate tectonics on a microscopic scale.

■   ■   ■

Geologists call these small, mobile blocks of rock "terranes"—a deliberate misspelling of the word terrain meant to signify a collection or region of rocks believed to have a similar history of origin and movement. Terranes that have moved are referred to as exotic terranes. Those that may have moved are referred to as suspect terranes. Today almost everything in Alaska south of the Denali Fault Zone is thought to be exotic.

On the wall of Plafker's office at the USGS is a large geologic map of Alaska. Different terranes are represented by different colors and at the moment, Plafker's map is a collage of fifty different colors that looks like a piece of abstract art. In the future, Plafker says, geologists may be able to trim that number down to just six or seven different terranes as they unravel the links between different pieces and packages of rock that now seem distinct. Deep-sea sediments surrounding the volcanic rocks of an ancient island arc, for example, may be part of a collection of rocks that moved as a single block: island, seafloor, and all. But at the moment the relationships are so complicated by folding and faulting that geologists are still struggling to understand them.

That notion of movement and uncertainty has changed the face of Alaskan geology. Plafker explains: "In Alaska when we started out we would just map out a little area and worry about that particular place and that was just great. It was simple. Now things have changed to the point where almost every little thing you look at you have to think, 'Well, how does this fit into the whole context of North American geology?' You have to concern yourself with the slightest detail: whether what you are looking at is really a piece of something local in Alaska or a piece of Baja California or maybe something even more distant. It's certainly been a revolution in the way of thinking about rocks. Of course the rocks don't change. That's the nice thing about them. It's about the only real factual thing we have."

•   •   •

Tents, stoves, sleeping bags, and rock drills were piled in the hallway in front of Jack Hillhouse's office at the U.S. Geological Survey in Menlo Park. In a few hours he would be heading off to Nevada to look for signs of moving rocks and continents. Hillhouse is one of the USGS's leading paleomagnetic experts—a paleomagician to his colleagues—capable of deciphering whether the rocks underneath your feet, for example, originated in Palo Alto or Peru. Using techniques similar to those used by scientists studying the patterns of magnetic anomalies on the seafloor, Hillhouse has helped trace the paths of islands and mountain ranges around the earth, the details of exotic terranes.

The origins of the earth's magnetic field are as much a mystery as the origins of gravity. While scientists know that the magnetic field comes from the liquid portion of the earth's core and that it is some kind of dynamo, details deeper than that are hard to come by. Apparently that field is also somehow related to the earth's rotation. While in line with the earth's magnetic field wanders, on average it stays near the poles, the earth's axis of rotation. The only other planets that have magnetic fields are those which rotate and have a metallic core. Both Mars and Jupiter have magnetic fields. The moon, while it has no magnetic field now, apparently had one in the past. Whatever lies behind its existence, however, the earth's magnetic field has been a useful tool for modern geologists, guiding them through the maze of plate tectonics as surely as it guided ancient mariners through the seas.

As molten rocks cool, the orientation of the earth's magnetic field is locked inside them. Particles of iron and other magnetically susceptible minerals inside the molten rock act like tiny compass needles, orienting themselves in line with the earth's magnetic field. Not only does the declination or direction of these tiny particles point to the North Pole, their inclination or tilt from horizontal also records the rocks' location with respect to latitude. The earth's magnetic field curls around it, diving almost straight into the earth at the poles. Near the equator the lines of force from its field are almost horizontal to the earth's surface. As one approaches the poles, however, they become more and more vertical, eventually pointing straight down into the earth. The fact that rocks could record this inclination of the earth's magnetic field as well as the orientation of its poles has enabled paleomagnetic specialists like Hillhouse to trace the paths of rocks and mountain ranges around the globe.

•   •   •

After getting his Ph.D. from Stanford in 1975, Hillhouse headed a few blocks down the road to the USGS to go to work. Originally he was supposed to work on the stratigraphy, or distribution and deposition of rocks, from the Quaternary era, the most recent 1.8 million years of time. When the geologist he was working under, Davey Jones, became interested in the Wrangell Mountains of southeast Alaska, however, Hillhouse suddenly found himself trying to unravel the paleomagnetic details of rocks more than 200 million years old.

As coincidence would have it the first rocks he went to work on were some samples brought back from the Wrangell Mountains by George Plafker, Ed MacKevett, and Richard Dole in the early 1960s. Dole, the co-founder of the USGS's paleomagnetic laboratory, had been hoping that the greenstones they had picked up in southeast Alaska would help him unravel the details of the earth's magnetic field during the Permian some 260 million years ago. When he analyzed the rocks in the laboratory, however, the results made no sense at all. They suggested that this stretch of Alaska had once been several hundred miles farther south. The results were so out of step with the conventional wisdom of the day that Dole decided they were better off forgotten and put them on the shelf.

A little more than a decade later Jones put Hillhouse to work on the same rocks. Laboratory techniques had improved dramatically over the intervening years, but when Hillhouse tested the rocks he came up with the same results as Dole had more than ten years before. The data, Hillhouse said, suggested that the Wrangell Mountains had moved a lot. But how much was uncertain. "We could only determine that the latitude was low, fifteen or twenty degrees. But we didn't know whether it was north or south." Apparently the Wrangell Mountains had drifted to southeast Alaska from as far away as Mexico and possibly as far away as Peru. The only thing that was certain was that by fifty million years ago those rocks had collided with the coast of southeastern Alaska.

After his initial lab work in 1975, Hillhouse spent the next six years from 1976 to 1981 working in Alaska on a regular basis with Jones and a collection of other mobility-minded geologists: Sherm Grommé, Norm Silberling, and Peter Coney—all of whom would eventually play a major role developing the idea of exotic terranes. They worked from hand to mouth. "Money was tight, as it often is at the survey (USGS)," Hillhouse recalled, but there were advantages to being part of the largest collection of professional geologists in the world.

When they needed money to keep them going they went from door to door at the survey's sprawling campus of offices in Menlo Park, "tin-cupping," as Hillhouse called it, begging for money from the survey's more established scientists in order to get themselves up to Alaska. Once up in Alaska, that process shifted to begging for helicopter time to get up into the mountains and do their fieldwork.

Paleomagicians like Hillhouse do not merely knock off a few handfuls of rock with a sledge hammer. They carefully drill a series of holes in the rock taking out cylindrical core samples whose location and orientation are precisely recorded. In the six years he spent working in Alaska, Hillhouse left small cylindrical holes the size of a 12-gauge shotgun shell scattered all over the Wrangell Mountains. In time Hillhouse's work and that of the rest of the group would add a whole new layer of complexity to plate tectonics. It was no longer enough to worry about the age of rocks, one also had to think about where they had been and how far they had traveled. The Wrangell Mountains, they found, had come from at least as far away as southern California. Everything was suddenly relative. It was quantum mechanics for geology, and the Wrangell Mountains of southeast Alaska were right at the center of the story.

In time the term Wrangellia was used to collectively refer to the exotic, displaced rocks of British Columbia and southeast Alaska. The Wrangell Mountains, Hillhouse says, became the classic illustration of this new perception of small-scale plate tectonics not because they were the first area studied, nor because the paleomagnetic evidence was so strong, but because so many facets—the rocks, fossils, faults, folds, and paleomagnetic record—all came together to create a body of evidence that was virtually undeniable. "Often when you have a good paleomagnetic story it doesn't have good geology surrounding it. People say, 'Oh that's impossible,'" Hillhouse said. "But there were real good geologic reasons for Wrangellia being out of place." Like the features of the Atlantic ocean that "proved" plate tectonics—the fit between continents, the pattern of magnetic anomalies on the seafloor, and the correlation of rocks on opposite sides of the ocean basin—it was the combination of things that was so important.

• • •

When David Howell started college the plate tectonics revolution was just beginning. It was an exciting time to be studying geology and Howell was looking forward to being part of it. Ideas and theories changed almost overnight. "We had all these uncertainties," he recalled. In 1969, however, all that changed when he was drafted for the Vietnam War. "When I came back plate tectonics was well entrenched and seemed to explain everything," he said. "I can remember thinking, 'My God. In ten or fifteen years we're going to solve about every problem there is.' And I was depressed."

When he started to work on his doctorate degree at the University of California at Santa Barbara, the topic of his dissertation was broad: basin rocks from the middle Eocene of both southern California and northern Mexico—the same fifteen- to twenty-million-year-old basin rocks that held Los Angeles's phenomenal oil deposits. In loose terms his field area covered several thousand square miles. The topic was enough to keep most university geology departments busy for several decades, let alone a single graduate student. For Howell, however, it was a formative experience—it taught him to think regionally and on a large scale. And after studying the complex faulting and folding of these basins and finding that these relatively young rocks had often been displaced by several hundred miles or more, he began to believe that perhaps the revolution he had been hoping for had not passed him by after all.

"I got all excited in the late 1970s when I realized that the simplistic model of plate tectonics wasn't explaining anything. We had all these other aspects. It was much more dynamic," he said. It was an idea he never forgot.

The problem, he says, was really one of scale. On a global scale the simplified theory of plate tectonics works quite well. "If you're at sixty-two thousand feet the geology is pretty simple and straightforward," he said. "This plate is going this way and that plate is going that way. If you get down on the ground and start wandering around you can see that this is indeed the case. But what happened to this particular piece right here is not so certain. Did it go this way or that way? Is it opening or closing? Did it rotate? When we really looked more carefully at the stratigraphic relationships we realized that there wasn't a one-to-one correlation between plate tectonics and local geology."

Instead of merely thinking about the age of rocks, one had to realize that they were often taking shape and moving at the same time. Often several processes were taking place at once. Subduction might form an island arc, but later that same arc might be broken up by faulting and spread out over several thousand miles of coastline. If you didn't have a flexible perspective of rocks these kinds of features and processes were virtually impossible to comprehend.

Today Howell is a geologist at the U.S. Geological Survey in Menlo Park and one of the world's leading thinkers on the evolution of continents and ocean basins. Over the past ten years he has played a key role in developing the realization that the earth's rocks are much more mobile and active than anyone had previously thought. He has been a leading proponent of the idea of suspect terranes, the possibility that small pieces of rock and seafloor have moved as readily as continents and ocean basins.

At an Audubon Society picnic at Bodega Bay, Howell and Peter Coney coined the term "tectonostratigraphic terrane." The difference in spelling with -*ane* instead of the more traditional -*ain*, as previously mentioned, was meant to emphasize the special geologic significance of the term. Not merely a terrain in terms of the physical landscape, but a terrane in terms of its history of movement and the processes that created it. It was a term they created out of necessity. "We had this problem of belts and fragments and blocks and zones and provinces: all these words. Geologists are constantly coming up with terms because we're so hamstrung trying to communicate kinematics and geographic features with words," he said. "It's really a visualization. You have to have a strong left brain to visualize things, but writing is a right-brain thing. Well we're horrible at it, but it's because we're trying to convey what are really substantive concepts, but the nomenclature always gets in the way."

While the concept of terranes was originally limited to southeast Alaska and British Columbia in the 1970s and early 1980s, today geologists in North America have come to believe that everything west of central Nevada is suspect terrane. While there is an overwhelming body of geologic evidence to suggest that this western edge of North America was only recently added to the continent, one of the other key pieces of evidence centers around the geochemistry of the region's volcanic rocks. Like paleomagnetism, the story lies in the details and in this case, details center around the ratios of two strontium isotopes found in volcanic rocks.

A host of radioactive elements are found naturally in the earth's crust, usually in small concentrations. In the case of strontium, there is a specific ratio between two of its naturally occurring isotopes, strontium 87 and strontium 86 (the number refers to the number of protons and neutrons inside their nuclei) that signals whether a rock's origin is oceanic or continental. As it turns out, rocks from a mid-ocean ridge that have had no contact with continental rocks at all have a strontium 87/86 ratio of .0702, while those that have come up through continental rocks have a ratio .0706 and often higher. It is an extremely delicate test, but also an extremely accurate one. "Geochemists," Howell said, "can contour it on a map." By measuring the strontium ratios of basalts and other lavas they can track what Howell calls the "cryptic edge of North America," the boundaries of the continent's ancient core or craton. "Everything west of .0706 is suspect terrane."

With the term suspect terrane now in use by geologists around the world, Howell finds himself fighting intellectual battles on both sides. With the rocks freed from their solid roots, some geologists have tried to move mountain ranges and pieces of coastline wildly across the face of the earth. At the same time, others stubbornly insist that aside from the large-scale features of continental drift, the rocks have hardly moved at all. "The term suspect terrane doesn't mean that everything came from China," Howell explains. "All it means is that there is some uncertainty with regards to its history and movement."

While the western edge of North America is suspect, Howell believes that only a small portion of it, perhaps only 10 percent, may actually prove to be as exotic as the Wrangellia terrane of southeast Alaska and British Columbia. At the same time, he believes it is important to keep that possibility of motion in mind.

"Most paleomagnetic studies indicate that there has been at least one thousand to two thousand kilometers of northward—poleward—translation in the last two hundred million years," Howell said. "Many geologists, because they're so conservative and worked in one area and think that they've got it all tied together, say 'My God, that's just not reasonable,' if someone tries to tell them that something may have moved one thousand or two thousand miles.

"But of course it's reasonable. If you take a very common rate of plate motion, even a slow rate, say just three centimeters per year and keep at it for eighty million years that's twenty-four hundred kilometers of movement. That's an airplane ticket. It's just a problem we have with relating to the consequences of very slow motions for a very long time. Most of us have a hard time grasping fifty to one hundred to two hundred million years. How long is that? It's really hard to understand how long one hundred million years is. You just have to play with these slow rates over and over again in your head until you finally realize that two thousand kilometers of movement is nothing."

• • •

It was a clear day in late September. The monsoon season, as Jim Monger jokingly refers to the wet Vancouver winter, had not yet arrived. Monger is a geologist with the Geological Survey of Canada. From the windows of his office in a high-rise office building on the edge of downtown Vancouver, you can see Mount Rainier 175 miles away. A dormant volcano, its conical shape rises nearly 3 miles into the sky. From here in British Columbia it seems to float over the soft, velvety layer of smog that marks Seattle. Other volcanoes dot the view to the south as well: Mount Baker and Mount Adams and the peaks of the Cascade Range further east. Between these volcanoes and the tight knot of the Olympic Mountains along the coast, however, the landscape is remarkably flat. A few thousand years ago ice from the wall of high coastal peaks that border the coasts of Alaska and British Columbia reached as far south as Puget Sound and planed the landscape almost flat.

From the other side of the building the view was entirely different. North of Vancouver mountains rise right up from the water's edge without prelude or preamble, the start of a chain of high coastal peaks that runs all the way to the tip of the Alaskan Peninsula more than a thousand miles away. There is no flat coastal plain here, only the mile-high granite peaks of the Coast Ranges and the volcanic cindercone of Mount Garibaldi just outside of town. On its flanks you could see fields of snow and ice. British Columbia is the start of a different world.

"You cross a fairly major structural boundary here—into rocks which you don't see farther south," Monger explained. "Unfortunately it happens right near the international boundary. People think, 'Oh it's just the geologists.' But it isn't really. The geology does change in this area and the change from Franciscan-type rocks to Wrangellian-type rocks happens right at the southern edge of Vancouver Island. To the south the rocks are young and typically related to the Franciscan Formation in one way or another and are often less than thirty million years old. To the north they are Wrangellian and not infrequently three hundred million years old."

North of Puget Sound the western edge of North America has been built up by pieces of the ancient seafloor that have often drifted for hundreds and thousands of miles. While the Pacific Ocean is several hundred million years old, the seafloor off the West Coast of North America is no more than fifty million years old. Those ancient rocks have all disappeared, subducted beneath the edge of the continent or smeared along its edge.

If you want to know what that ancient seafloor looked like, you need look no further than the maze of islands and oceanic plateaus scattered across the South Pacific. Unlike the Atlantic, the floor of the Pacific is asymmetric. The spreading ridge that runs through it lies not in its middle, but along its edge. Off the coast of Asia the seafloor is some one hundred million years older than it is off of North America. Features like those that collided with the coast of North America to build the high peaks of Alaska and British Columbia still fill the South Pacific—chains of volcanic islands and coral reefs, a rich tropical sea.

"Over the past two hundred million years more than ten thousand miles of oceanic crust has been thrust under the edge of the continent," Monger said. The western edge of the Pacific has gone through the Pacific "like a big wind-screen wiper," he explained, scraping islands, seamounts, and submarine plateaus from the floor of the sea.

The process of accretion, however, did not happen in a single instant or involve merely the collision of a single terrane or island chain. It spanned several million years and several dozen terranes—fragments of continents and seafloors that drifted across the face of the earth to collide with the continent.

■   ■   ■

If you could go back some six hundred million years in time to the start of the Paleozoic, you could stand on the western edge of North America and watch the land take shape. Here in Canada, however, you would not be standing in British Columbia, but several hundred miles farther east in Calgary, according to Monger. The rocks that would build the coasts of British Columbia and Alaska were still several hundred or even several thousand miles away in the Pacific. Some may not have even appeared yet on the earth's surface but were still taking shape as islands and oceanic plateaus. For the first four hundred million years almost nothing would happen. The edge of the Pacific was as placid and passive as the Atlantic is today.

By two hundred million years ago, about the time that the first dinosaurs appeared, the first terranes would begin to arrive, appearing over the edge of the horizon and then slowly colliding with the edge of the continent to build up high mountains—the Cassiar, Monashee, Kootenay, Yukon-Tananan, Slide Mountain, Quesnellia, Cache Creek, Stikinia, Alexander, and Wrangellia terranes. Their arrivals would come at various times: the Alexander Terrane some seventy million years ago; the Wrangellian Terrane just fifty million years ago. It is important to distinguish between the age of the rocks and their time of arrival.

While these rocks were colliding with the continent, their roots, the deeper rocks below, were often being subducted beneath it. From time to time volcanic lavas and bodies of molten granite would rise to the surface, stitching the arriving blocks of terrane together with flows of volcanic rocks and molten bodies of granite. Today British Columbia has one of the largest belts of granitic mountains in the world, part of the same granitic rocks that make up the Sierra Nevada Mountains of California.

Ice would be the last to arrive, appearing some three million years ago as the climate gradually cooled. You could watch it slide out of the snow fields and ice fields nestled in the high peaks and spread out along the coast, carving fjords, deep bays and U-shaped valleys. By ten thousand years ago that ice would begin to retreat back up into the mountains, leaving sheets of bare rock and loose stones behind. Plants would begin to move in. The transition from bare rock to forest would happen so quickly that the trees would seem to spring up right out of the rocks. Mats of lichen and moss would cover the rocks, preparing the way for the flowering plants and low scrubby trees that would evolve into a dense coastal forest. Freed from the weight of the overlying ice the land would begin rising up out of the sea, like the freshly worked dough for a loaf of yeast bread left near a warm oven. Looking north you would see a chain of small islands stretching all the way to Alaska, where the rivers of ice still run to the sea.

• • •

Mendenhall Glacier is less than fifteen miles from the center of Juneau, Alaska. A narrow stream of ice reaching down from the Juneau Icefield, its terminus is more than a mile wide and two hundred feet high. It reaches almost all the way to the water's edge, separated from Auke Bay and the sea by a narrow *U*-shaped valley carved by its moving ice. Readily accessible by road, it is a popular day trip for tourists who flock to Juneau each summer on the armada of cruise ships that ply the fjord-laced coastline of Alaska.

"To visitors it looks pretty stable, but to people who live here it's really quite dynamic," said Rita O'Claire, as we stood near the small visitor's center overlooking the glacier. "For the past ten years it has been retreating at a rate of ninety feet per year. When I first moved here the ice reached all the way to that waterfall," she added, pointing to a stream of water cascading down the side of a high cliff nearly a quarter mile from the front of the ice.

What is captivating scenery for tourists, is an outdoor laboratory for Rita O'Claire, a professor of biology at the nearby University of Alaska at Juneau, a place to bring her students to see glacial plant succession firsthand. The shape of the landscape here has as much to do with the movements of plants as the movements of rock and ice. When the ice retreats the plants move in, changing the face of the landscape. At the close of the ice ages plants reshaped the glacial landscapes of Massachusetts, New York, Ohio, and other eastern states. Here in Alaska those processes are still taking place.

At the head of the Mendenhall Valley the retreat of the Mendenhall Glacier has left a narrow glacial moraine behind. The glacier itself is bordered by a small lake. The shore of the lake is bare silt and gravel. Less than a half mile away, however, what was barren ground less than thirty years ago has been transformed into a forest of sitka spruce and hemlock. Following the winding trail that leads from the edge of the lake and into the neighboring forest you can see several decades of plant succession in the space of a few hundred yards. As we walked down the trail across the moraine, O'Claire identified the plants and trees, describing how they had shaped the landscape.

• • •

The rock and silt left by a retreating glacier is one of the most barren landscapes on earth, devoid of both soil and nutrients. Its surface is alternately hot and cold—covered by snow and ice during the winter and baked almost dry during the long summer days when the daylight is almost endless. The first plants, however, arrive almost as soon as the ice leaves—tiny lichens that anchor themselves to the sides of rocks and begin to grow. They materialize out of thin air, carried by the wind and the rain. Lichens are simple plants, produced by the combination of algae and fungi. They live symbiotically, the algae providing food and energy from the sun and the fungi anchoring them both to the rock. Algae are carried to the moraine by rainwater; the spores of the fungi are carried by the wind. Scientists dragging fine nets behind planes have collected spores of fungi at ten thousand feet. "If they can get up to ten thousand feet, they can travel around the world," O'Claire says. The reaction between algae and fungi that creates a lichen can occur in the space of just three hours. With more than eighteen thousand species of fungi and forty genera of algae, the possibilities are almost endless.

A few yards away from the lake the ground is already covered by low scrubby willows and alders. But lichens are still very much a part of the landscape and here and there O'Claire picks up small stones and points out the lichens: small splotches of color—lime-green, yellow, rust-red, and black—that covered their sides like a chipped layer of paint. Here and there were reindeer lichen, map lichen, and daisy lichen. They seemed to grow almost everywhere, even on top of one another. Pushing aside the leaves of a low willow trailing across the ground we found them covering the rocks beneath it. "No fresh surface lasts for long out here," O'Claire said.

Lichens are pioneers, the first plants to arrive on the barren landscape. They need no soil to grow in, but get the nutrients they need right out of the air, fixing nitrogen out of what is readily available in the atmosphere (the air we breath is 78 percent nitrogen). In fact lichens are such efficient gatherers of things floating in the air that the results can be almost deadly. Widespread atmospheric testing of nuclear weapons in the 1960s, for example, released huge amounts of radiation into the atmosphere. In the Arctic lichens and mosses growing on the tundra collected so much of that radiation that they caused a dramatic increase in cancer among the Inuit and other native peoples of the region. In the late 1980s fallout from a massive leak at the Chernobyl nuclear power plant in the former Soviet Union was also collected by mosses and lichens. When reindeer, herded by Laplanders and other native groups, grazed on those plants, they too became contaminated by radiation. Ultimately, tens of thousands of animals had to be slaughtered. Aside from the these man-made disasters, lichens are invaluable plants in this icy landscape.

Once lichens begin to establish themselves, they begin the slow process of building up soil. While conventional wisdom says that they accomplish this task by breaking down rocks with secreted acids and other chemicals, O'Claire dismisses that idea with a laugh. "Baloney," she said. "That may be true in areas with limestone, but not here in Alaska where we have hard silicate rocks." Instead of breaking down rocks, O'Claire says, lichens build up the ground piece by piece, collecting bits of windblown dust and silt to create a paper thin layer of soil. That collection of debris is not much, but it is enough to pave the way for mosses that make up the next phase of plant succession.

Like the lichens, mosses were plentiful here near the lake as well, covering rocks here and there like a layer of green felt. As we headed down the trail, O'Claire called out the names of passing varieties like someone seeing old relatives at a family reunion: "Fern moss!—Look, club moss!" "Mosses are really nothing more than low-lying plants," O'Claire said, hefting a grapefruit-size chunk of granite covered with lichens and club moss. Telling me to look at the moss closely, O'Claire pointed out the tiny clublike heads that gave that variety of moss its name. Up close they looked almost like a carpet of tiny trees. Small stalks of club moss seemed to sprout up right out of the rock, ending in a tiny club-shaped head. The analogy is particularly fitting. Three hundred and fifty million years ago in the Carboniferous era these tiny club mosses were as tall as trees. Today they are barely two inches tall.

Because of their greater thickness, mosses are even more efficient gatherers of soil and nutrients than the lichens which precede them, trapping bits of dead lichens, sand, and silt in their soft green surfaces. Eventually they change almost every aspect of the land they settle: acidity, nutrient levels, soil thickness, even temperature. It can be as much as twenty degrees (Fahrenheit) warmer inside a carpet of moss than on the bare rock surrounding it. That warmer temperature makes it easier for the windblown seeds of flowering plants like fireweed and lupine to germinate.

Alongside the trail you could see how successful those flowers had been at taking root in the mosses below. Here and there we passed knee-high thickets of fireweed and lupine. There had been a light rain that morning and drops of rain had gathered in the center of the lupine's clustered leaves where they glittered like small, single diamonds. By the time these flowering plants arrive, O'Claire said, the plant succession from bare rock to evergreen forest is nearly half-complete. Like the lichens and mosses, they too fix nitrogen in the soil, building it up even further with the debris derived from the decay of their own leaves and stalks.

• • •

Trees follow closely on the heels of the fireweed, O'Claire says. But it is interesting that while fir trees are the dominant trees in Alaska, the first trees to arrive are not pines, but deciduous trees like willow, alder, and cottonwood. Willows are the first to arrive, but they rarely do well right away: exposed on the bare rock and blown by the wind, they crawl across the ground like a vine. You could see them scattered across the rocky ground of the moraine. As we huddled over one studying its low shape, O'Claire moved its branches aside and pointed out the litter of leaves and twigs piled around its trunk. "Look at all the leaves it's catching," she said. "If it didn't, it wouldn't have anything to grow in."

A few yards ahead the trail disappeared into a thicket of fifteen- and twenty-foot-high alder trees. The trail was suddenly rough and bumpy. "Notice how all the tree roots are right up at the surface. There's no need to go down deep here. There isn't any soil below. You might as well be up where the action is," O'Claire said. Fingering the yellow, shriveled leaves of a tree, she added, "A lot of plants are trying to tell you something: they're trying to tell you that they're starving to death."

"Alder is a really key guy up here," O'Claire said as we walked on through the trees. They have nitrogen-fixing bacteria attached to them, leather lichens growing on the sides of trunks and twigs. Where they overhang streams you can see a surge in nitrogen levels in the water when their leaves fall. Their leaf litter also helps build up the soil quickly, and once the alder establish themselves, the cottonwoods and willows begin to grow rapidly.

Their success, however, is short lived. Like the lichens and mosses they are soon overshadowed by other plants that take advantage of the changes they have made to the landscape and the layers of soil they have built up. While the cottonwoods and alders thrive, the sitka spruce that will overtake them are already waiting patiently beneath them—biding their time before they reach up to the sun.

Spruce trees arrive relatively early on the scene after the retreat of a glacier, but it takes several decades for other plants to prepare the ground for them. Back among the lichen and moss-covered rocks closer to the lakeshore we had found small spruce trees sticking up out of the ground—little more than small nubs an inch or two high. Wind plays havoc with the small trees. Chips of windblown ice and snow slice off their tops and small branches. Once the willows and cotton-

woods begin to climb over the landscape, however, the spruce are typically well-established too. Relatively shade tolerant, they can remain
beneath the leafy canopy of the trees that precede them for decades,
growing slowly and persistently.

Once they finally rise above the cottonwoods and willows, however, they begin growing rapidly, shading out the leafy trees below
them in a matter of years. The surge of growth is so rapid, O'Claire
says, that you can see it in the trunks of cut spruce trees—dozens of
tightly spaced rings and then a sudden surge of growth marked by fatter and fatter rings as the trees race toward the sun. It can take them
forty to fifty years to dominate the forest.

As we passed into a dense grove of forty- and fifty-foot-tall sitka
spruce, the day seemed suddenly darker. In contrast to the scrubby
plain of lichen and moss-covered rocks that had bordered the lake, few
plants seemed to grow here beneath the spruce trees. Sounds were
muffled by the needles carpeting the ground. With the arrival of the
spruce, O'Claire said, the cycle is nearly complete. But while the
ground underneath the trees seemed barren there is what O'Claire
calls a whole "underground economy" going on within the soil. Like
the fungi that link up with algae to form lichens, certain fungi also live
within the roots of trees under a similar plan of mutual assistance.
While the fungi fix nitrogen and phosphorous in the soil for the trees,
the trees in turn provide the fungi with sugars and vitamins. For the
fungi the role of a single cell of algae has been taken over by a massive
tree.

Even now the forest is not quite finished developing. For all their
imposing height, spruce are still not the dominant tree in the forests of
southeast Alaska. That distinction belongs to the hemlock, a tree even
more shade tolerant than the sitka spruce. Hemlock have not yet dominated the forest here near the Mendenhall Glacier, but they are already
visible. Looking under the sagging boughs of a spruce, O'Claire found
one lurking in the shade. "Ah there's one," she said. "A rotten, sneaky
hemlock tree."

Sitka spruce could dominate the forest here forever except that
they do such a good job of shading the forest floor that they shade out
their own seedlings—ultimately bringing on their own demise. Hemlocks, however, need even less sun than the spruce. So while the spruce
slowly die off, the hemlocks begin to rise up and take over the forest.
The tree O'Claire had found was barely four feet tall, but a full-grown
hemlock can be as much as one hundred feet tall with a light lacy
foliage that resembles that of a redwood tree. Unlike the forests of
spruce, cottonwood, and alder that preceded it, the hemlock does not
bring about its own demise. Once the hemlock trees take over, other

shade-tolerant plants move into the forest floor to complete this stable, climax forest: dwarf dogwood, lace flower, and feathery moss, the archetypical coastal rain forest of British Columbia and southeast Alaska.

The progression from bare glacial moraine to coastal rain forest typically takes from two hundred to three hundred years. In theory such forests could dominate the land here entirely given enough time, but in reality storms and disease continue to topple stands of trees here and there, starting the cycle of plants over again.

While this climax forest is the product of a two-hundred-year progression from simple mosses and lichens to tall stately trees, it is something of a paradox that the most dominant plants of the coastal rain forest in terms of total biomass are not the hemlocks, O'Claire says, but the layers of lichens and moss that cling to their uppermost branches of the trees. After starting the whole process by building soil and capturing nutrients on bare rock, they climb up into the tops of the trees, hanging down from branches and twigs, drawing energy from the sun and raining it down on the plants below.

By the time we finished our short walk through the moraine and the neighboring forest it was nearly 8:30 P.M., but in early August the day was still bright, even with the low, wet clouds that hung overhead. As we headed back to the parking lot, O'Claire wandered off the trail looking for a rare Norfolk pine. Originally from Norfolk Island off the eastern coast of Australia and often used today as an ornamental shrub in the southeastern United States, its presence here on the edge of a glacier in Alaska is something of a mystery. "I like this tree," O'Claire said. "It's reminding us that we don't know everything."

•   •   •

Out on the Taku River the fish wheel turned, spinning in the current like a waterwheel. It looked almost like a child's overgrown toy. Giant net baskets more than six feet high dipped into the gray silty water of the river. Mounted between two floats in the river, there is no mill or generator attatched to its spinning axle. It seems to spin without any rhyme or reason. But every third or fourth turn of the wheel seems to catch a salmon. As the basket reaches the top of its arc, the salmon falls down toward the wheel's axle where a carefully contrived system of chutes and boards channels it to a holding pen alongside the floats. Taggers then pull the salmon out of the pen with a net and dump it on a table where they try to hold the thrashing fish still and weigh and tag it before tossing it back into the river to continue its trip upstream. The work is hard. The fish are two, three, and almost four feet long in some cases—thirty pounds or more of solid, agitated muscle. Some leap right off the table back into the river as soon as they are pulled out of the holding pen. A day of wrestling with salmon can leave your arms black and blue and aching.

I was out on the river for the day with Dave Sterritt and Steve Elliott of the Alaska Department of Fish and Game. The tagging station at the fish wheel was one of several operated by the department in an effort to track salmon populations on the Taku River. From late spring until early fall, the Taku River is a highway for fish—salmon and steelhead trout heading back up into the mountain streams to spawn.

Salmon once reached as far south as Mexico on the West Coast of North America. Today, however, they are found no farther south than northern California. They are extremely sensitive to pollution and disruption. Two hundred years ago the richest salmon runs in North America were concentrated in the wild, wooded rivers of the Pacific Northwest. When the salmon headed up the coastal streams, they flooded the land with wealth and food. Along the coast, and even several miles inland from it, the lives of the native peoples of the Pacific Northwest—the Chinook, Makah, Kwakiutl, Salish, Haida, and Tlingit—were centered around this seasonal surge of life from the sea. They built some of the most sophisticated and elegant societies in North America, a world of elaborate potlatch feasts, totem poles, and beautifully carved dugout canoes.

Today, however, those rich runs in the south have been crippled by dams and clear-cut logging. Here in southeast Alaska and northern British Columbia, however, the ruggedness and remoteness of the landscape has kept it almost untouched. While fisheries in the south depend on hatcheries and carefully timed releases of water from irrigation dams to keep the fish alive, the runs here in the north are still wild. Salmon follow the rivers and streams up into the mountains to spawn just as they have for several thousand years.

The Taku River and its tributaries run inland for more than one hundred miles, across the border and deep into northern British Columbia, all the way to the arctic desert that lies on the far side of the mountains. The river slices through the heart of the peaks via a canyon more than one thousand feet deep. From the high, dry plateaus that lie beyond the mountains you can see the headwaters of the Taku, Stikine, and Yukon Rivers stretching out below you. Here to the south you can see ice: tongues and fingers of glacial ice that reach down from the Juneau Ice Field through valleys and cuts in the mountains toward the river.

On the day of our trip, however, the surrounding peaks and ice were all hidden from view by low, wet clouds. We had flown out that morning in a seaplane from Juneau through broken clouds that had become steadily thicker as the day wore on. (A helicopter would have to come out to bring us back into town that afternoon.) The land seemed low and wet. Less than two hundred years ago it had all been covered with ice. Forests of cottonwood and sitka spruce now lined its banks.

• • •

We stayed at the fish wheel for the better part of an hour, watching the taggers at work before heading back down river in our flat-bottomed aluminum boat to Yehring Creek, a small side stream on the Taku where the department had set up a base camp and built a weir to trap the migrating fish for tagging.

As we turned into Yehring Creek the water was suddenly clear— as if an invisible wall separated it from the milky waters of the river. In the deep pools near the mouth of the creek you could see the bottom fifteen feet below. Up above the weir, however, the shallow reaches between pools were less than a foot deep in places and pocked with fallen logs and boulders. While common sense suggested a slow and careful trip upstream, we had to keep the small boat moving quickly to keep as much of its hull as possible out of the water. We shot up the stream at nearly full speed, weaving between rocks and stumps and ducking beneath the branches of alders and willows that overhung the stream.

Along the bank we spotted the half-eaten carcass of a salmon—its bright pink flesh spread like a filet amid the bright green grass. "Grizzly bears," Sterritt said over the whine of the engine. "Looks pretty fresh. He must have just run off when he heard us coming."

Farther upstream a maze of marshes and small lakes fed into the stream. A low beaver dam of sticks and mud blocked the way, but Elliott wanted to look around. It would have been simpler, perhaps, to get out in the mud and push the boat over the dam, but not nearly as much fun. Elliott pushed his baseball cap down on his head, loosened the tilt lock on the jet outboard, and then gunned the edge. We hit the low dam at full speed and went sailing through the air like a skier taking off from a jump landing with a thump in the water on the other side. Elliott smiled. Here and there in the shallow waters of the lake you could see the small fry of coho salmon barely two inches long.

"There are about two million coho salmon in southeast Alaska," Elliott said. "That's a large number of fish, but they're spread out between nearly three thousand streams. Half of those streams are so small that you and I could jump across them. Also, most of them are only five to seven miles long. Without their connection of the sea, most of those streams couldn't support fish any larger than three or four inches in size."

After hatching, the small coho fry spend up to two years in streams and small lakes of southeast Alaska. When they finally leave for the sea they are barely three inches in size. When they return they are more than thirty. What happens in between, Elliott said, "is a big black box."

Biologists do not know exactly where the salmon go once they leave the stream or exactly what they feed on in the open ocean. Fish from southeastern Alaska have been caught as far away as Japan, although most stay closer to home. They seem to swim out into the Gulf of Alaska and then travel with the prevailing currents of the Alaskan gyre, which carry them clockwise through the North Pacific in a large circle. By summer, Elliott said, the fish are in "what we refer to up here as the southern latitudes," off the coast of southern Oregon and northern California, feeding in the rich upwellling currents centered around Cape Mendocino. By fall they are back in Alaska, generally slightly to the north of their native streams, where they start swimming southward along the coast, eventually heading inland to spawn. Different species stay out in the ocean for different lengths of time: pink salmon for just a season, coho typically for a year and a half. King salmon can stay out for as long as six or seven years. The odds against their return are high. From the eggs that hatch in the streams each spring, less than one in a thousand returns.

• • •

In late September I went back to the Taku River with Dave Sterritt to spend a few days at the department's base camp on Yehring Creek. In the morning of the day we were scheduled to leave it had been foggy with a light rain, a not unusual state of affairs for Juneau, Alaska, and we had doubted that we would be able to get into the air. By midafternoon, however, the sky had suddenly cleared, and at 4:00 P.M. we were speeding down the runway in a DeHavilland Otter seaplane laden with food and supplies.

As the plane left the ground, Mendenhall Glacier came into view. From the air you could see how it rose up from the Mendenhall Valley toward the fields of snow and ice cradled within the high peaks. In the summer we had hardly been able to see the ground at all—catching only glimpses of green trees and gray water between the thick wet clouds. That day, however, the air was sharp and clear. As we headed up the Gastineau Channel, the deep fjord that borders Juneau, toward the mouth of the Taku River the water below was almost turquoise and as opaque as paint—colored by fine particles of glacial flour.

Taku Glacier near the river's mouth was clean and white. It ran right down to the water's edge, poised, it seemed, to jump right across it and seal it off from the adjoining fjord like a sliding door. Small icebergs had calved off its front and drifted out into the fjord. Unlike the fjord, the river water was gray and laden with silt. Where it met the blue water of the sea it spread outward like a fan. Up ahead you could see the wide curves of the Taku River, winding back into mountains. Jagged peaks nearly two miles high rose up to either side. In the bright, sharp light you could almost count the stones on scree slopes that cascaded down their sides. In the summer the land below had been green and wet. Now you could see the signs of the approaching fall. Tall stands of cottonwood trees alongside the river had already turned bright yellow and the grasses of the marshes beyond them were touched with streaks of gold. Down below the salmon were running, more than two hundred thousand coho, heading back up into the mountains after a year and a half at sea.

Fifty miles upriver we circled the base camp. You could see it below among the alder and sitka spruce bordering the creek: a narrow weir that ran across the stream like a small dam, and the white canvas roof of the wall tent. The plane landed right on the river, its twin floats skimming the water's surface, sending out a rooster tail of spray behind. The transition from air to water was sudden.

We taxied up to a small grassy point and began to unload our gear, the seaplane's engine choking and coughing as it idled. In less than five minutes the pilot was roaring down the river with the throttle wide open and rising back up into the air. As the plane left, Jerry Owens and Suzanne Crete—who more commonly goes by the nickname of Jarbo—came shooting down the creek in a flat-bottomed boat with a jet outboard engine to ferry us upstream. Mountains seemed to hem us in on all sides.

Ice was running low at the camp so after we settled in we decided to make a run up to Twin Glacier Lake, heading back down the creek and out into the fast, wide, gray waters of the Taku. We shot the rapids of the small stream that led up into the lake at full speed, weaving back and forth between its banks which were lined with tall cottonwood trees. Turning the corner the lake suddenly opened up in front of us—a flat, still reach of water ringed by high peaks and glaciers, its surface filled with drifting icebergs. We moved slowly through the water, picking up twenty- and thirty-pound chunks of glacial ice floating in the water. Here and there larger icebergs shaped like ships plied the surface of the lake. Others floated by like whales, their tops barely above water. Balance was precarious. When we chipped a melon-size chunk of ice off of one larger iceberg, it quickly began to roll, capsizing in a matter of seconds. The ice was unbelievably cold. A few seconds of lifting left fingers and hands numb for minutes.

We headed back to camp just before sunset. In the clear pools of water near the mouth of Yehring Creek hundreds of salmon shot across the bottom like the shadows of leaves blown by the wind.

• • •

After dinner we sat inside the tent playing cards and Yahtzee by the light of a kerosene lantern. For the past week a pack of black wolves had been feeding on salmon near the mouth of the creek. We listened regularly for the howls of wolves, hoping to have a chance to drift downstream and see them at work. From time to time messages come in over the shortwave radio, a kind of communal telephone here in the bush: requests for a new boat engine, a problem bear down near St. Petersburg—news and gossip from more than a dozen remote camps scattered along the nearby coast and neighboring river drainages. Around 10:00 P.M. the still, cool of the night was broken by a rattling and pounding down at the weir, as if a moose or bear was trying to break through its metal poles to head upstream. "Sounds like salmon to me," Jerry announces. We all head out the door and down to the weir, picking our way across it in the starlight, out to the holding pen that lies in its center, the gate that leads upstream. Inside (the gate is shut until tomorrow morning so we will have a chance to tag the fish as they pass through) are six fish, their sleek sides rippling in the current, ramming the metal slats of the weir with their heads, searching for a way upstream.

The next morning, however, all but one of the fish have vanished, having moved back downstream to wait for a few more days before running for the shallower, gravel bottomed water upstream.

With no fish to tag we decide to drag a large net through a deep pool just above the weir and see how many fish were still lurking inside it. While Sterritt and I manned the lines along shore, Jarbo and Jerry waded out into the stream. The pool was just twenty feet wide and no more than six feet deep. When we pulled the net into shallow water there were more than seventy fish inside it, a collection of twenty- and thirty-pound-coho and sockeye salmon. After we tagged and checked the fish we tossed them into the water beyond the net. They were in no mood to be counted or weighed and thrashed wildly about in the water. Standing in the shallow water enclosed by the net you could feel them nudge your legs and shins, almost strong enough to knock you off your feet. Two and three feet long, they seemed strangely out of place in such a small stream, but their size, Sterritt assured me, was typical for salmon in southeast Alaska. When the fish return from the sea, they bring the richness and fullness of life in the ocean with them.

• • •

The lives of salmon are one the most dramatic stories in nature. After traveling thousands of miles in the open ocean for months or even years, they return to their native streams, leaping over rocks and up waterfalls to reach the small streams and gravel bars where they were born. Here on Yehring Creek they are only a few miles from the ocean. In places, however, they migrate through rivers and streams for hundreds of miles. Farther south in the Pacific Northwest, for example, salmon travel up the Columbia River and into the Salmon River, swimming all the way to its headwaters in Sawtooth Mountains of central Idaho, more than eight hundred miles from the sea and nearly a mile high in the sky.

No one really seems to understand how salmon find their natal streams after spending a year or more at sea, but there appears to be some "hierarchy of capabilities," according to Dave Gibbons with the U.S. Forest Service in Juneau, built around their sense of smell and their perception of the earth's magnetic field. During my first visit to Juneau I had spent a morning in the offices of the Alaska Fish and Game Department in Douglas across the water from Juneau talking about salmon in a roundtable discussion with both Gibbons and the Fish and Game Department's Steve Elliott.

Salmon, Gibbons explained, are highly sensitive to magnetic fields. If you take a tank of young salmon and subject them to a magnetic field, they will line up in the orientation of that field like "tiny compass needles." Apparently, he said, that sensitivity to magnetic fields functions as a built-in directional device when the fish are swimming in the open ocean to bring them back to the general area they hatched in. At that point smell seems to take over, guiding them to a particular stream.

As an undergraduate at the University of Washington in Seattle, Gibbons worked on an experiment with researchers who plugged the noses of young salmon ready to head out to the ocean to see how it affected their homing ability. The fish, it turned out, got back to Puget Sound all right, but after that, he said, "we ended up with salmon all over the place."

Heredity also seems to play a role, but that aspect of the salmon's homing ability is even less well-understood than that of magnetism and smell. Relocated fish, those that were spawned in a particular stream and then moved to another, are often confused. Gibbons offered some examples: "If you take eggs from stream A and transplant them in stream B, after they go out to sea they will not return to stream A but will go back to stream B. No one knows quite how or why this works. But as you move them farther and farther away from the stream they were originally deposited in, they have a harder and harder time finding their way back." Wild salmon, he added, have much better homing capabilities than those that were raised in hatcheries.

Like most things in nature, however, the homing capabilities of salmon, even wild salmon are not foolproof. Some fish do stray, but that too is put to good use—offering a way for the fish to begin colonizing new streams. A few years back Gibbons got a call about a steelhead trout that had migrated up the concrete channel of the Los Angeles River. Children playing along the bank had not known what it was and, surprised by the large fish, had stoned it to death. Perhaps it was drawn to the river by some ancient memory embedded within its genes from the days when salmon reached as far south as Mexico.

Whatever force enables them to travel back to their native streams, the urge to migrate is almost undeniable once the fish reach maturity. Where fallen logs and shallow water block a stream you can often see the salmon piling up behind the blockade, as irresistible and persistent as the waters of a rising flood—waiting for an opening that will grant them a way into the land and back up into the streams where they were born.

• • •

Two days later four people from the tagging station at Canyon Island stopped by to capture some sockeye salmon from the upper reaches of Yehring Creek for some laboratory work being done back in Juneau. Jerry, Jarbo, and I waded upstream with them to help out, walking alongshore through thickets of alder to avoid the deeper pools. Our group was spread out along the stream, those in the lead carrying a .306 rifle for bears. In the height of the salmon season the possibility of a run-in with bears is a very real threat, although our group and its collective noise cuts down on the risks considerably. Here and there we saw their prints in patches of sand and mud scattered along the banks. Their paws were the size of baseball mitts. You could see the sharp curving reach of each claw.

A mile and a half from camp the water was barely ankle deep and the stream bed was lined with gravel and small fist-size stones. Where the creek divided to pass around a small alder-covered island, the shallow water was filled with more than one hundred sockeye salmon. They shot upstream like rockets when we drew closer. You could hear the stones rattling beneath them. The water was so low that the backs of the largest fish protruded halfway out of the water. Their bright red sides and olive green tops looked almost surreal, as if they had been hand-painted.

We split up into groups, some of us wading through the water in a line, herding the salmon downstream like cattle. Others waited below with a gig, trying to spear fish as they passed by. The fish were quick and elusive. With each miss of the gig you could hear the musical clang of its tines on the rocks below. The best of our group proved to be Clayton from Canyon Island, a full-blooded Tlinglit from British Columbia. Thrust after thrust he never seemed to miss, spearing a fish each time right on the center of the tines. "Must be in my blood," he said with a broad, tight-lipped grin.

All of the fish we caught had already spawned. Most seemed like they had been healthy enough to live for a few more days or even weeks. On the sides of the swift-moving shallow stream, however, there were others that seemed nearly dead, covered with clouds of fungus and mold, their fins rotted away. Some had gone blind, their eyes milky with infection or even fallen out from rot. The sight was disturbing and unnerving. It is a long way from the Gulf of Alaska to this small creek: so much work and effort for so little gained. Its seems almost extravagant and pointless. Even here, however, there is purpose to things.

■ ■ ■

As salmon begin to migrate upstream, their whole body chemistry changes in preparation for the chance to spawn and perpetuate themselves. Body structure changes. In sockeye and pink salmon the males develop large humps on their backs and a long pointed jaw or kype. They change color as well: sockeye which are blue-backed silver fish in the ocean become bright red and olive green. Silver salmon, the coho, turn gun-metal gray with broad pink stripes running down their sides. The migrating fish stop eating—everything becomes focused on getting upstream. They strike at hooks and lures not out of hunger, but annoyance. For nourishment they begin reabsorbing their scales to stay alive, using them up like hidden reserves of fat. In some salmon even their immune system seems to shut down, leaving them susceptible to a host of diseases and fungi.

Once ready to spawn, the female—in the case of coho—hollows out a crater in the gravels of the streambed, a foot to a foot and a half deep, and then lays her eggs. The male then swims by and releases his milt, fertilizing the eggs which are then buried and hidden in the gravel. Their young fry will not emerge from the gravel until the following spring.

Contrary to popular perception, the salmon do not die immediately after spawning. Most survive for a few days or weeks, but they do not return to the sea. Biologists, Dave Gibbons told me in Juneau a few months before my second trip to Yehring Creek, have tried to keep salmon alive after spawning, taking them back to salt water—even force-feeding them, but nothing seems to work. After the salmon have run upstream, Gibbons said, "the clock has apparently run down."

But while the death of the salmon seems like a needless extravagance, it is actually critical to the health of the stream and even the surrounding forest. The schools of spawning salmon that run up through the coastal streams feed a host of other animals—bears, eagles, wolves, minks, and seals. Like the forests, the coastal streams in Alaska are nutrient-poor, their richness comes from the sea. In the spring the nutrients released by the carcasses of decaying salmon will foster a surge of productivity in the stream that will help carry their young fry to the sea.

••••

That night the sky was sharp and clear. Standing on the edge of the creek in the starlight you could look up and see the northern lights, a shimmering curtain of green and white light dancing across the sky. By mid-morning a heavy overcast had settled in and the air smelled like rain or even snow. In the sharp cold weather, the creek was edged with ice. The water had been falling steadily for the past few days and after the overnight freeze it had dropped another two feet, leaving the weir almost enitrely out of the water. With the cold weather the snowmelt and small streams higher up in the mountains that normally fed the creek had already started to freeze. With the rapidly dropping water, there was a chance the salmon still waiting to swim upstream would not be able to make it. Sterritt looked out over the low water and the almost-dry weir and sighed. "Now it's a race between ice and fish."

With no fish to tag we decided to take a trip upriver and do some wandering around. We stopped in at Jerry's cabin on Canyon Island in the middle of the Taku River to say hello to his wife Jan and see their small son Zach. The Owens live on the river year-round. In the summer Jerry works in the field for the Fish and Game Department. In the winter he traps for furs, running a trapline spread out along the river and up the scattered sidestreams that stream down out of the mountains. In mid-winter the temperature can drop to twenty below and the river that surrounds their island home becomes a frozen highway for packs of wolves and moose traveling through the mountains.

Dave, Jerry, Jarbo, and I went for a walk on the backside of the island along the edge of the river. On the far shore you could see the white shapes of mountain goats scrambling across the top of Kluchman Peak more than a mile above us. Bald eagles and osprey flew by fishing for salmon, some with fish firmly locked in their claws. For a half an hour we watched two grizzly cubs try to wade across the river. To the south you could see blue and white tongues of ice reaching through gaps and valleys in the walls of the surrounding mountains. They seemed to rise over the river like bright, low-lying clouds.

The water was low here on the Taku as well and on the backside of the island; the low water had exposed a wide sheet of sand. The vanished current had marked it with dunes and ripples. Walking along the edge of the narrow stream left by the falling water we could see the tracks of wolves and bears. Here and there were deep pools filled with hundreds of salmon, their sleek shapes weaving back and forth. Scattered among the clean gravel you could see the bright orange of salmon eggs, the size and shape of a pea—bright points of life.

• • •

By midafternoon we were back at the camp on Yehring Creek and light rain had started to fall. Sterritt and I were supposed to fly out that afternoon, but with the thick clouds closing in we made plans to stay another day. Suddenly there was the unmistakable sound of an approaching plane, quiet at first and then skimming the treetops only one hundred feet above the ground. In a matter of minutes we packed up our things and hopped into a boat and speeded through the shallows down toward the mouth of the creek. By the time we reached the river the seaplane was already waiting and tied up on shore.

We took off heading upriver at full throttle, barely clearing the trees, before banking away in a sharp turn to avoid the wall of the mountains ahead. We flew back into town across the Juneau Icefield, skirting the tops of the highest peaks that rose above the fields of ice and snow like small islands. Over the past few thousand years that ice has carved the landscape here. It has only recently retreated back up into the peaks, opening a pathway for salmon into the heart of the mountains. Like the high peaks that line the coast of southeast Alaska here, the salmon have come out of the sea.

Three months after my last trip on the Taku River, I got a letter from Dave Sterritt giving me news of this and that and the comings and goings of different people I had met on the river. He wrote that the day after we left the river the water had begun to rise. In Yehring Creek more than three hundred salmon had shot through the weir in a single day. Now the eggs were all there, hidden in the gravel and waiting for spring, when the ice would break and start the cycle all over again— life heading back to the sea.

THE ALEUTIAN ISLANDS

UNANGAN, *WE THE PEOPLE*

WE HAD BEEN AT SEA FOR ALMOST FOUR WEEKS, TRAVELING BACK AND forth across the North Pacific from the Alaskan Peninsula to the edge of Siberia, a distance of more than a thousand miles, mapping the floor of the sea. For weeks the only sign of land during this time had been a series of small orange dots on the ship's radar screen—the tracings of islands.

After a month at sea we were finally heading back toward land. At first, I had been fooled by a raft of high clouds that looked almost like islands. But now in the rose-colored sky you could suddenly see them, the serrated shapes of peaks on the edge of the horizon. Off the port bow the Islands of the Four Mountains seemed to rise up out of the Pacific like small, perfect pyramids, part of the Aleutians, an arc of volcanic islands that stretches across the North Pacific like a chain.

The sun did not set until nearly midnight. In the long, almost Arctic twilight the islands were lit by the light of a full moon and surrounded by a soft, red sky, growing steadily in size as we drew closer. Rivers of water seemed to run through the sea, shimmering with the patterns of ripples and waves.

The Aleutians reach across the North Pacific like a series of stepping stones leading out from the coast of Alaska all the way to Siberia, separating the shallow waters of the Bering Sea from those of the North Pacific. Their names are almost magical—Umnak, Tidáglda, Akutan, and Unalaska—the last an Aleut name meaning "where the sea breaks its back." For several thousand years these islands were home to the Aleuts, a native people whose lives were as closely tied to the sea as those of the seals and whales they hunted.

Even after more than twenty centuries of habitation, the islands are still remote. Of the chain's more than two hundred islands, only six are inhabited. Some have never been mapped. Twenty thousand years ago during the peak of the ice ages the islands sat just offshore from the land bridge that linked Asia to North America, a pathway for peoples and animals to the new world. Scientists call this lost link between worlds Beringia. While it is commonly referred to as a land bridge, Beringia was actually a vast subcontinent more than one thousand miles wide in places, an arctic steppe or plain inhabited by elephant-like woolly mammoths and primitive camels and horses. Today it lies beneath the Bering Sea.

Since early June I had been traveling on the *RV Farnella,* a British research ship leased by the U.S. Geological Survey to map the floor of the North Pacific. We had left Dutch Harbor in the first week of June after a few days in port taking on supplies and checking the ship's sonar gear. At first the days had been clear and bright. The tundra-covered slopes of the hills were bright gold and the grass had not yet come back to life. Ravines and crevices were still filled with drifts of snow, as if winter were not all that far away. The passage of time was deceptive. At 10:00 P.M. it seemed as bright as midday. From our anchorage in the middle of the harbor we could watch the comings and goings of fishing boats and seaplanes. Bald eagles gathered on the high cliffs of black volcanic rock alongshore like pigeons in a city park.

We left Dutch Harbor in a spitting rain. All hands that were not tied to duties below were up on deck to watch the islands slip out of sight. For the next three days the weather was rough, as a summer storm blew through with fifty-knot winds and twenty-foot seas that sent the ship plunging through the waves like a horse run wild. Up on the bridge, twenty or more feet above the deck of the ship, you could look out over the wet hills of the waves and watch streamers of spray blow across the bow in wet sheets. Albatrosses and shearwaters flew through the troughs between waves, rising and falling over the crests with subtle shifts in their wings. The storm hardly seemed to faze them at all. From time to time messages came in on the radio from nearby ships. A hundred miles away a 350-foot freighter broke up in the waves and went down with all hands. Our ship, which had seemed massive in port, suddenly felt small and vulnerable.

The day after the storm the air was bright and clear. The waves were still high but already regular and rolling, a deep navy blue, their crests topped with bright white foam. You could see across the open water for miles.

For the next three and a half weeks the sea was suprisingly calm, broken only by passing banks of clouds and fog and occasional rainstorms. Calmness is a rarity in the North Pacific, which has some of the worst marine weather in the world. Some nights the swell was so smooth and regular that it would rock you to sleep. Other times it was cut by crossing waves and irregular patterns of swell that sent the ship pitching and rolling, toppling chairs and throwing open unfastened drawers.

At sea the ship was a small, floating village, under the command of the ship's master, John Cannon, and its two senior scientists, David Scholl and Andy Stevenson from the U.S. Geological Survey. The crew numbered just thirty-five, drawn from Great Britain, Ireland, and the United States. Carrying our own food, water, and fuel, we were free and self-sufficient. On board was a small library, gym, galley, mess, and laundry. Communications with the outside world were sporadic, limited to bulletins of news from the BBC, typed up by the ship's first mate, and occasional radio messages and telephone calls relayed by satellite. The ship ran twenty-four hours per day with shifts of scientists and crew members working around the clock as we sailed back and forth across the North Pacific at a slow and steady eight and a half knots, towing our seismic and sonar gear behind us.

My own schedule was as regular as factory work, running from 8:00 to 4:00 every day, seven days a week. I worked in the lab helping keep track of the computers and sonar gear gathering data from the floor of the sea. Meals were served at a table by the ship's steward, and there were hours of time after work for reading and thinking. After dinner I often spent the evening on deck or standing at the rail outside the bridge watching the sea. The evenings were typically sharp and clear, with light that lasted until well after midnight. Looking over the water you could see the patterns of ripples stirred up by the wind, shaped like cat's-paws. Sunsets were slow, the clear sky colored by masses of almost solid color: first gold, then yellow, orange, and red that slowly faded to black. For days we would see nothing but water, then suddenly ships would appear: freighters and container ships traveling between Japan and the United States and Canada along the great circle route. After a lifetime of looking at flat map charts, it is hard to remember that the earth is round. The shortest route from west to east is not due east, but north along a curving arc that skirts the edge of the Arctic.

When the rains came you could see the clouds approaching from several miles away—a wall of purple and silver floating over the sea and fringed with a gray curtain of rain. You could see gusts of wind move across the water and hear the patter of raindrops as the squalls drew closer.

On other days we traveled through banks of fog and mist. Some early mornings, the water was almost flat calm—as smooth and still as the surface of a mirror. As sunrise approached, small puffs of wind would begin to blow, sending gently rolling ripples across the water that caught the diffused light at odd angles, transforming them into ribbons of color, a hundred shades of purple and gray that danced across the surface of the sea.

Life in the sea was well-spaced but often highly concentrated. On these still mornings we would drift through the fog and mist and come across small flocks of puffins roosting on the open water. As the ship drew closer they would paddle to get out of the way. Others would fly, running across the surface of the water to gather speed, each footstep leaving a ring of disturbed water behind. As they gathered speed and circled the ship you could see their bright orange-webbed feet trailing out behind them like small square flags. On bright, clear days we were often followed by solitary watchers: black-footed albatrosses from the Antarctic that followed in our wake, soaring for hours on their eight-foot wings only a few hundred yards behind us. From time to time we were joined by gulls and kittiwakes that stopped to rest on the fantail of the ship.

At times the contrasts were striking. After days of almost empty water, the sea would suddenly be filled with thousands of birds, flocks of shearwaters, puffins, and storm petrels following a school of small fish or a cloud of plankton. One day we passed through the midst of a flock of twenty to thirty thousand storm petrels, dancing across the surface of the water with their webbed feet, feeding on krill. While the birds moved over the water above, a pod of perhaps as many as a dozen fin whales rose up from below, the small knucklelike fins on their backs just breaking surface as they came up for a breath before diving back down below to feed. An hour later you could still see the birds: a black cloud stretching across the horizon.

Other days were punctuated by the appearance of solitary sea lions or whales. They would often appear without any sign or suggestion. I would spend hours watching an empty sea, then a humpback whale would suddenly leap out of the water—head, flukes, and tail all clear of the sea for a few fractions of a second before it went crashing back down. Spouts of water would appear on the edge of the horizon and then vanish as the whale dived down below.

• • •

The scenes up on deck, however, were merely a sideshow, a break from the work going on below deck. Half a dozen pieces of sonar gear and measuring devices trailed out behind the ship, sending out pulses of sound. Down in the lab, rows of computers tracked those waves of sound as they bounced off the seafloor to create graphs and pictures of the rocks and sediments below. The sound was sent out at different frequencies and the waves that came back told different stories. Some measured the depth of the bottom, others revealed the structures of rocks several thousand feet below the surface. The instruments were so sensitive that you could see clouds of plankton, tiny microscopic floating plants, just below the surface.

At the center of all these instruments was Gloria, a sophisticated sonar device on loan from the British Institute of Oceanographic Science, along with a team of British scientists trained to operate it, capable of mapping the seafloor in swaths some fifty miles wide. It is one of only two such devices in the world. Like the other sonar gear, it sent beams and patterns of sound toward the seafloor and then listened for the echoes that came bouncing back. But rather than a simple graph or chart, computers then filtered and deciphered the incoming waves, transforming them into black-and-white pictures that looked almost like photographs. As they were pieced together in the chart room, the emerging mosaic of images became a panoramic picture of the seafloor—almost as if the ocean had been suddenly emptied and photographed by a plane flying overhead.

In these images of the seafloor you could see how the Pacific Plate ended, diving down into the Aleutian Trench, a deep gash on the seafloor. It was all as clear and straightforward as the diagram in a textbook. As the plate neared the trench its surface was creased with ripples, like a throw rug caught by the edge of a frequently used door. The trench was there as well, a giant scar running across the seafloor. Watching the sonar gear was even more dramatic. As the ship passed over the trench you could watch the bottom plunge from ten to fifteen to more than twenty thousand feet deep as the seafloor sank out of sight. This is where La Paz, Los Angeles, Santa Barbara, and Santa Cruz are all heading: up to the North Pacific and down into the Aleutian Trench on the back of the Pacific Plate.

The Aleutian Trench is a subduction zone, the place where the moving Pacific Plate is being thrust back down into the earth. The rift zone that runs through the Gulf of California to drive Baja California away from the coast of mainland Mexico is one end of the spectrum of plate tectonics—the beginnings of an ocean basin and the start of forces that drive the earth's plates around the globe. This deep trench on the floor of the sea is the other—the place where those mobile rocks dive back down into the earth to be reshaped and reformed, rising to the surface in the form of volcanos and bodies of hot, molten granite.

On charts and maps its size is almost manageable: a thousand miles long and more than four miles deep, a comfortable collection of numbers to keep in the back of your mind. On dry land, however, its features would defy both imagination and perception: a chasm four times as deep as the Grand Canyon stretching all the way from Los Angeles to Denver.

• • •

"We're really a satellite," Dave Scholl told me one morning as we sat monitoring the computers. "It's all remote sensing. We're just bombarding these rocks with sound. We're listening to them, trying to understand what they're telling us." As a seamount appears on the tracing of one of the sonar charts, Scholl pulls at the paper, spreading it out on the floor of the lab to study the patterns of folds and faults in the rocks. "See that?" Scholl says, pointing to a domelike mound of rock that seems to rise above the seafloor. "That's the plug. Pull it out and the whole Pacific Ocean will go right down the drain."

During the cruise, Scholl and I shared one of the ship's small cabins. We would often stay up until long after midnight discussing everything from the origin of the Pacific to religion and art. Now in his fifties, Scholl is still taut and lean, a dedicated runner on land. His hair is sandy colored and his face is dotted with freckles. Even when sitting still he seems like a bundle of nerves and energy, joking and laughing, running on 220v while those around him are forced to get by on 110v. On the government's ranking system for federal employees that runs from GS-1 to GS-15, Scholl is a GS-16, a recipient of the Distinguished Service Award for his work as a marine geologist in the employ of the federal government. He is also one of the world's leading authorities on the evolution of the North Pacific, an intellectual jack of all trades, equally at home in both geology and geophysics.

Scholl grew up in Torrance, California, and became interested in the ocean while surfing and spear fishing. He was curious about the rocks he found on the floor of the sea. When Senior Day rolled around, a chance for local high school students to visit area colleges, he told the people at USC that he was interested in marine studies and they sent him to talk to Ken Emery. Scholl didn't know it at the time, but Emery was one of the world's leading marine geologists. "I asked him stupid questions and he answered them with a great deal of understanding and got my curiosity going. He told me they had a ship that I could go out to sea on and that I should get into geologic studies and they would take it from there." It was a chance meeting, but for Scholl the future was set. To this day he is still not sure what Emery saw in him at the time other than enthusiasm. "I was a typical California high school product. I couldn't write a sentence. I wasn't sure that verbs were necessary," he says.

At USC Scholl worked hard both at his studies and at odd jobs to earn enough money to keep him in school. "I was a poor, skinny kid who never locked his door. Nobody was going to steal from me. I didn't have anything." He subsisted on day-old bread and bulk cheese and margarine. Also peanut butter. By turning the space heater inside his room on its side, he learned that he could make cheese-melt sandwiches.

The ship Emery had told Scholl about was Allan Hancock's *Velero*. Life on board the ship was unlike anything Scholl had ever seen before. Hancock often traveled with them, bringing his personal chef along. "It was the first time I had ever tasted food like that," he recalled. During their short trips at sea, Scholl would gorge himself on the rich food, eating enough to tide him over for the next few days. In the ship's lounge he heard Tchaikovsky for the first time and tasted his first glass of sherry. Poker games with Hancock and others proved to be a good way of making some much-needed spending money and Scholl got very good at it. At a few dollars per hand, he could win enough to keep him going for a week or two back on land. To this day Scholl can still not relax while playing poker.

Science, however, was what the ship was all about and on his first trips out Scholl stayed awake the entire time. "I didn't go to sleep," he says. "I simply stayed up for days because there were so many exciting things happening on that ship, things I had never seen before. Geologic things. Biologic things. I couldn't sort them out. Which was more important, the fish or the bottom?" At night the ship would stop running and they would put lights over the side. The clarity of the water was remarkable. Scholl sat on the deck for hours watching the squid and flying fish rising toward the surface.

Spending time on the ship gave him an entirely new perception of science. "For the first time I began to perceive that you could learn about and contribute to knowledge about how the earth evolved by framing a problem and going out and getting information," Scholl said. "Basically you could become a detective and solve a mystery by going out and getting evidence about what the hell had happened here. It was something under your own control."

After finishing his master's degree at USC, Scholl went to work in the Arctic, doing a detailed geologic study of the site for a proposed deepwater port on Alaska's northwest coast. Part of the Atomic Energy Commission's Project Plowshare—a giddy and ill-conceived program for finding "peaceful" applications for atomic weapons (and a guise for testing them)—the port was to be blasted right out of the coast with a nuclear bomb. To Scholl and his co-workers the project was laughable. Camped out on the edge of the Arctic Ocean they used to spend the almost endless arctic nights joking that the AEC really wanted to build a large turning basin for the kayaks of the nearby Inuit villages.

But while the goals of the project were ridiculous, working in the Arctic proved to be a formative experience. Scholl developed a taste for working in the far north and studying remote areas. Soon after receiving his Ph.D. from Stanford, Scholl joined the U.S. Geological Survey (he was first hired on by Parke Snavely) in the early 1960s. Before long he was working in the Arctic and in the Bering Sea on a regular basis.

Instead of studying the research of others, he worked from a clean slate. On land and at sea the Arctic was an open book. As far as geology was concerned almost no one know what lay beneath the surface, much less how it all had formed. And once the theory of plate tectonics began to take hold, those few existing ideas would change dramatically, often due to discoveries unearthed by Scholl and his colleagues.

While others working in Europe or the lower forty-eight (as the rest of the United States is known to those in Alaska) fought to protect their scientific turf and keep others out, Scholl was constantly bringing people in, looking for partners to work on problems that he could never tackle alone—the evolution of the entire North Pacific, for example, as opposed to the evolution of a single mountain range. In the mid-1970s he was appointed head of the U.S. Geological Survey's branch of pacific marine geology where he put his restless energy to work pulling political and bureaucratic strings for funding and ships to explore the seafloor. Being at sea and finding out what the rocks had to say was what it was all about as far as Scholl was concerned, and once off of dry land he had an uncanny knack for picking the right spots and the right projects. He knew where to find a new frontier.

In the Bering Sea he discovered the Navarin and St. George Basins. After looking at the magnetic anomalies in the Bering Sea he suggested that it might in fact be a piece of trapped oceanic crust. His work helped redefine science's perception of the North Pacific and the Bering Sea. "Basically Dave is one of the greatest explorers of modern times," explained Tracy Vallier, one of Scholl's fellow scientists at the USGS. "He found opportunities to get on ships. But what was more important is that he knew where to take the ship. And when he saw the data he knew how to interpret it. He went to the Navarin Basin and looked at the thick piles of sediments and said, 'Boy, those are deep.' He went over the deeper parts of the Bering Sea and studied the bottom and said, 'Boy, it's got to be old.' He went across the Aleutians and was able to put things together. He was in the right place at the right time and he was in the right place at the right time because he wanted to be there. He found ways of getting things done. No one in the field does a better job than Dave in being able to put rocks and geophysics and old terranes together and making sense of them."

• • •

Computers and sophisticated sonar gear gave scientists a new way of looking at the earth, allowing them to see broad pieces of it as a coherent whole. Water is such a perfect medium for conducting sound that ships can travel at the surface and send out beams of sound and map the features and structures of the seafloor—dozens of square miles a day. (On land it is a tedious process of drilling holes and setting off charges of dynamite to shake the rocks below.) There is no need to drill holes or search for an outcrop. The rocks are all there underneath your feet, waiting to be heard. Until the widespread availability of sonar and seismic surveys, that three-fourths of the earth's surface that lies beneath the seas was virtually unknown. While rocks on land had clearly been faulted and folded and pushed, the spreading ridges and subduction zones that were the key to understanding all that motion were located on the floor of the sea. After their discovery it didn't take geologists long to realize that the earth was much more complicated and mobile than they had ever imagined.

Although Scholl was right on the edge of the plate tectonics revolution, he was not an early convert. While other leading geologists of the day, like Bud Menard and J. Tuzo Wilson, spent their time studying midocean ridges, Scholl spent his time studying subduction zones in Alaska off the Aleutian Islands and along the coast of South America, where things were not so clear. "If I had had any brains at all I would have gone out and worked on the ocean floor where crust is being made. It's much easier and much better behaved," said Scholl.

"Rigid plate tectonics, as it was called in the 1960s," he went on, "believed the earth behaved as if you had plates made of stainless steel that bounced off of one another and maybe threw out a spark or two like a volcano when they collided. Of course they're not at all rigid as they come together. They're plastic and deformable and shearable, but nobody really appreciated that back then. I was confused by the whole thing because when I went to look at the edges where the plates were supposed to collide and scrape off, I couldn't find the rocks that were supposed to be there. Information was compelling from the center of the oceans, but it was not so clear if you worked along the edges."

Scholl was skeptical. "I didn't seem to be following the new religion very fast. While everybody else was thinking about rocks you could see, I was worrying about rocks you couldn't see."

Plate tectonics had arrived in the early 1960s, but subduction— which dictates that as the seafloor spread out from midocean ridges it collided with the edges of continents—was still unheard of at that time. According to the theory of rigid plate tectonics, the rocks along-shore should be the same age as those on the floor of the sea, with older rocks being located progressively further inland. But when Scholl went to South America he found that there was no correlation at all between the age of the rocks alongshore and those in the ocean offshore. While the rocks offshore were six to eight million years old, those onshore were three hundred to six hundred million years old—rocks from before the time of dinosaurs even, ten to twenty times older than anything seen on much of the West Coast of the United States.

The more Scholl thought about it, the more he believed that the reason the gradual progression of older and older rocks was nowhere to be found was that the middle-age rocks had all been subducted out of sight—forced back into the mantle. "When you go to where the rock is being disposed of and you don't find it, you've got a problem," he said. "Somebody's been putting money into the vault and then you come by to count it and the vault's empty. It's a robbery!"

There had to be, Scholl reasoned, a recycling of things: rocks came up out of the earth as midocean ridges and volcanoes and were then eroded and carried back down into the earth. It was a radical idea at the time, but Scholl and a number of other geologists were begin-ning to believe that it was the only explanation of the features they were finding on the floor of the sea. "Most geochemists at the time were under the impression that the mantle was a one-way process: things got soaked out of the mantle and made volcanoes and the volca-noes eroded and made mud and everything stayed on the surface of the earth. So the mantle was a one-way system. Everything went out and nothing went back."

Like the faint patterns of magnetism that played such a key role in convincing scientists the Wrangell Mountains of southeast Alaska had traveled for hundreds of miles, small details would play a key role in confirming the idea that subduction was recycling the rocks found on the surface of the earth. This time, however, the key to the story was not magnetism, but Beryllium 10—a radioactive isotope geochemists unexpectedly found in the volcanic rocks of the Aleutian Islands.

Beryllium-10 is produced from atmospheric fallout. With a rela-tively short half-life, it all but disappears in the space of only a few mil-lion years. Scientists, therefore, were surprised to find Beryllium-10 in the lavas erupting from volcanoes in places like the Aleutians. What it seemed to suggest was that Beryllium-10 had been deposited on rocks and sediments near the surface that had then been thrust down into the earth. Long before the Beryllium-10 had time to decay, those same rocks had then been transformed into magmas that rose back to the surface to fuel volcanoes. Not only were rocks going back into the mantle and being recycled, they were also doing it quickly.

"The mantle is not a closed system," Scholl said. "There are inputs—i.e., plate tectonics—and it's happening faster than anyone thought. Before, geochemists had thought that the mantle was slowly being transformed into continental crust, but apparently there are inputs. What we're looking at here in the Aleutian Trench is the process of rocks going back into the mantle. We can see the material just getting tucked away and disappearing out of sight. The circulation pattern in plate tectonics is like tomato soup—it boils up over here but sinks down over there and it goes round and round and round!"

• ▪ •

By morning the small points of peaks had become islands. As we sailed past Kagamil Island, Mount Cleveland seemed to tower over the water, its cinder cone rising up through a skirt of clouds. The snow on its flanks seemed almost blue in the bright light, colored by the surrounding sea and sky.

The waters near the island were filled with life. Dolphins followed our path, riding the pattern of waves left in our wake. Flocks of gulls skimmed over the surface of the water. Here and there were seals and sea lions. Overhead a laysan albatross circled and soared in the currents of air, here on the edge of the Arctic, away from his nesting grounds on Midway Island in the Central Pacific. An immense gray and white bird with an eight-foot wingspan, you could see the slight hook at the end of its yellow beak as it flew alongside. A few yards away from the boat a forty-foot log floated by—driftwood from the mainland more than three hundred miles away. There are no trees in the Aleutians, only low scrub, tundra, and grassland. In the air you could catch the faint, almost imperceptible smell of land.

In late afternoon we pulled the sonar gear out of the water, lashing it down on deck. The ship was strangely quiet and still. The work that had been going on nonstop for four weeks suddenly came to an end as we headed back toward Dutch Harbor. We sailed through Akutan Pass, threading between rocky, tundra-covered islands, and we spent the night with our engines softly idling out near Egg Island just within sight of Unalaska Island. Everyone on board seemed to be awake and on deck that night watching the land and waiting. A bank of low gray clouds had moved in after dinner, a soft silver mist that kept the islands shifting in and out of view. As the sun set the light pierced through the lower edge of the clouds, making the sea shine like a burnished metal.

The next morning at 7:00 we sailed past Priest's Rock into the open arms of Unalaska Bay. Hillsides that had been covered with snow when we left four weeks earlier were now full and green, a quilt of tall grass and tundra, reaching all the way from the water's edge to the rocky slopes of the high peaks and volcanoes. It was the Fourth of July, and the harbor was filled with ships: deep-sea trawlers, crab boats, and a collection of research ships and fishing boats from the United States, Japan, Korea, and Russia. Others, from the U.S. Coast Guard and the National Oceanic and Atmospheric Administration, were also in port on full dress parade, their hulls decked out from bow to stern with flags and pennants. After so much time at sea, the short hour's wait while we cleared customs seemed to last for weeks. As I looked around the bay I felt like I had never seen land so lovely. When I walked off the ship I kept waiting for someone to call me back. For all that, leaving the ship was like leaving a warm cocoon. For almost a month it had been our home shelter out in the sea. I had never known that one could grow so attached to mere metal and machinery. For the next few hours I found myself wandering regularly back to the ship to reassure myself that it was still there.

Most of the ship's scientific crew were flying out that day, heading off to San Francisco or London. A new group was already waiting onshore to take our place. In three days the *Farnella* would be back at sea continuing our task of mapping the seafloor in front of the trench. Scholl and I, however, decided to stay on the islands for a few days and look around, gathering samples of rocks and seeing the land we had been traveling on the edge of for the past four weeks.

After we dropped a crew off at the airport with their mountain of baggage and data, I walked down to the shore of Makushin Bay on the edge of the airport. The ground felt good under my feet. After the constant noise of the ship—the sounds of engines, blowers, and sonar gear—the silence was almost overpowering. Laying back on the cobble beach and looking up at the sky I could hear the soft hiss of the waves running over the rocks.

• • •

Later that day Scholl and I climbed to the top of Mount Ballyhoo on Amaknak Island in the middle of Unalaska Bay, the small island that contains Dutch Harbor. The center of the Bering Sea's profitable fishing industry, Dutch Harbor has the ambience of a restless frontier town. Money seems to come and go in spurts here. When the fishing is good, deckhands on a profitable crab boat can make several thousand dollars a week. Others on giant factory ships catching pollock and hake to make surimi, the artificial crab of supermarket shelves, routinely clear as much as $10,000 a month. The work, however, is long and hard and often dangerous, a succession of sixteen-hour days that run back to back for weeks on end, broken only by a few hours of sleep before getting up to clean a steady stream of crab and fish.

Weather in the Bering Sea is rough. Fall overboard without a survival suit on and you can die of exposure in less than five minutes in the cold water that hovers just a few degrees above freezing. Ships disappear in the storms and high winds that blow through on an almost weekly basis—breaking up or simply flipping over and heading straight for the bottom, taking down everyone on board. It is hard, dangerous work and after a few weeks at sea the deckhands come ashore to blow off steam and cash. On the Fourth of July things seemed to be running at a feverish pitch. A sign on the back of the motel room door seemed to capture the ambience in a simple, eight-word warning: NO DRUGS OR FIREARMS ALLOWED IN THE ROOMS.

A winding dirt road behind the airport led up the side of the mountain, past the remains of bunkers and ammunition huts dug right into the side of the slopes. During the first half of the Second World War the Aleutians were the site of a bitter struggle between the United States and Japan—a deadly and dangerous war on the edge of the Arctic. The battle was and still is often overlooked. During the war more than fifty thousand troops were stationed here and several thousand more were spread out across the chain all the way to Attu more than four hundred miles away.

The islands were the first U.S. territory seized by the Japanese in the Second World War. In 1942 the Japanese bombed Dutch Harbor and seized the western islands of Attu, Agattu, and Kiska. From bases in the Aleutians the Japanese hoped to hit the U.S. mainland—from Dutch Harbor, Seattle and its invaluable Boeing Plant would have been within reach of the Japanese bomber squadrons. Later these islands would be the first lands recaptured, but not before months of a bloody and often frigid struggle spread out over a front nearly a thousand miles long. The weather claimed more causalities than did combat. Ships would disappear at sea in the midst of heavy gales. Airplanes became lost in the thick, cold fogs and flew into the sides of volcanoes and mountains. When U.S. troops retook Attu Island the outmaneuvered and outgunned Japanese troops elected to die rather than surrender. After bayonetting their wounded in their hospital beds, they retreated to the top of a ridge and then turned to fight leading a series of suicidal charges with shouts of "Banzai" against the advancing U.S. troops. Others held hand grenades to their chests. By the battle's end more than twenty-three hundred Japanese were dead.

After the Japanese pulled out of the Aleutians, U.S. forces stayed on, guarding against the remote possibility of another invasion. While armies in Europe and the South Pacific captured headlines, troops in the Aleutians sat on the islands and waited. The land seemed harsh and inhospitable. Most felt like they had been sent off into exile. Winter covered the islands with drifting snow and sheets of ice. Sudden blasts of wind—williwaws—blew through mountain passes and channels between islands at one hundred miles an hour, toppling tents and even buildings. It seemed to rain almost constantly. There were no trees on the islands, and so firewood and coal had to be shipped in from the mainland. The joke among those who skirted the rules was: "So what are they going to do? Send me to the Aleutians?"

After D-Day the pullout was swift and unceremonious. Plates were left on mess hall tables with the food still on them. On Tanaga Island soldiers drove jeeps out onto the runways and drained them of oil and let their engines run until they froze up. Twenty years ago when Scholl first came to the islands you could walk through the abandoned buildings and find jackets and coats still hanging from hooks on the walls, as if their inhabitants had simply vanished into thin air.

Today the bunkers and buildings that litter the hillsides have been weathered to a silvery gray by the wind and winter snow. They look almost like driftwood. Some have been abandoned. Others, however, have been pressed into service as homes and storage sheds or stripped of their wood and plumbing for homes closer to town.

■ ■ ■

Halfway up the mountain we left the road that had led up from the airport and began walking straight up the grass-covered slopes. The flowers were in full bloom and we waded through knee-high stands of lupine and carpets of small yellow and white flowers. The salmon berries were already starting to ripen. Here and there we surprised birds: ptarmigans and snow buntings and rosy finches, natives of Siberia that had come to the islands for the brief, green summer. On top of the peak a collection of sailors and fishermen sat in the sun taking in the view. Some had carried folding chairs and six-packs of beer up the steep slopes.

From the top of the peak you could see the maze of small islands and inlets that made up the harbor below. To the north, sheer cliffs of loose rock tumbled down to the Bering Sea more than one thousand feet below. To the east you could look across Akutan Pass and see the Akutan Peak, a cone-shaped volcano. To the west was Makushin Volcano, a sixty-six-hundred-foot cinder cone on Unalaska Island rising right up from the water's edge; plumes of steam rose off its crest. To the east and west were others—including Shishaldin, Okmok, Kagamil, Carlisle, Yunaska, Kiska—a chain of active volcanoes reaching across the north Pacific in a fiery arc.

The volcanoes here are part of the ring of fire, a chain of active volcanoes strung around the rim of the Pacific. Scattered among the islands of the Aleutian chain are more than eighty large mountains believed to be volcanoes. Forty-six of them have been active since written history began here with the arrival of Russian Orthodox priests in the early 1700s. The thick deposits of volcanic ash scattered around the islands, as well as the Aleuts' own collection of legends and oral history, suggest that they were active for thousands of years before that. On Unalaska Island, for example, Aleut legends say Makushin Volcano and neighboring volcanoes on Akutan and Umnak became involved in a dispute over which was the largest volcano. Instead of spears, Aleuts hurled lava and molten rocks through the air at each other, killing the animals and burying the islands under a layer of fiery ash.

Eruptions here are often explosive. In the early 1900s an eruption of Katmai Volcano sent five million cubic meters of volcanic ash into the air—1.5 times as much ash as from the legendary eruption of Krakatoa in the South Pacific in 1883. Since then the volcanoes have been active many times. Makushin Volcano had major eruptions in 1930, 1938, and 1951; like Okmok on nearby Umnak Island, it regularly steams and smokes.

The Aleutians are some of the most remote terrain in the world. Unlike the Arctic and Antarctic, there are no scientific research stations here and few ongoing scientific studies. The almost incessant cover of clouds and rains makes the islands all but impossible to study from the air or by satellite. The storms that blow across this stretch of the North Pacific keep them hidden from view. Reports of eruptions, accordingly, are haphazard, made by bush pilots flying between the islands' remote settlements, or by air crews on the regularly scheduled flights between Dutch Harbor, Adak, and Anchorage. Other reports come by word of mouth, from fishermen working in the Bering Sea— at times the clouds suddenly part, exposing a mountain whose sides are covered with sheets of fire.

As volcanic islands, the Aleutians are often erroneously compared with the Hawaiian Islands. Unlike the Hawaiian chain, however, the Aleutians are not created by a hot spot on the earth's surface and fed by lavas that spring from deep inside the earth; they come from the subducting rocks of the Pacific Plate that slide into the Aleutian Trench. Unlike the relatively clean rocks of the Hawaiian volcanoes, those in the Aleutians are contaminated by seawater and sediments thrust down into the trench along with pieces of the seafloor. As a result, eruptions here are not smooth and regular but explosive—sending clouds of ash and steam and flaming stones into the air.

Until the Second World War almost nothing was known about volcanoes in the Aleutians other than that they were active. When an eruption of Okmok Volcano in 1944 nearly forced the evacuation of Fort Glen on Umnak Island, however, the army suddenly took a profound interest in the area's geology and gave a $1 million grant to the U.S. Geological Survey to study the islands' volcanoes and see if there was any way to predict when future eruptions might occur. Heading up the research was geologist R. R. Coates, one of the leading field geologists of the 1940s. Coates was not able to offer the army any help on predicting eruptions—there was nothing much to say in that regard except that the chain's volcanoes were remarkably active and the army should be prepared to lose its bases at any time.

Coates, however, did come up with some interesting geological ideas about the origins of the Aleutians. The islands, he suggested, were bordered by a large fault. Sediments and basalts were carried down into the earth along that fault and then granitic and volcanic rocks were "sweated out of the rocks" (Coates had no clear idea why this occurred) and then rose to the surface, fueling volcanoes. In essence he believed that subduction was taking place here, but the theory of plate tectonics was still twenty years away and few people, perhaps not even Coates himself, understood the implications of his work.

For thirty years Coates' ideas would be all but forgotten. Later, Scholl and others would bring them back to life, weaving them into the fabric of plate tectonics. "He was like a guy playing a piano in a forest," Scholl said. "He was a really great piano player, but he only played in the forest so nobody knew how good he was." Today these volcanoes are not seen as the end of something but as the beginning. Their flows of lava and ash represent nothing less than the creation of a new continent.

· · ·

The next day, Dave Scholl and I, along with Herman Karl and Tracy Vallier, the two chief scientists who would be heading up the *Farnella's* upcoming cruise, went walking along the warren of roads that leads into the hills behind the town of Unalaska. We spent the day looking at rocks and the remains of huts and gun emplacements from the Second World War. Following a steep ravine back into the hills, we surprised a rare Steller's sea eagle, its black wings bordered with streaks of white.

Near the ruins of an abandoned officer's club we spent some time studying a low cliff of green sandstones shot through with small pebbles and stones. It is axiomatic in geology that what goes up must come down, and while lavas have been building volcanoes here for the better part of fifty million years, erosion has also been wearing them down, building up thick piles of sandstone like these.

Vallier broke off a fist-size chunk of rock and pointed out the cobbles and stones inside it. Coating them was a thin, reddish-brown layer—rust, he explained. Underwater, things may decay and corrode, but rust requires exposure to open air. The small rings of rust suggest that these rocks were deposited in open air, on dry land. And that is a key point here in the Aleutians—the islands may be sandwiched between the Bering Sea and the North Pacific, but their rocks are more continental than oceanic.

While fossils, faults, and folds and patterns of remnant magnetism suggest that almost everything else on the West Coast has moved, the Aleutians have been firmly in place since their appearance some fifty-five million years ago. Their arrival predated the beginnings of the San Andreas Fault, related not to the movements of the Pacific Plate but to an all but vanished piece of oceanic crust known as the Kula Plate.

Over the past two hundred million years the West Coast of North America has been bordered by a collage of plates and spreading ridges. For a brief period of time extending roughly from eighty million years ago until sixty million years ago, much of the western edge of North America was bordered by the Kula Plate. Like the Pacific Plate, it too was heading north, but almost twice as fast—perhaps by as much as twelve centimeters per year instead of the five or six centimeter-per-year clip of the Pacific Plate today. While the Farallon Plate slid eastward beneath the edge of what is now California and Mexico to create Franciscan-type rocks of Coastal California and the Sierra Nevada farther inland, the Kula Plate was speeding towards Alaska and the Arctic. Subduction was taking place on the northern edge of the Pacific as well—not beneath the Aleutians, but farther north along the northern rim of what is now the Bering Sea—as the Kula Plate slid into a trench on the seafloor.

Fifty-five to sixty million years ago, however, that subduction ground to a halt. That seems to have happened, geologists believe, when a chain of oceanic islands or a cluster of oceanic plateaus became stuck in the trench, backing it up like a sink with a plugged drain. To the south, however, the Pacific Plate was following close on the heels of the Kula Plate and heading north as well. When subduction ground to a halt, the subduction zone simply jumped farther south, opening up a new gash on the seafloor—the Aleutian Trench. Faulting and folding built up a bench or shelf on the seafloor. Later, volcanoes would poke through its surface to complete the construction of what is known today as the Aleutian Islands.

Behind the island arc and its shelflike base, the Bering Sea is being slowly filled in with sediments from the Yukon, Kuskokwin, and

Anadyar Rivers, the streams that drain Arctic Alaska and Siberia. Today those sands and silts on the floor of the Bering Sea are nearly a mile thick. In seismic profiles, "it looks almost like a bathtub filling up with sediment," Scholl said. As for the floor of the sea, "that may well be one of the largest pieces of exotic terrane in North America," he added. The floor of the Bering Sea is now made up of a land-locked fragment of the Kula Plate. As the sediments slowly fill it in, that fragment will rise up out of the sea.

While the rifting that opened the Gulf of California represented the beginnings of a new ocean basin, what is happening here in the Aleutians is nothing less than the making of a new continent. In fact, as recently as fifteen thousand years ago, that parts of that emerging continent were already above water, exposed by the drawdown of sea level that accompanied the ice ages—a conduit between Asia and North America that brought plants and animals as well as humans to the new world. The Aleutian Islands are not the end of the world but a link between the old world and the new.

• • •

Scholl and I stayed on the island for two more days, traveling around in a well-worn rental truck, following gravel roads back into the hills, stopping from time to time to walk in the mountains and look for fossils and rocks. The days were gray, marked by a low, wet sky. In the mornings the tundra was often wet with the mist from a soft rain.

On a cliff just outside of Unalaska we found the bones of a Steller sea cow, a manatee-like animal hunted to extinction by Russian sailors and trappers less than fifty years after their arrival in the early 1700s. Scholl was pleased and excited, his face lit by a broad mischievous grin. "I don't hunt," Scholl said with a laugh. "But I do like hunting for fossils. You don't have to kill anything. The animals are already dead."

He then held two fossils in his hands like puppets. "You know what these guys were saying back then?" he asked. "'The Russians are coming! The Russians are coming! Eat! Eat! Eat! We'll all be dead in fifteen million years!'"

Most days we traveled around the outskirts of Unalaska, the small village opposite Dutch Harbor. Although Dutch Harbor was created by the military and now thrives on the heady influx of dollars from the profitable crab and fishing fleets of the Bering Sea, Unalaska is more than three hundred years old—the site of one of the first Russian Orthodox churches in Alaska. A native village created by the Russians, it is still home to most of the island's remaining Aleuts.

Less than one hundred fifty years ago the islands were part of the Russian Empire, the start of territories in North America that included all of Alaska and forts and settlements as far south as Fort Ross in northern California. Even today Unalaska's links to Russia are plain to see. Signs at the tribal health clinic and the tribal office in the center of town are written in both Russian and English. Most of the town's native residents are fluent in both languages, and walking down the streets you are as likely to hear Russian as English. Down at the edge of the bay onion-domed spires of the Church of the Holy Ascension of Christ rise up above the water's edge. Graves in the churchyard were well kept, surrounded by white picket fences and small green plots of grass. White wood crosses have a slanted bar at their base—symbolizing Christ's agony on the cross.

Outside of town the roads are dusty, even after the rain. As we head south on the dirt road that skirts Unalaska Bay I break out into an uncontrollable coughing fit from the dust. "Don't worry," Scholl said with a fabricated look of concern on his face. "That dust is all from Miocene volcanics. The Miocene was a happy time. The earth was getting warmer."

Scholl has spent a great deal of his professional life trying to understand how the Aleutians were built up, but he can already see their demise. Like so many things in geology the collision of the Pacific Plate with the trench is not entirely head-on but angular, and as one heads farther west in the arc, the direction of movement slowly changes. While plate movements here near the center of the change are nearly perpendicular, farther west toward Attu they are almost parallel to the trench—westward toward Asia. Plate movements there are pulling the islands apart, sending the far islands shooting off toward Siberia. Whether the Russians will accept them, however, is a matter of conjecture.

Eventually the collapse of the Soviet Union may do as much for geology as it promises to do for economics. As far as the religion of plate tectonics is concerned, most Russian geologists are determined skeptics. In physics Russian scientists are right on the cutting edge of science, but in matters of geology they still lag decades behind in a kind of politically induced time warp. As far as physics is concerned, "it's pretty hard to argue with the reading of a meter," Scholl explained, "but when it comes to geological things, you can get philosophical. The databases are sort of difficult to interpret and kind of spooky."

The break between U.S. and Russian scientists in geology is large, almost hard to comprehend. Six months after Scholl and I were in the Aleutians, a team of Russian geologists visited the U.S. Geological Survey in Menlo Park, and their head scientist refused to even look at Scholl's data from our cruise over the Aleutian arc. "He didn't want to be tainted," Scholl later told me. "To him science and political philosophy are the opposite sides of the same coin. Plate tectonics is a Western idea. Therefore it's Western political philosophy and he can't accept it. He's a good Christian, from the old school of Soviet geology, which taught that the Pacific Ocean used to be a continent. It's mythical, OK? And it sank out of sight. Everything goes up and down. Not sideways. He can't handle sideways. It's kind of weird. When you present him with some newer data, including all the things you can add into the pie, he won't even consider it. It's sort of like talking to the Pope about, 'let's examine whether Christ was God or not.' It's not to be examined. It doesn't work. We don't talk about those things. That's already a given."

• • •

We had lunch at Summers Bay a few miles north of Unalaska. Afterward we spent time looking for artifacts in the small patch of sand dunes near shore. Near the base of the dunes were lead-cored .45- and .50-caliber bullets, relics from the Second World War. Higher up in the dunes were broken pieces of murex shells, small pieces of charcoal, and the bones of salmon and seals.

At their peak the Aleuts numbered no more than twenty thousand, but the sites of their villages and hunting camps litter the islands, leaving behind a rich treasure trove for archaeologists today. Like the Egyptians, they apparently mummified their dead, burying them in sea caves and mounds along the coast along with their kayaks and clothing. The bones here are probably the remains of midden or shell pile from an ancient hunting camp. The bones and charcoal are all old, weathered, and shot through with holes. How old, however, is impossible to say. Perhaps only a few hundred years, perhaps as many as several thousand. To an archaeologist these deposits in the sand would be nearly useless; the layers of bones and shells have been repeatedly winnowed and sorted by the wind. The scene, however, was enough to keep us occupied for several hours, walking through the sand looking at bits of bone and carved stone and thinking about what might have been.

From the top of the dunes you can sit and look out across Unalaska Bay to the Bering Sea. To the west are the high cliffs of Priest's Rock, which stands near the entrance of the bay. Even from here, several miles away, you can see how the cliffs' black rocks are shot through with veins of light gray rock that look almost like streaks of lightning. It was images like these that first taught Scholl to think about the lives of rocks and how they had changed. In the end those thoughts would shape his life.

Scholl's oldest memories of rocks are from when he was a small child, driving through the San Gabriel Mountains with his parents and noticing the veins of quartz and pegmatite that shot through the rocks in the heart of the mountains. "I didn't know what those were in those days," he says. "I was just stunned how one rock could penetrate another like that. It was a simple naive thought: How do you do that? Here was this massive rock. It was obviously something that was not soft and it has these funny bands of color going through it. It just struck me. How is that possible?

"It's not much of a question, but it made me think of rocks in a different way. Seeing that they were complex, that they had a history. Something had happened and then something had happened to them. It wasn't that they were all made at once and dropped on the seafloor or wherever they were. It was a very profound moment. I could never quite put it into words."

His family had a hard time understanding. When he told his grandmother that he wanted to go to college and study rocks she told him simply: "I don't think you should study things that have always been." She had this static view, Scholl said, that rocks were just made by God and left there.

Scholl, however, couldn't stop thinking about it. His mind was always full of questions and most of them were centered around rocks. "We were conservative, fundamentalist Lutherans and I was basically being told that eternal damnation was where I was headed for asking these questions about how things were related in time. I was confused by such stupid silly issues, but at the time they were very major for me. I was a very religious kid and I was beginning to have problems with catechism and all. Here was the question in my mind: 'Well, did they, Peking and Java Man, exist before Noah or after? Because if they existed before Noah they were made in the image of God, and, hey, God must be a pretty ugly guy, you know?' And if they were after Noah, then I thought, 'My God! What happened to Noah's children? Why did they go back into the caves and live like that?'

"Things in my mind said that there had been a change. It wasn't just a fixed thing anymore. How do you relate this to a Christian tradition, these ancient people that lived in caves or whatever they were? How do you put them into the context of the mythology of the Old Testament? Well, you can't of course. But I was being told that you had to and finally there was no other way. Either I was going to deny Peking Man or I was going to deny the Church. I believed in Peking Man. I just couldn't stop asking the questions and ultimately I had to leave the Church because there was no room for me."

· ■ ·

As recently as fifteen thousand years ago much of what is now the floor of the Bering Sea was dry land, a subcontinent known as Beringia, a link between Asia and North America. While much of

Canada and the United States was covered with ice, Beringia was an ice-free Arctic grassland, similar in many respects to the Arctic steppes of present-day Siberia. We think of Asia as a world away, but the link between these two lands is still close: at the Bering Straits just fifty-three miles of open water separates Alaska from Siberia.

The climate of this submerged subcontinent was characterized by warm, dry summers and wet winters with heavy snows. Familiar animals like reindeer, caribou, horses, and mountain sheep roamed the plain, but fossil evidence suggests there were other more exotic animals as well: steppe antelope, camels, woolly rhinoceros, bison, and woolly mammoths, a collection of large animals that make the area sound almost more African than Arctic. Some paleontologists have suggested that the area was a kind of Arctic Serengeti that teemed with wildlife, but it is hard to accurately judge their numbers based on fossil collections today. Others suggest that the animals were no more plentiful than today but traveled in large herds, leaving much of the area open or abandoned, visited only occasionally. From the Arctic these animals eventually migrated south to populate North and South America.

Whatever their true population numbers, the appearance of so many different types of large animals is unusual because the area supports almost none today. Today almost all of those giant mammals—the woolly mammoth and rhinoceros—are extinct and many scientists believe that people who followed these herds across Beringia may have played a role in their extinction.

Humans are the most recent arrival to Beringia, but it is not clearly understood exactly how or when people migrated across the land bridge. Like the land itself, the archaeological sites that hold the answers to that riddle are all underwater. Whereas archaeological sites in the southern United States and central Mexico suggest that humans may have been present in North America as early as 50,000 to 125,000 years ago, the oldest artifacts in the Arctic are just 11,500 years old. The only certainty about early humans in the Americas is that to date there is no sign of a Peking Man or a Java Man. However, one of the clearest links to the roots of early humans in the Americas lies through the Aleutian chain within the history and culture of the Aleuts.

• • •

"I have come to believe there are two kinds of people in the world," Lydia Black said with a laugh as her voice faded in and out over our long-distance phone-link from San Francisco to Fairbanks. "Those who hate the Aleutians and those who love the Aleutians. You must be one of those strange people who love the Aleutians."

Black, a Russian-born anthropologist now working at the University of Alaska at Fairbanks, is one of those strange people as well. She first went to the islands in 1975 and has been working there regularly ever since, studying Aleut culture and history. Today she is one of the world's leading authorities on the Aleuts. After a series of long distance phone calls from California, I finally traveled to Fairbanks to spend a day with Black talking about her work among the Aleuts.

There has traditionally been a strong split between Russian and U.S. anthropologists on the history of the Aleuts. While U.S. scholars tend to view the Aleutians as an isolated corner of North America, the Russians approach it as an extension of Asia and traditionally view its natives from the larger context of Asian history. Black shares that view, and together with Harvard professor William S. Laughlin she has been a champion of the idea that the Aleuts represent much more than an evolutionary cul de sac. "The cul de sac idea is out, it's just out," she said emphatically. "The Aleutians were not a cul de sac but part of a prehistoric pathway to the New World."

Culturally and physically the Aleuts are closer to the people of Asia than they are to the Indians of North America. While they have skeletal similarities to Inuit or Eskimo peoples of the Arctic and are often lumped together with them, the differences between their languages today are on par with the split between Russian and English. Language, blood type, and dental characteristics all resemble people from the eastern edge of Asia along the Chukchi Sea and the Sea of Okhotsk.

While other scientists believe that Asian contacts with the Americas ended with the submergence of the Bering land bridge fifteen thousand years ago, Black's ideas about the relationship between the peoples of the Aleutian Islands and those of Asia are as fluid as Scholl's perceptions of their rocks. Instead of seeing them as living in isolation for the last twelve thousand years, Black sees the Aleuts and other seafaring peoples of the North Pacific as traveling back and forth along the Aleutian chain, venturing between the Old World and the New World, sporadically exchanging ideas, materials, and culture. "The Japanese current," she says, "wasn't just invented in the twentieth century."

While much is made of the 180-mile gap between Attu, the westernmost island of the Aleutians, and the Komandorskiye Islands, which reach out from Siberia, Black points out that Attu is clearly visible from Bering Island in the Komandorskiyes and from there it is just ninety miles to the mainland. The Aleuts, she explains, were seafaring people and often traveled out of sight from the islands.

Scattered artifacts found on the Aleutian Islands tend to suggest that there were some periodic links to Asia, possibly even a primitive trade route: embossed disks of copper and bronze that seem to be from China or Japan as well as small pieces of amber and iron found in ancient burial sites—neither of which are found in the rocks of the islands. Ancient Japanese legends speak of seafaring people traveling in skin-covered boats (kayaks) who visited their islands.

Not only did the Aleuts possibly travel back and forth across the Pacific, keeping contact open between Asia and North America, their mode of travel—boats—may also be a further picture of how early humans came to the New World. In the Arctic today boats are still a principal means of travel where and when airplanes are not available. Rivers are highways through the tundra. Early humans, Black says, didn't trudge across the Bering land bridge, they came by boat or log, following the coastline and rivers eastward, floating on logs or perhaps simple rafts. The perfection of the kayak achieved by the Aleuts came later, but they were so well adapted to this ocean world on the northern rim of the Pacific that their culture and peoples survived intact for more than ten thousand years—longer than any other known civilization on earth.

••••

From U.S. eyes the Aleutians have always been the end of the world, but to the Russians they were the gateway to *Bolsha Zemlya*, the New Land. The islands themselves were seen as the remains of a primordial continent that had sunk out of sight to create the Pacific. While European eyes looked westward across the Atlantic to the new world, Russian eyes looked eastward across the Pacific.

In the early 1600s Russian fur traders began moving east, traveling from river to river across Siberia, following one up from the coast and then walking east to the next and riding it back down to the sea. By 1680 they had reached the Pacific coast of Siberia and the edge of the Bering Sea. Along the shore there were rumors of a new land just south of Kamchatka. Their imaginations ran wild. Some suggested that one could travel through the polar seas to India. Scholars in Moscow believed that the seas in the Arctic were ice free north of eighty degrees latitude due to subterranean volcanoes. Beneath all those ideas was just one inescapable fact: knowledge of the New World across the Pacific reached no further than the Spanish explorations along the coast of California. North of thirty-eight degrees every chart was blank.

In 1733 the Dutch navigator Vitus Bering was put in charge of what was perhaps the most ambitious voyage of exploration the world had ever known. Empress Anna of Russia was interested in expanding her empire eastward; once there, Bering was expected to begin settling the land as well. After exploring the nearly seven thousand miles of land that lay between St. Petersburg and the continent's eastern shore, Bering was expected to settle and develop the Pacific coast, founding elementary schools and a network of post offices with bimonthly mail service while introducing cattle raising to the area. To promote maritime trade he was instructed to build a network of lighthouses, seaports, and shipyards along the coast, using his spare time to set up rope works, iron mines, and foundries. After accomplishing that he was to load his men onto ships and look for the land that lay just offshore.

Bering finally left St. Petersburg in 1734 with more than ten thousand people, a small army of cooks, drivers, carpenters, artists, and cartographers. The journey east took more than seven years, not surprising since east of Moscow the land they traveled was a roadless wilderness, virtually as unknown as the face of North America that had greeted Columbus nearly two hundred forty years earlier. Communications with the royal court in St. Petersburg were sporadic, and Bering was routinely criticized for both falling behind schedule and overspending his budget.

Upon reaching the edge of the sea that now bears his name, Bering's own instincts were to look for the land that was rumored to lie offshore (no one at this time had conceived that it was part of North America) by sailing east. But although he was several years and several thousand miles from St. Petersburg, political pressure and influence still made itself felt. When Bering announced his plans to head east, his cartographer, an effete and well-connected Frenchman by the name of Louis de la Croyere who took snuff and wore a white powdered wig throughout the duration of their wilderness trek, told Bering that his duty was to search for Gama Land, a large island that his highly imaginative and highly inaccurate maps showed lying to the south-southeast.

Bering eventually gave in. De la Croyere was influential in St. Petersburg and not a man to be easily crossed. It a was a fatal mistake. Instead of reaching Alaska within a week they spent more than four months floundering about in the Central Pacific without seeing any sign of land. South of Kamchatka the nearest land was New Zealand. De la Croyere, for his part, complained regularly of seasickness. By June the crew was on the verge of mutiny.

On July 16 they had their first sight of land after more than four months at sea—Mount St. Elias on the coast of southeastern Alaska, towering more than three miles above the water. As the boat neared the shore of Kayak Island, a young naturalist on board, George Steller, who managed to argue his way onshore with a watering party, leaped out and ran ahead through the surf, becoming the first white man to leave footprints on Alaskan soil. A decade earlier, Steller had gone to St. Petersburg with the dream of reaching the New World. His sprint through the surf was a moment he had been waiting for almost all of his life.

After ten years of preparation, he was given just ten hours to explore the land. He wasted no time, collecting plants as he walked inland to the base of a cliff now known as Steller's Bluff, while his assistant, Lepchukin, shot birds for him to identify. Walking around the island, he stumbled across a small village or campsite, suggesting that the area was not uninhabited. Steller noted the bone utensils filled with smoked salmon and the sweet grass being soaked to make liquor and deduced that these people were somehow related to the tribes occupying Kamchatka on the opposite side of the Pacific. That afternoon he spread his samples out on the beach and began writing, identifying more than a dozen different species of plants and birds—the salmon berry, the black crow berry, the whortleberry, and the Steller's jay. The bird was unusual, unlike anything he had seen in Siberia or Kamchatka. To Steller it was positive proof that they had finally reached America.

From Mount St. Elias they sailed west along the Alaskan coast and into the Bering Sea, where they discovered the Pribilof Islands. Then they headed back toward Kamchatka along the Aleutian chain, discovering the islands of Unalaska, Kiska, Attu, Aggatu, Umnak, and Adak. By late fall, however, they were still several hundred miles from Kamchatka and the crew was now stricken with scurvy. Bering was so sick that his second in command, Ivan Khitov (later remembered as "Khitov the Incompetent" by the crew), took over command of the ships and nearly killed them all. They spent the next several weeks sailing headlong into full gales. The crew wanted to head back to their port in the Sea of Okhotsk, but when they sighted Bering Island, Khitov insisted it was Kamchatka. As they tried to land, the ships broke up on shore.

The crew wintered over in dugout huts, suffering from scurvy and near starvation and tormented by foxes. While the crew collected sea otter and sea lion pelts, Steller worked on natural history, writing a definitive paper on fur seals and identifying the spectacled cormorant, a survivor from prehistoric times that would be extinct in less than one hundred years, as well as the Steller sea lion and the Steller sea cow, a kind of Arctic manatee that lived in the shallow waters near shore. For Steller and the rest of the crew the sea cows were a godsend—a docile animal that weighed almost eight thousand pounds with meat that tasted like beef. Hunted, they put up almost no defense at all and seemed so unconcerned about their safety that Steller wondered if they could actually see and hear. In less than fifty years they would be hunted to extinction.

Bering died over the winter, and the next spring the crew began building a new ship out of the wreckage that remained of their old *St. Paul*. Construction, however, was not without its problems. One officer, Lieutenant Ovstin, objected to the idea. They could not, he insisted, break up one of Her Majesty's ships without permission. They would be destroying government property. Ovstin was overruled. When the makeshift ship was finished they loaded up their treasure trove of furs and set sail for the mainland. They quickly ran into a storm and nearly sank—the crew was forced to throw their precious furs overboard to keep the ship afloat. Khitov, it turned out, had carefully hidden his away.

Those hidden furs would eventually reshape the face of Alaska. When they finally reached the mainland, Khitov's secret hoard was enough to convince others that they had found a rich new world teeming with foxes, seals, and sea otters just waiting to be harvested. The next spring a group of fur trappers sailed to Bering Island and came back with the pelts of more than five thousand sea otters, fur seals, and foxes. A stampede of princes, thieves, and fortune hunters would sweep out into the Aleutians and the Bering Sea, building rafts and barges along the coast of Kamchatka and setting sail for the New World.

• • •

Contact between the Aleuts and Russians were confrontational from the start. The first meeting between the two occurred on Umnak Island when Bering sent a longboat ashore for water. Natives came out in boats to meet them. They had straight black hair and lithe slender features, reminding the crew of natives they had met along the coast of Kamchatka and Siberia. When the Russians tried to trade beads and clothing with them, however, the Aleuts would have none of it—they wanted knives. When Bering's men tried to leave, the Aleuts attempted to take them hostage, a move that was stopped only when one of the crew fired a gun over their heads, scaring them off.

That initial contact set the tone for the future. For the next forty years contact between the Russians and the Aleuts would often be deadly for both, although the Aleuts would eventually lose out in the end. The warlike Aleuts and the Cossacks and Siberians who made up the bulk of the fur trappers flooding the area swapped brutalities. Aleuts would attack an unsuspecting camp and kill dozens while keeping others alive for days of torture. The Russians, in turn, repaid the atrocities with interest, killing an entire village on occasion in retribution for a single death. One seaman killed eight Aleuts with a single shot. He had lined them up front-to-back in a row and then shot the first man in line. He had wanted to see how many men a single bullet could kill.

By the 1760s the Aleuts and the Russians were in a state of open warfare. In 1763 Aleuts in the eastern islands sank four ships and murdered 188 sailors. Out of an initial crew of 200, just 12 survived. In 1766 the Russians finally brought the Aleuts to heel as Ivan Soloniev (remembered as "Soloniev the Destroyer") attacked the Aleuts at Unalaska, destroying their sea kayaks and breaking their spears and harpoons—essentially destroying their ability to survive. By 1781 the Aleuts were virtually extinct.

News of the atrocities, however, reached the royal court in St. Petersburg. When Russian Orthodox priests arrived in the Aleutians in the 1820s conditions for the Aleuts began to improve rapidly. Unlike the Catholic priests and monks who had been sent to California, the Russian priests sought to convert the natives by example, not force. They were, according to Ivan Veniaminov, the first archbishop of the Aleutians, instructed to "speak only when asked and to remember at all times they were guests in someone else's house." The faith spread by word of mouth. The priests devised alphabets for Aleut, Tlingit, Yupik, and Alutiq and translated the Bible into native languages. For their part, the Aleuts took readily to the heavy symbolism and elaborate rituals of the orthodox church. Conversion of the islands was rapid and complete and the Aleuts came to enjoy the same status as free peasants in Russia, while those with a putative ancestor in the male line could apply for creole status, which entitled them to government education and opened the doors to a host of bureaucratic jobs within the government.

The Russian settlement of North America was far different from that of western Europe. While the Europeans landing on the Atlantic coast gradually drove the natives westward and ultimately onto reservations, native peoples were an integral part of the Russian Empire—a reflection of Imperial Russia itself, a multicultural society pieced together out of literally hundreds of different ethnic groups. While the Europeans excluded, the Russians included.

At the height of the Russian presence in North America there were just 595 Russians on the Pacific coast, and 495 of those were located in a single town—Sitka. The rest were spread over several thousand miles all the way from Fort Ross in northern California to the Aleutians. By the 1850s the Aleut world had been transformed from a subsistence economy based on what could be gathered from the sea to one based on wages and money. There were Aleut bureaucrats, teachers, and clergy. After nearly one hundred years of struggle they were an essential part of the Russian Empire.

After Russia's defeat in the Crimean War in 1856, the threat of British expansion in Alaska and British Columbia was suddenly very real. Faced with the possible loss of their colonies in North America, Russia sold Alaska and the Aleutian Islands to the United States in 1867 for two cents an acre.

When the U.S. flag was raised over Alaska, the Aleut world changed in an instant. Aleuts became Indians. In place of the creole status that had been offered by the Russians they found the color bar. They had no rights or protection under the law and their property was seized and requisitioned by the arriving whites. The sale had stipulated that the native peoples would be afforded the same treatment as they had enjoyed in the past, but those provisions were conveniently ignored. For the Aleuts, who had finally adjusted to life with the Russians, it was like starting over. Both Russian and Aleut languages were suddenly illegal. Depression and alcoholism took a heavy toll. Protestant missionaries rushed to the area to convert the "heathens." Methodist missionaries, in an amazing display of ignorance and arrogance, announced that the Russian Orthodox Church was not a Christian Church. In the Pribilof Islands the manager of the local fur company objected to the Aleuts under his employment "wasting their money" on church candles. When Peter Trimble Rowe, the first Episcopalian bishop of the Aleutians, was asked if he recommended that funding for Aleut schools be increased, he replied that there wasn't any need. "The children," he said, "would all be dead in two to three years anyway."

During the Second World War the few remaining Aleuts on the islands were forcefully relocated to southeastern Alaska and housed in warehouses. Those under Japanese control fared no better—many were sent to Japanese concentration camps. When they came back to the islands after the war they found their homes and villages destroyed. Unlike in Western Europe, there was no battlefield cleanup in the Aleutians and little or no money for reconstruction. Some, like the residents of Adak, were not allowed to return to their island at all. It had been permanently requisitioned as a military base. After a history spanning ten thousand years, the Aleuts were teetering on the brink of extinction. Today there are just six native villages and some two thousand Aleuts. Finding a place in the sea proved far easier than finding a place in western civilization.

• • •

The year after my trip on the *RV Farnella* I went back to the Aleutians to spend more time on the islands. After a month and a half of shuttling back and forth along the coast of southeastern Alaska on a succession of planes and ships, I spent my first few days back on Unalaska Island simply catching my breath. I set up a small camp near the edge of a mountain lake in the tundra and spent the next few days watching and walking. From a nearby ridge you could look out over the town of Unalaska Bay and see the towns of Unalaska and Dutch Harbor and watch the sun set over the Bering Sea. Days were clear, but at night layers of fog drifted in from offshore, slowly covering the maze of inlets and channels that ran through the bay below. The tundra was soft and green and I spent several hours one day just lying on my back watching eagles soar overhead. From time to time I would walk into town to buy food and supplies and to look up Lydia Black's various Aleut friends to whom I passed on messages.

• • •

EVERYTHING IS SOMEWHERE, said a small sign on the door of Nikki's Place in Unalaska. IF YOU CAN'T FIND IT, ASK. IF WE CAN'T FIND IT TOGETHER THEN WE MAY HAVE TO ACCEPT THAT NOT FINDING EVERYTHING IS ONE OF LIFE'S GREAT MYSTERIES. That being said, you can get almost anything you want at Nikki's Place: fishing lures, imported coffee, Japanese dictionaries, Russian literature, wood block prints, and original watercolors. Island life requires a certain amount of self-sufficiency and innovation. At the time, however, you could not get a copy of Ivan Veniaminov's book on life in the Aleutians. It was, sadly enough, out of print, said the owner and sometimes operator of Nikki's, Abi Woodbridge, but she had (true to form) been pestering the publisher to reprint the book. Maybe next year. When I told her I was working on a book that would mention the Aleutians, she was interested.

"Have you been out on the water?" she asked. I told her about my time on the *Farnella,* but she shook her head and said, "Not a big boat. A small boat. A skiff. You don't really know the islands until you've traveled around them in a skiff. The Aleutians are the most beautiful place in the world."

The next morning we were carrying cans of gasoline, food, and fishing poles down to her fourteen-foot skiff tied up at the town pier in Dutch Harbor. The next day would be the opening day of halibut season, and the fishing boats alongside the dock were busy baiting their hooks and preparing their lines.

After tying our packs to the hauling post in the center of the boat we headed out of the harbor, sliding through the channel out into Captain's Bay, drifting past Hog and Amaknak Islands and out into the open waters of the Bering Sea. It was like opening the door to a new world.

• • •

For a while I sat huddled in the bow against the wind, but after twenty minutes I gave in and stood up in the bow, wrapping the bowline around my arm and leaning back against it like a water skier, taking up the bounce and skip of the boat over the waves with my knees and inhaling deep lungsful of air. There was a soft, rolling swell to the sea—you could feel it as the boat rose and fell. Outside the mouth of the bay there were thousands of shearwaters floating in the water. As we passed by they rose up into the air and curved off over the sea in a gentle roll, rising and falling like leaves blown by the wind.

There was a high overcast in the sky and the water was as black as ink. Ripples in its shiny surface caught the light of the tundra-covered slopes alongshore, sending ribbons of green water dancing across the surface of the sea. Cliffs alongshore were sharp and square, almost as if they had been cut by a knife. In their sides you could see columns of basalt and the roseate patterns of dykes and rings. Waterfalls cascaded two hundred and four hundred feet down the sides of coastal cliffs. In places, the regular blocky pattern of the basalt seemed to rise up from the water's edge like the seats of an amphitheater. Others were hung with carpets of moss and flowers and looked almost like hanging gardens.

Farther up the coast toward Reese Bay we floated through a small sea cave at the base of some high lava cliffs. The cave opened back up onto the sea like a tunnel. In the clear dark water you could see crabs scurry across the rocks below. It was low tide, and the sides of the cave were carpeted with purple-blue shells of mussels and knots of brightly colored starfish. Tucked away on ledges inside were puffins and cormorants that flew off in alarm as we drifted through the cave.

Offshore from Reese Bay we tied up in a bed of kelp like a sea otter, tying strands of kelp to the bow. Puffins from a nearby seastack circled overhead. Out in the kelp there was a gentle roll to the swell, but in the still air you could hear the sea break on the beach. The waves were almost ten feet high. On shore you could see a grassy ridge running along the edge of the beach—the remains of a longhouse from an Aleut village that once bordered the bay.

• • •

When Bering and his men wintered on Bering Island they nearly starved to death in the midst of one of the richest marine mammal habitats in the world. Unlike the Aleuts, they were not at home in the sea. They traveled across the water almost like astronauts in a space ship, dependent on food and water gathered on land to survive. On the Aleutians, however, dependence on the sea was total. "Here," Ivan Veniaminov said of the Aleutians, "man is not at risk only with respect to air and water. All else depends on chance."

The islands are beautiful, but they are also stark and barren. East of Unalaska Island there are no terrestrial animals, no native fox, bear, or even ground squirrel. The islands serve only as a temporary resting ground and nesting place for seabirds and marine mammals like seals, sea lions, and sea otters. The ground is treeless.

The Aleuts called themselves simply *Unangan,* which meant "We the People." They believed that they came from the west where the land was rich and the weather was warm. Here in the islands they had no wood to build boats and fires, aside from the driftwood that washed in from the mainland. There was also no clay or flint, which meant that the Aleuts could not make pots or fashion anything more than simple tools.

In spite of all these limitations the Aleuts used what they could and built a life of remarkable richness and complexity. Snares for catching seabirds were made out of thin strips of baleen from whales. Seagull tendons were used for fishing lines and leaders, driftwood was carved and shaped into spears and darts. Kelp and sea lion intestine was used to make braided ropes. From head to foot they fashioned their clothes from the fabrics and materials they gleaned from the sea lion. The primary concern on the northern edge of the Pacific was not keeping warm but keeping dry, and Aleuts fashioned *kamleikas*— anorak-like rain parkas made out of sea lion gut. Boots were fashioned out of the animal's esophagus, while flippers saw duty as soles, their naturally rough surface providing traction on the slick, wet rocks alongshore. Water bottles were fashioned from the sea lion's peri-cardium, the heart's outer sheathing. Almost nothing went to waste. Their tools, Veniaminov wrote, were all of the simplest kind. "Their function did not extend beyond the needs of subsistence and defense. They had no machine or complex tools of any kind, tools which could increase the power of a man or accelerate his activity—except the wedge. Everywhere their principal machine and tool was patience."

Their clothing was often breathtakingly beautiful, decorated with cormorant feathers, braided hair, and gut-on-gut appliqué. The *baidarka*—or sea kayak—however, was their crowning achievement. Wooden boats were too heavy for traveling far in the rough seas found to either side of the Aleutians, but the *baidarka* could move through the water as smoothly as a seal. It was an aphorism among the Aleuts that it was impossible to tell if the *baidarka* was made for the Aleut or the Aleut was made for the *baidarka.*

The *baidarka* was as sophisticated as a modern sea kayak, only the materials were different. It was built out of sea lion skin stretched over a frame of wood and bone. On board were a host of spears and darts and a throwing board fastened to the deck. Inside were sealskin floats to improve buoyancy, a wooden pump for bailing out the boat at sea, and a gut spray skirt that could be fastened around the waist to keep the craft watertight. The Aleuts' skills in handling the boats and their spears were so remarkable that they were hired by Russian and U.S. fur traders to hunt seals and sea otters in Alaska, the Pacific Northwest, and California. Aleuts in their *baidarkas* performed for the royal court in St. Petersburg and entertained crowds at the St. Louis World's Fair in 1902, giving a performance of their skill in the pool of a fountain.

The hunt was the key to life in the Aleutians and the *baidarka* was what made it all possible. At sea in their skin-covered boats the Aleuts ranged as freely across the water in search of sea lions as the Sioux or the Blackfoot roamed across the Great Plains in search of buffalo and antelope. Hunting was best in the open sea, and they frequently traveled several miles from shore, navigating by the aid of weighted floats. They weathered storms lashed together in a group. Their endurance was remarkable. It was said they could paddle for sixteen to twenty hours without stopping.

Hunting was highly ritualized and often mystical. Animals were thought of as human spirits. Sea otters were transformed human beings created to assuage the grief of parents who had lost their children. One could draw them close by wearing beautiful clothes and hunting hats or having a well-adorned kayak. The animals were relentless judges of character, avoiding not just those whom they considered lazy or immoral but those who, for example, had been cuckolded or had an unchaste sister. With nothing more than spears or darts to hunt with, skill was imperative. One had to know the habits of sea lions and sea otters—know where they hauled out and where they were likely to travel in search of food—as well as how to sneak up on them unawares. The Aleuts sought not domination over the world around them but harmony with it.

• • •

From the perspective of our own highly structured and industrialized world it is easy to envy the freedom and independence of the Aleuts, their intimate feel for the natural world around them. Primitive life, however, held its own costs, many of them startlingly familiar. The almost constant wet, cold weather caused severe arthritis. The Aleut developed a kind of tennis elbow from constantly throwing spears and darts. In late winter, food was scarce and March was known as *quixmagnix*, the month when they gnawed straps. Hunters out at sea could travel for days without seeing a single animal. After eighteen hours of nonstop travel they would land on the beach so crippled with cold and exhaustion that they were virtually unable to move and had to be lifted out of their kayaks and warmed up by an oil lamp placed under the edge of their *kamleika*. Others riding out storms at sea in their thin, fragile boats had nervous breakdowns or went completely insane with fear. The sea was relentless. There were two classes of people: the quick and the dead.

Contrasted against all this, however, is the beauty of a brightly colored bent wood hunting hat decorated with geometric designs, glass beads, and sea lion whiskers, or a parka studded with puffin beaks and iridescent cormorant feathers. Aleut society was often surprisingly sophisticated: they practiced acupuncture and bloodletting and had a system of counting that reached to more than ten thousand. Disputes within villages were often settled with singing duels. Life on the edge of the sea did not exclude a sense of beauty and elegance. Art was a part of life itself.

•   •   •

Henry Swanson remembers standing on Adak Island a few years back and thinking, "This is all mine." The island was deserted. He was the only person on the tundra-covered island near the western end of the Aleutian chain. He had leased it from the federal government for just twenty dollars per year for raising foxes. Originally the islands had no foxes on them, but Swanson and other trappers transplanted them, bringing them out in boats and releasing them, a practice started by the Russians. They would let the animals reproduce for a season or two and then come back and start trapping.

The only other animal life on the islands aside from a few seals and sea lions was seabirds. The foxes and birds had their litter at approximately the same time, Swanson recalled. It worked out well for the foxes, but not so well for the birds. Money was good. Fox pelts brought $50 to $75 apiece. If you were lucky enough to have an island, the only real risk was other trappers. "We'd steal foxes from one another," Swanson recalled. "I'd even help myself to a couple," he said with a chuckle.

I spent a rainy afternoon in Unalaska talking with Swanson in his small wood-frame house, a simple one-bedroom home salvaged from the Second World War. The wood walls were thin and drafty, but the roof was watertight. Henry kept the oven on with its door hanging open to ward off the chill.

I had come by to talk with Henry about life in the Aleutians and also about a walk I was planning to take through the hills behind town to Makushin Bay—a cross-country trek of about fifteen miles. Henry liked the idea and was sorry he couldn't make the trip himself. "Two to three years ago I used to just run up those hills," he said, "but I don't get around as much as I used to." I didn't feel he needed to make excuses. At ninety-two, one is entitled to sit back and take life more slowly.

His eyes were still bright and blue. Patches of gray hair covered the side of his head. He was so thin he hardly seemed to fill his clothes. At the time, he was the oldest living person in the Aleutians and perhaps Alaska, and over the past ninety-plus years his eyes had seen a lot of history. He spent most of his life in the Aleutians or at sea—the two are so closely intertwined that it is hard to separate them.

At fourteen Swanson was part of the crew on the ship that made the last legal sea otter hunt in Alaska. He says they had twenty-four men on board—all Aleuts—and hunting parties of twelve *baidarkas*. The year was 1910 and by that time the islands were pretty well stripped clean. In the five months they spent cruising around the Aleutians and then to Kodiak Island, they took only fourteen sea otters. They sold them in London where their pelts brought $2,100 apiece. Sea otter, Swanson says, was and still is the official fur of the rich.

The Aleutians, for as long as Henry could remember, always had a boom and bust economy. Over the years one line of work has supplanted another as styles and fads—and the animals themselves— came and went: sea otter hunting, fox trapping, cod fisheries, whaling stations, crab boats, and, more recently, factory ships and surimi boats, which turn pollock and hake into artificial crab for Japan and Korea. The money has almost always been good, even during the Second World War when Henry, although too old to enlist, was made a chief warrant officer and put in charge running supply ships out to remote outposts in the islands; he made $1,500 a month. Earlier, in 1917, Swanson had been a member of the crew of the first troop ship to carry U.S. soldiers to Europe for the First World War.

The same boom and bust cycle also applies to nature. The tundra-covered slopes seem barren, but some years the salmon berries can get as big as your thumb, Henry says. One day the sky is still and empty; the next it can be filled with hundreds of thousands of birds. One year the krill that washed up on the bare cobble beach that borders the village was more than a foot deep. Out in the bay the water was full of whales. They could sneak up on you unexpectedly. One day when Henry was taking a boat over to Akutan Island there were suddenly whales all around him, hundreds of them, and he could see their spouts extending off to the horizon in all directions. The tide was turning, the water flowing out of the Bering Sea through the narrow passes between islands, and as Henry slipped through Akutan Pass there were twenty- and thirty-foot standing waves in the channel. "You don't go over them when they're that big," he said matter-of-factly, "you go through them." The whales were trying to shoot through the pass as well, bucking the head tide. "They would dive down for a couple of minutes and then come back up swimming as hard as they could. They only moved about five feet." They looked almost like giant salmon swimming upstream to spawn.

• • •

Two times each year the world's entire population of Pacific gray whales, some sixteen thousand animals, swims through Unimak Pass on their way to and from their summer feeding grounds in the Bering Sea. The pass, which lies between Unimak and Akun and Akutan Islands, is less than twenty miles wide, and during the late spring and early fall the procession of whales traveling through it seems almost endless.

In the summer the shallow waters of the Bering Sea are transformed into some of the richest waters in the world. Sunshine in the Arctic summer is almost endless, and after a long winter locked in ice and darkness the water is filled with nutrients and ready to explode with life. Like the gray whale, fish, crabs, seals, and walrus thrive here during the summer. Other whales are drawn here as well, and the waters near the islands or even farther north above the Arctic Circle can be alive with fin, bowhead, and humpback whales traveling alone or even in large pods of several dozen animals.

In the early 1700s, before widespread whaling decimated their numbers, there were times, according to the Aleuts, when Unalaska Bay was so full of whales that it was unsafe for a *baidarka* to travel across the water. Today, protected by an international ban on whaling, the numbers have begun to come back.

Whales are the largest living animals on earth. Washed up on a beach they look almost prehistoric—massive mounds of flesh with the size and weight of a large house, their tails and flippers, or flukes, larger than a human. In the water, however, their movements are soft and seductive, rising to the surface and then slowly arching their backs as they dive. Like us they are warm-blooded mammals. They nurse their young on milk and must come to the surface to breathe—although they are so well adapted to life in the ocean that they can stay submerged for more than an hour. When ice forms early on the Bering Sea, however, they can suffocate beneath the ice, unable to break through to the surface to breathe. Studying the massive flukes of a whale you will find that its bone structure is almost identical to that of a human hand.

Here in the North Pacific the lives of people are closely tied to those of whales, hunted by the Tlingit of British Columbia and southeastern Alaska as well as the Inuit and Eskimos who lived farther north around the fringes of the Bering Sea. Aleuts hunted whales as well, but while the Tlingit and Inuit hunted with harpoons with groups of men in large wooden boats, whale hunting for the Aleuts was a solitary quest. They stalked the whales at sea in their *baidarkas* using poisoned spears and darts empowered by magic spells and contact with the dead.

While the Aleuts depended primarily on the sea lion and the sea otter, whales were coveted prizes. A single whale could supply a village of fifty with food for more than a month. Its large size made it useful for other things as well. Whale ribs could be used as poles and beams to support huts and houses. Shoulder blades were fashioned into tables and chairs. Almost nothing was wasted.

Whale hunters were a breed apart in Aleut society, and those who hunted whales offered themselves to the service of evil for the sake of the community. Poison was fashioned from the roots of plants, but other ingredients were considered essential as well—fat from the tissue of corpses, for example, or blood from the menstrual flow of a woman.

The poisoned spears were not actually used to kill the whale but to paralyze or cripple its flukes and fins. Success depended on paddling up to within a few feet of the whale and to just the right spot for the throw. Damaging the flukes interfered with the whale's ability to maintain its balance. Eventually if the throw was good and the poison was strong enough the whale would be unable to stay upright at the surface, and it would drown. After spearing a whale the hunter would return to shore and then go off into isolation for four days, laying on his stomach. On the fourth day he would roll over on his back, symbolizing the whale's death as it rolls over in the water. If all had been done correctly—the poison, the hunt, and the ritual—they believed the animal would die and wash ashore.

Using a small skin-covered boat to chase an animal that could weigh twenty to forty tons took courage and skill. Whale hunters were admired and feared within the village. It was said that those who hunted whales died young and hopelessly insane.

• • •

My next to last day on the island, I went for a long rambling walk through hills behind Unalaska, walking to Beaver Inlet and over the ridge of the mountains that lead to Summer's Bay on the Bering Sea. It was a bright, clear day with rafts of low, flat clouds in the sky. Near the wet ravines that cut through the hills the tundra was carpeted with flowers—blooms of pink, white, yellow, and red, speckled here and there with the deep purple of small, wild irises. For most of the walk I followed the worn gravel roads left from World War II over the hills, but from time to time I wandered off through the steep tundra-covered slopes, stopping here and there to gather handfuls of salmon berries or explore promising canyon.

Climbing to the top of a high ridge I could suddenly see Summer's Bay spread out below. Ridges of high green hills led down to the edge of the bay, framing the metallic gray waters of the Bering Sea that stretched out to the north. The walk down to the mouth of the small bay took nearly an hour. A small stream ran down the center of the valley in a tight, narrow canyon, its path bent into a series of sharp S-shaped curves as it tumbled over boulders and cobbles of granite. Bald eagles perched on rock walls of the canyon, taking wing from time to time to circle over the tundra.

Down at the edge of the sea I climbed up on top of the same sand dunes that Dave Scholl and I had explored a little more than a year earlier. There was a low shelf of clouds hanging over the Bering Sea, but a shaft of sunlight streaming through the clear sky beneath them caught the massive triangular shape of Priest's Rock, giving its black rocks an almost golden glow. The sun was bright and warm, but the wind off the water was cool, as if the wind carried with it a touch of the coming winter. Along the narrow beach two fishermen were fishing for salmon, casting their lines out into the sea.

Behind my back the Pacific Plate was slowly carrying the edge of the continent northward. In time pieces of Mexico and California will arrive and dive down into the trench, rising up behind it in sheets of fiery rock. Pieces of the seafloor are headed here as well: submarine canyons, spreading ridges, and deep basins filled with oil. By the time they arrive here they will already have been reshaped dozens or even hundreds of times—cut into sharp points that drift along the coast or worn away into sands that are blown by the wind into high dunes. Others will be cut by glaciers or seared by vents on the seafloor. Looking at the rocks alongshore they suddenly seemed alive—as if I could see their movements and histories in a single glance—all the features I had seen in the past two years of travel.

In a few weeks the ice would be forming on the Bering Sea and the gray whales would begin heading south, passing through the islands on their way back to the warm coastal lagoons of Baja California. Like the rocks, they too are a culmination of things. Their massive shape is the product of more than 4.5 billion years of evolution, a process that began with single cells of bacteria that flourished on the fringe of volcanic vents on the seafloor and gradually developed into the rich variety of plants and animals that fill the land and sea today .

As the whales head south, the rocks beneath them are heading north. This is not an end here, but a beginning. In time a new continent will rise up out of the sea where the whales now feed, built out of oceanic islands and small pieces of continent that have fallen down into the earth only to rise up again in sheets of fire and molten rock—new land, the makings of a new world.

Like the whales, the lives of rocks are marked by cycles. They have been in motion for millions of years, shaping and reshaping the face of the earth. There is no clear beginning or end, only change and endless cycles of rock and water.